Lithium-Ion Batteries: Design, Preparation, Reaction Mechanisms of Electrode Materials, and Battery Life Evaluation

Lithium-Ion Batteries: Design, Preparation, Reaction Mechanisms of Electrode Materials, and Battery Life Evaluation

Guest Editors

Zhenbo Wang
Tingfeng Yi
Gang Sun

MDPI

Basel • Beijing • Wuhan • Barcelona • Belgrade • Novi Sad • Cluj • Manchester

Guest Editors

Zhenbo Wang
School of Chemistry and Chemical Engineering
Harbin Institute of Technology
Harbin
China

Tingfeng Yi
School of Materials Science and Engineering
Northeastern University
Shenyang
China

Gang Sun
College of Materials Science and Engineering
Shenzhen University
Shenzhen
China

Editorial Office
MDPI AG
Grosspeteranlage 5
4052 Basel, Switzerland

This is a reprint of the Special Issue, published open access by the journal *Batteries* (ISSN 2313-0105), freely accessible at: https://www.mdpi.com/journal/batteries/special_issues/C4A5156883.

For citation purposes, cite each article independently as indicated on the article page online and as indicated below:

Lastname, A.A.; Lastname, B.B. Article Title. *Journal Name* **Year**, *Volume Number*, Page Range.

ISBN 978-3-7258-6119-4 (Hbk)
ISBN 978-3-7258-6120-0 (PDF)
https://doi.org/10.3390/books978-3-7258-6120-0

Contents

batteries

MDPI

Article

Synthesis and Performance of $NaTi_2(PO_4)_3$/VGCF@C Anode Composite Material for Aqueous Sodium-Ion Batteries

Bo Ding [1,2,*], Mingzhu Li [1,2], Fuzhou Zheng [1,2], Yangzhou Ma [3], Guangsheng Song [3,*], Xiulong Guan [1,2], Yi Cao [4] and Cuie Wen [5]

1 School of Materials and Chemical Engineering, Bengbu University, Bengbu 233030, China; julielimingzhu@163.com (M.L.); zfz617578876@163.com (F.Z.); guanxiulong2002@163.com (X.G.)
2 Anhui Provincial Engineering Laboratory of Silicon-Based Materials, Bengbu 233030, China
3 School of Materials Science and Engineering, Anhui University of Technology, Maanshan 243002, China; yangzhou.ma@outlook.com
4 Enpower Energy Technology (Anhui) Co., Ltd., Maanshan 243000, China; yi.cao@enpower-energy.com
5 School of Engineering, RMIT University, Melbourne, VIC 3001, Australia; cuie.wen@rmit.edu.au
* Correspondence: www024151@163.com (B.D.); song_ahut@163.com (G.S.)

Abstract: This study combines self-prepared $NaTi_2(PO_4)_3$ (NTP) with commercial vapor-grown carbon fiber (VGCF) using a solid state calcination, then coats it with carbon to synthesize the composite anode material $NaTi_2(PO_4)_3$/VGCF@C (NTP/VGCF@C). The microstructure and electrochemical properties of the composite material were then analyzed using microstructure analysis and electrochemical testing equipment. Single phase NTP shows nanoparticles with a polyhedral structure, and there is good contact at the interface between the nanoparticles and the VGCFs. The carbon coating formed on the NTP particles displays a nearly 6.5 nm thick layer of amorphous carbon. From the coin-cell battery performance measurements, after 850 cycles, the composite material NTP/VGCF@C exhibits an excellent retention rate of 96.3% compared to that of the pure NTP material when the current density is 200 mA/g. As a result, the composite material and lithium manganate (denoted as LMO) were assembled into an LMO-NTP/VGCF@C aqueous sodium-ion soft pack full battery system. The full battery shows an initial capacity of 31.07 mAh at a rate of 0.5C, and a reversible discharge capacity retention rate of 95.8% after 480 cycles, exhibiting a good long-cycle stability performance.

Keywords: aqueous sodium-ion battery; composite material; electrochemical performance

1. Introduction

With the increasing demand of the large-scale energy storage market, a rechargeable energy storage battery has been developed and researched. Currently, the lithium-ion battery is the most widely used chemical energy storage system, but it cannot meet the needs of large-scale application, mainly due to the shortage of high-cost lithium resources [1–6]. In contrast, sodium is abundant on the earth and is also much cheaper than lithium; thus, lithium-ion batteries may be replaced by sodium-ion batteries in large-scale energy storage applications [7–10]. However, sodium-ion has a significant impact on the deintercalation of sodium ions due to its large ionic radius (1.02 Å) [11,12]. Therefore, it is particularly urgent to develop suitable materials for the development of the sodium-ion battery (SIB). At present, hard carbon is the most widely used anode material for SIB since it exhibits good Na^+ storage performance. However, hard carbon easily forms sodium dendrites when the working voltage is lower than 0.1 V, thus causing a safety concern [13,14]. However, TiO_2 is also widely studied as a metal oxide anode materials for use in sodium-ion batteries, since under the condition of deep discharge, the theoretical capacity of TiO_2 is up to 335 mAh/g. However, TiO_2 is a semiconductor with low conductivity, leading to its low electrochemical activity. It is necessary to combine TiO_2 with highly conductive materials to achieve its purpose [15–19]. Therefore, it is necessary to develop a new type of electrode material suitable for sodium-ion batteries.

It is well known that NTP contains Na super ionic conductor (NASICON), with an open three-dimensional frame structure which facilitates ion diffusion. It consists of TiO_6 octahedrons and PO_4 tetrahedrons that are connected to each other through oxygen atoms [20–22]. The reversible sodiation and desodiation between $NaTi_2(PO_4)_3$ and $Na_3Ti_2(PO_4)_3$, i.e., the redox reactions of $Ti^{3+}\leftrightarrow Ti^{4+}$ via a two-phase reaction mechanism, can be represented by equation: $NaTi_2(PO_4)_3 + 2Na^+ + 2e^- \xleftrightarrow{\text{charge-discharge}} Na_3Ti_2(PO_4)_3$ [23,24]. However, due to its poor electronic conductivity, the NTP anode material cannot play its proper role in the sodium-ion battery [25]. In order to improve the electronic conductivity, several methods have been tested: (1) introducing conductive agents to increase the conductivity of the material [26–29]; (2) enhancing the conductivity or hydrophilicity of materials with doping elements [30,31]; and (3) constructing nanostructures and carbon-coated structures to improve the stability and electronic conductivity of materials [32]. For example, A sol-gel method was reported to embed nano-NTP particles into a mesoporous carbon matrix to artificially improve electronic conductivity [33]. The sodium-ion anode material showed excellent electrochemical performance at the current density of 4 A/g and the high rate of 20C. The presence of the interconnected porous carbon material effectively improved the electrochemical performance of the sodium-ion battery anode composite material. In addition, a three-dimensional (3D) porous structure containing fluoride $NaTi_2(PO_4)_3$@C (F-NTP@C) was created to improve the storage capacity of Na^+. After 2000 cycles at 10C, the capacity retention rate was reported to be 75.5%. The addition of F-ions and the three-dimensional porous structure design effectively enhanced the electronic conductivity and Na-ion dynamics [34]. Since graphene possesses both astonishing surface area and conductivity, the $NaTi_2(PO_4)_3$-graphene nanocomposite anode material for aqueous sodium-ion batteries was synthesized using the solvothermal method and subsequent high-temperature calcination [35]. Highly crystalline $NaTi_2(PO_4)_3$ nanoparticles were reported to be uniformly distributed on the surface of the nanographene, with good electrical conductivity. This type of nanocomposite exhibited excellent high-rate and cycling stability performance in a 1 mol/L Na_2SO_4 aqueous electrolyte; it showed high specific capacities of 110 mAh/g at 2C rates, and the capacity retention rate was 90% after 100 cycles. In addition to graphene, a carbon nanotube (CNT) was mixed with $NaTi_2(PO_4)_3$ and graphite by mechanical ball milling to prepare a composite material [36]. The composite showed the best electrochemical performance when 6.2 wt.% graphite coated on $NaTi_2(PO_4)_3$ and 15 wt.% CNT were used as the conductive agent, and the initial specific capacity reached 130 mAh/g at 0.1C rate. Carbon coating has been reported to be an effective strategy for improving NTP electrochemical properties [37]. The $NaTi_2(PO_4)_3$/C composite material coated with double carbon layers using the soft chemical method yielded reversible specific capacities of 133 and 64 mAh/g at 0.5C and 50C, respectively, showing excellent cycle stability and rate performance. To our knowledge, the addition of nano carbon fiber into NTP anode material has not been previously reported.

At present, the organic electrolytes used in the sodium-ion energy storage battery are highly inflammable and explosive, resulting in a safety risk. However, the aqueous sodium-ion battery is a safe energy storage technology because it adopts a salt water solution as an electrolyte, which does not easily ignite or explode; thus, battery combustion and explosion accidents can be basically avoided. It is noted that the narrow thermodynamic electrochemical window of water is about 1.23 V. This means that the voltage of the sodium-ion battery in water under normal conditions will not exceed 1.5 V. As a result, only a few aqueous sodium-ion battery systems have been studied. At present, the battery system matching the $NaTi_2(PO_4)_3$ sodium-ion anode material mainly includes $NaTi_2(PO_4)_3||Na_{0.44}MnO_2$, $NaTi_2(PO_4)_3||Na_3V_2(PO_4)_3$ and $NaTi_2(PO_4)_3||$Prussian blue cathode [38–43]. For the aqueous sodium-ion battery consisting of ($NaTi_2(PO_4)_3||$1 mol/L $Na_2SO_4||Na_2CuFe(CN)_6$) [44], the reversible specific capacity is 104 and 50 mAh/g at 2C and a 100C rate, and after 1000 cycles at the 10C rate, the specific capacity and capacity retention rate are 74 mAh/g and 88%, respectively. As for the aqueous sodium-ion battery consisting of ($NaTi_2(PO_4)_3$||1 mol/L Na_2SO_4 + 1 mol/L Li_2SO_4 ||$\lambda - MnO_2$), the discharge capacity of the full battery is 800 mAh at a current density of 50 mA, and the specific capacity reaches 89% of the theoretical capacity [45]. Another aqueous sodium-ion full battery consisting of $NaTi_2(PO_4)_3$||1 mol/L Na_2SO_4||$Na_{0.44}MnO_2$ reveals that the reversible specific capacity is 50% of the initial specific capacity after 1600 cycles [46]. Overall, the cycle stability and energy density of an aqueous sodium-ion full battery must be further improved in order to meet the requirements of existing large-scale energy storage. Therefore, the development of key materials is very important for the application of an aqueous sodium-ion full battery for large-scale energy storage.

In this paper, the vapor-grown carbon fiber (VGCF) was first added into the $NaTi_2(PO_4)_3$ anode material and then coated with carbon to synthesize an NTP/VGCF@C composite material. The microstructure and electrochemical performance of the composite material were investigated. Moreover, an aqueous sodium-ion soft pack battery system was assembled with a lithium manganate (LMO) cathode material (Daxiang Energy Technology Co., Ltd., Baiyin, China) and an NTP/VGCF@C anode material. This full battery showed a very good long cycling stability. Hopefully our findings can provide useful guidance for the industrial production of an aqueous sodium-ion battery.

2. Experimental Procedure

2.1. Sample Preparation

In this experiment, the analytically pure raw materials Na_2CO_3 (Chinese medicine chemical reagent, purity 99%), $TiO(OH)_2$ (Macklin reagent, purity $\geq$ 99%), and $NH_4H_2PO_4$ (Macklin reagent, purity 99%) were used for preparing the NTP phase as a first step, and then carbon fiber VGCF and citric acid were added as a second step to create a nanostructured composite NTP/VGCF@C material using the mechanical grinding + high temperature calcination + carbon surface modification method. Specifically, sodium carbonate, mettitanic acid, and ammonium dihydrogen phosphate were weighed and added to a stainless steel ball mill tank, according to molar ratio of 0.5:2:3. Then, 20 mL of alcohol containing 0.5% of the thickening agent polyvinyl butyral (PVB) was added into the stainless steel ball mill tank and mixed for 300 rpm for 2 h to ensure full grinding and mixing. The mixed solution was stirred and dried in a heated magnetic agitator to prepare the precursor material of carbon composite NTP for aqueous sodium-ion batteries. The precursor powder was pressed using a 20 MPa tablet press (size: diameter: Φ60 mm$\times$ thickness: 10 mm) and then transferred to a muffle furnace for high temperature calcination. The high temperature calcining process was as follows: pre-calcining at 350 °C for 2 h, then calcining at 800 °C for 9 h, with a heating rate of 3 °C/min, finally yielding the NTP sodium-ion anode material. The sintered titanium-phosphate anode material was crushed and sifted by the crusher, and the particle size was controlled below 20 μm. The NTP anode material, carbon fiber VGCF, citric acid, and thickening agent PVB were added to the alcohol solution according to the mass ratio of 90:0.5:8.5:1, nano-grinding and mixing were carried out in the sand mill for 5 h, and the grinding mixed solution was stirred and dried in the heated magnetic agitator. Then, the carbon composite NTP/VGCF@C aqueous sodium-ion nano-anode material was prepared by carbonization sintering in a tubular furnace under an argon atmosphere. The carbonization sintering process is described as follows: presintering at 300 °C for 1 h, followed by calcination at 650 °C for 3 h at a heating rate of 3 °C/min. The preparation process of NTP and NTP/VGCF@C sodium-ion anode materials is shown in Figure 1.

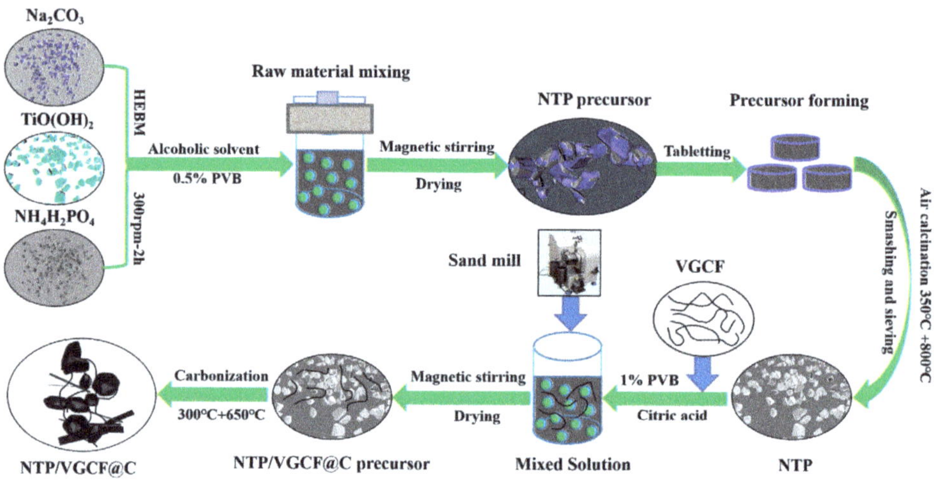

Figure 1. Schematic illustration of the preparation process for NTP and NTP/VGCF@C sodium-ion anode materials.

2.2. Battery Assembly

The prepared active material (NTP, NTP/VGCF@C, or LMO), conductive agent (Super-P), and binder (polyvinylidene fluoride (PVDF)) were mixed in an agate mortar at a mass ratio of 8:1:1, and an appropriate amount of N-methyl-2-pyrrolidone (NMP) solvent was added to prepare an evenly mixed slurry. The slurry was then evenly coated onto the copper foil using a doctor blade and dried at 80 °C for 24 h. A circular electrode disc of 12 mm in diameter (the soft pack battery pole size was 60 mm × 60 mm) was prepared using punching equipment as the research electrode and a sodium metal piece as the counter electrode. A total of 2025-type coin-cell batteries were then assembled in a glove box. The cutting NTP/VGCF@C anode electrode and LMO cathode electrode were assembled into an LMO-NTP/VGCF@C soft pack full battery. The electrolyte was 2 mol/L CH_3COONa and 2 mol/L CH_3COOLi, and the diaphragm was made of non-woven fabric.

2.3. Microstructure and Electrochemical Properties Analysis

The microstructure of the prepared anode materials was analyzed using a field emission scanning electron microscope, model TESCAN MIRA LMS. The microstructure and composition were analyzed by a transmission electron microscope (TEM), model JEM-2100F. Phase analysis was conducted using the Bruker D8 Advance X-ray diffractometer (Scientific Compass). The cyclic voltammetry (CV) and electrochemical impedance spectroscopy (EIS) were tested using the CS310H model electrochemical workstation tester, with the CV scanning speed of 0.02 mV/s. The electrochemical performance of the coin-cell batteries were tested using an electrochemical cycle system (Neware Technology Co., Ltd., Shenzhen, China). The prepared active material was used as a work electrode, while a thin sheet of sodium was used as a reference electrode. The voltage testing range of the coin-cell batteries was from 1.5 V to 3 V. For the LMO-NTP/VGCF@C soft pack full battery, and the voltage testing range was from 1.2 V to 1.9 V at a 0.5C rate, and the test temperature was maintained at a room temperature of 25 °C.

3. Results and Discussion

3.1. Phase and Element Analysis

Phase analysis was carried out on the synthesized NTP and its composite NTP/VGCF@C powder samples. These two XRD patterns are compared and shown in Figure 2a,b.

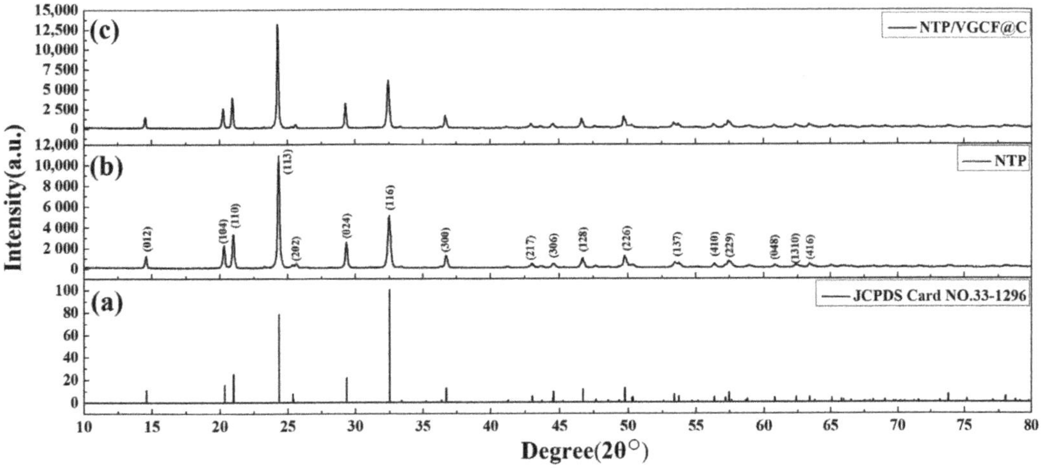

Figure 2. XRD patterns of: (**a**) JCPDS Card NO.33-1296; (**b**) NTP anode material and (**c**) NTP/VGCF@C anode composite material.

According to JCPDS Card NO.33-1296, in the diffraction range of 10°~80°, as shown in Figure 2a, the characteristic diffraction peaks of pure NTP appear at 14.52°, 20.27°, 20.9°, 24.25°, 25.54°, 29.27°, 32.42°, and 36.31°, etc., corresponding to the crystal plane index of (012), (104), (110), (113), (202), (024), (116), and (300), etc. This standard card pattern perfectly matches the measured whole pattern obtained from the synthesized NTP powder, as shown in Figure 2b, indicating that a single phase NTP material has been successfully prepared. By comparison, the XRD pattern of the composite NTP/VGCF@C (Figure 2c) is found to be very similar to that of the synthesized NTP material, indicating that the NTP exists as a main phase in the synthesized composite material. The phase structure of NTP does not change in the subsequent composite processing procedure, showing good structural stability. Although no characteristic peaks of carbon materials are reflected in the XRD pattern due to their minor amounts, the existence of VGCF nano-fiber and an amorphous coating layer can be confirmed by TEM.

The elemental composition and chemical valence state of the NTP/VGCF@C anode material were analyzed by XPS. It can be clearly seen from the XPS general spectrum in Figure 3a that the NTP/VGCF@C electrode material contains Na, O, Ti, P, and C elements. This indicates that the Na, O, Ti, and O elements come from the NTP phase, while the C element comes from carbon material, and the signals at 134.0 eV, 284.6 eV, 531.8 eV, and 1072.3 eV correspond to the P2p, C1s, O1s, and Na1s components separately. In the Ti 2p spectrum shown in Figure 3b, it can be seen that NTP/VGCF@C exhibits two peaks at 459.4 eV and 465.0 eV. These two peaks correspond to the characteristic peaks of Ti $2p_{3/2}$ and Ti $2p_{1/2}$ of Ti^{4+} in the NTP/VGCF@C anode material [47–49]. It is proved that Ti exists at the +4 valence in the NTP/VGCF@C anode material, indicating that no phase transition occurs in the subsequent process of the NTP anode material.

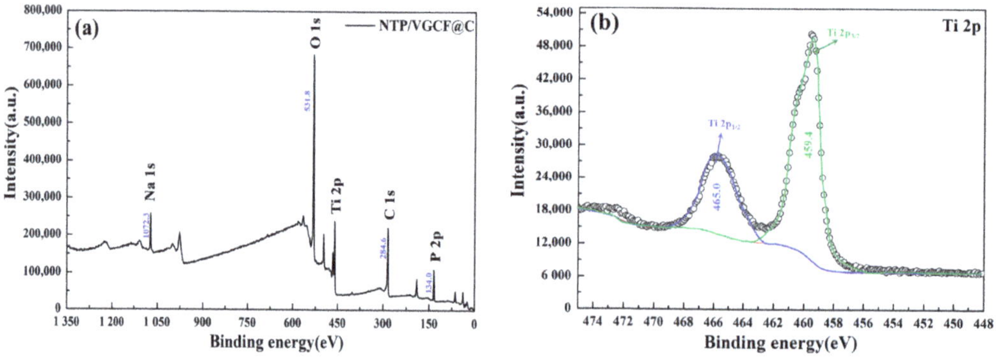

Figure 3. NTP/VGCF@C (**a**) total XPS spectrum, and (**b**) Ti 2p XPS spectrum.

3.2. Microstructure Analysis

Microstructure analysis was carried out on the prepared NTP/VGCF@C composite material. The SEM morphology, TEM morphology, EDS element distribution, and corresponding selective region electron diffraction (SAED) images of the prepared composite NTP/VGCF@C composite material are shown in Figure 4a–d.

It can be seen from Figure 4a that the prepared NTP/VGCF@C composite material shows polygonal particles ranging from 50–350 nm in size. The carbon fiber VGCF is evenly distributed among the particles, acting as a "bridge" connection between them. This is helpful for the construction of an effective conductive network. According to the TEM image (Figure 4b) analysis, the VGCF carbon fiber seems to be bound on the interface of the composite. The corresponding elemental distribution images are shown in Figure 4(b-1)–(b-5), and the qualitative test results of Na, Ti, P, O, and C in NTP/VGCF@C are 5.12, 26.2, 14.95, 39.08, and 14.16wt.%, respectively. The high amount of carbon and its whole area elemental distribution suggest that the carbon element mainly comes from the carbon coating on the NTP particles in the NTP/VGCF@C composite material.

High-resolution TEM (HRTEM) image analysis was conducted to obtain more microstructural information regarding the presence of carbon in the NTP/VGCF@C composite material. As shown in Figure 4c, there is a layer of amorphous carbon coated on the NTP particle with the interplanar spacing $d_{(104)} = 0.437$ nm. The carbon layer was measured to be about 6.5 nm. In addition, the carbon structure of the VGCF carbon fiber (002) plane (d = 0.341 nm) shows a stack of linear lamellae parallel to the carbon fiber surface. The diameter size of the carbon fiber is about 4.3 nm. From a higher magnification image (Figure 4(c-0)), it can be clearly seen that the VGCF lamellae structure is overlapped with the NTP lattice fringes, indicating a good contact at the interface between the carbon fiber and the nanoparticles in the NTP/VGCF@C composite material. This type of structure is helpful for increasing the interfacial strength of the connection between them, as well as the superiority of electron transport in the NTP/VGCF@C composite material.

The selective area electron diffraction (SAED) was used to obtain structural information for the aggregates of the NTP nanoparticles. The SAED pattern from the nanoparticle aggregates (Figure 4(d-0)) shows many discrete spots forming a series of rings, as shown in Figure 4d. Three discontinuous diffraction rings are selected and marked with a dotted line circle. According to the radius data obtained from these three circles, three d-values were calculated to be d1 = 0.210 nm, d2 = 0.170 nm, and d3 = 0.131 nm and indexed as (119), (137), and (336) crystal planes of the NTP phase. Overall, the synthesized NTP/VGCF@C composite material is composed of NTP, VGCF, and amorphous carbon.

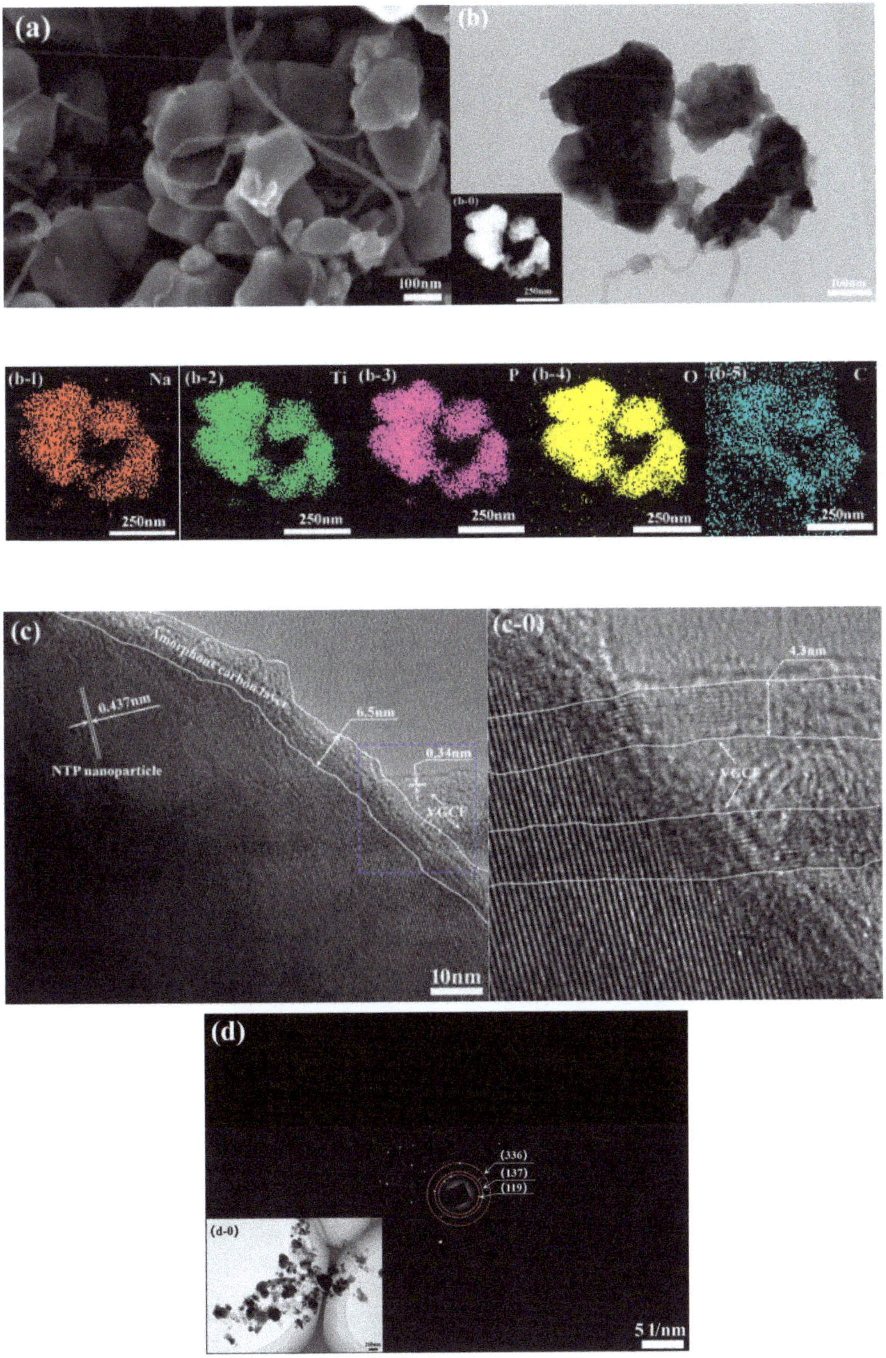

Figure 4. NTP/VGCF@C SEM morphology (**a**), TEM bright field image (**b**) and dark field image (inset, **b-0**), TEM-EDS scanning elements distribution (**b-1**–**b-5**), HRTEM lattice fringe images (**c**,**c-0**), TEM image showing a selected area of nanoparticle aggregates (**d-0**), and corresponding SAED pattern (**d**).

3.3. Cyclic Voltammetry and Electrochemical Impedance Spectroscopy Analysis

The cyclic voltammetry curves of the NTP/VGCF@C anode composite material are shown in Figure 5a. The Nyquist plots of the pure NTP and the NTP/VGCF@C anode composite materials are shown in Figure 5b.

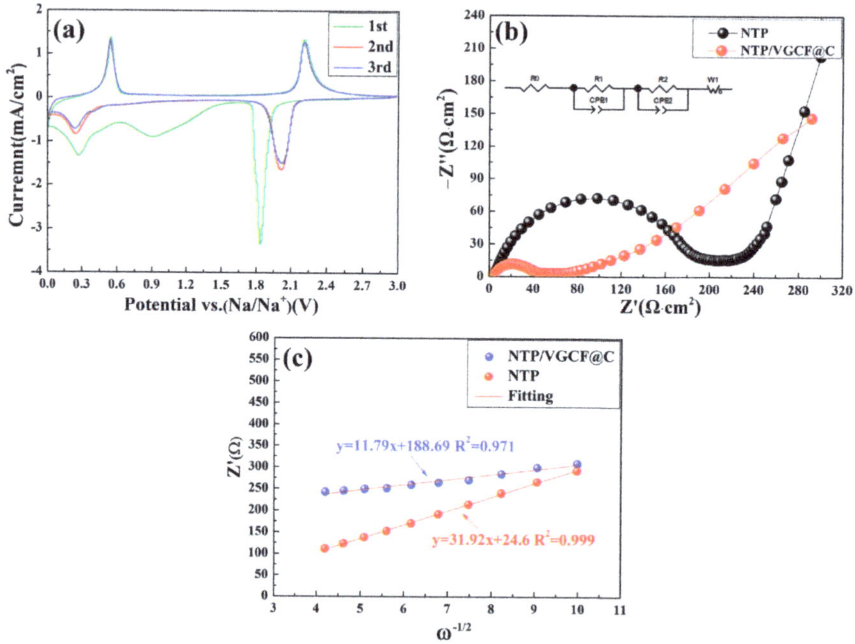

Figure 5. (**a**) CV curves of the NTP/VGCF@C anode composite material at the 1st, 2nd, and 3rd cycle; (**b**) Nyquist plots of the NTP and NTP/VGCF@C anode composite materials; (**c**) relationships between Z_{re} and $\omega^{-1/2}$ at low frequency.

From the curves of the first three cycles of cyclic voltammetry shown in Figure 5a, two broad peaks at 0.92 and 0.28 V can be seen during the initial discharge, but these are not present in subsequent cycles, when the potential range is 0.01–3.0 V vs. Na/Na^+, and the scanning rate is 0.1 mV/s. These irreversible peaks are attributed to electrolyte breakdown and SEI film formation, causing the initial capacity loss. In addition, there are two distinct redox reaction peaks at 2.23/1.85 V and 0.55/0.27 V. 2.23/1.85 V, which correspond to the reversible oxidation/reduction process of $Ti^{4+} \leftrightarrow Ti^{3+}$, while the redox peaks at 0.55/0.27 V correspond to the reversible oxidation/reduction process of $Ti^{3+} \leftrightarrow Ti^{2+}$. It should be noted that almost all CV curves overlap each other in subsequent cycles, indicating a good electrochemical reversibility. It can be determined from the Figure 5b that the SEI film resistance (R_{SEI}) and electrode/electrolyte interface resistance (R_{ct}) values of NTP/VGCF@C are lower than those of the NTP. The electrochemical impedance results show that the addition of carbon fiber VGCF and amorphous carbon has a great effect on the structure of the entire electrode material, providing a better electrode dynamic process. The ionic mobility of the NTP/VGCF@C was significantly improved because the VGCF formed a conductive network in the anode composite material, and the VGCF, acting as a "bridge", was attached by the amorphous carbon material between the active materials. The electrochemical properties of the materials can be improved. Moreover, the Warburg coefficient (Z_{re}) values of the NTP and NTP/VGCF@C are 31.92 $\Omega \cdot S^{-1/2}$ and 11.79 $\Omega \cdot S^{-1/2}$, respectively (as shows in Figure 5b) [50]. This indicates that the presence of the VGCF and the amorphous carbon coating layer has a significant influence on the electron transport in the composite material; the conductivity of the NTP/VGCF@C electrode was significantly improved. Therefore, the long cycle and rate performance of the NTP/VGCF@C anode composite material could be further improved.

3.4. Cycling Performance

The effects of pure NTP and its composite material on the cyclic stability and rate properties are shown in Figure 6a,b. The NTP/VGCF@C composite material is assembled as an anode, with LMO as a cathode, to create a 32 mAh LMO-NTP/VGCF@C aqueous sodium-ion soft pack battery system. The full battery cyclic performance is shown in Figure 6c.

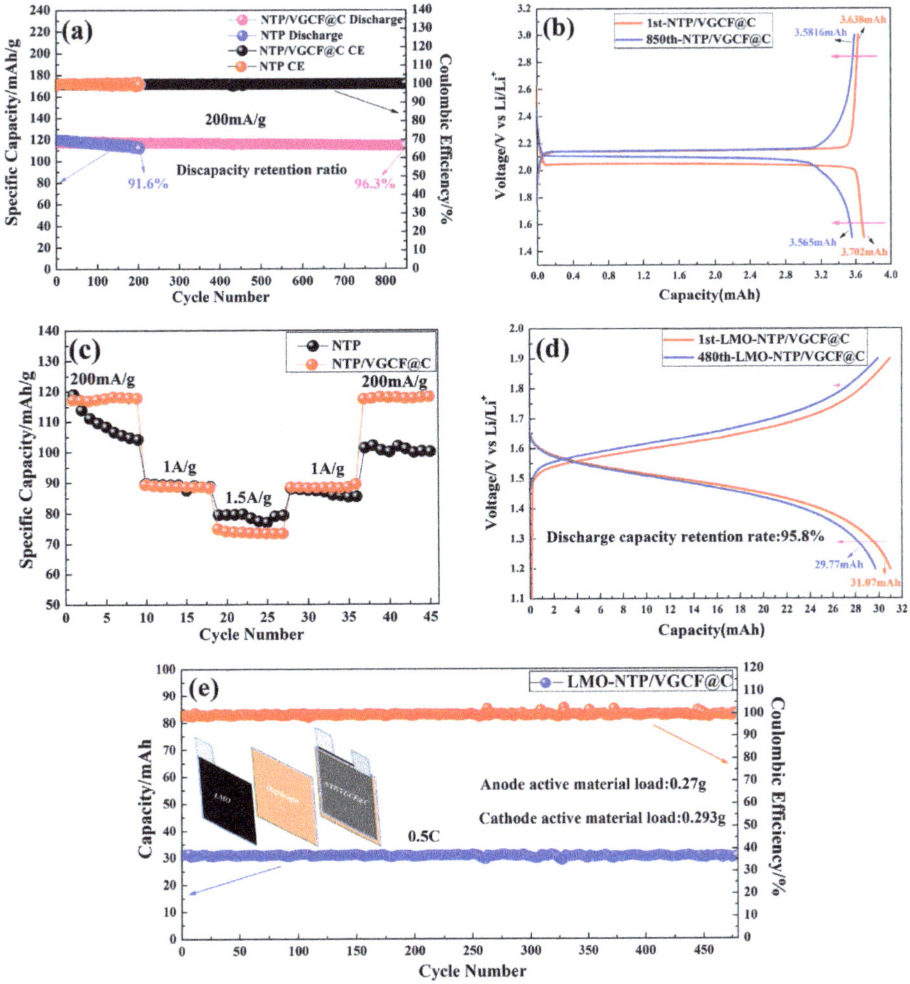

Figure 6. (**a**) Cycle performance of NTP and NTP/VGCF@C; (**b**) discharge/charge voltage profiles of NTP/VGCF@C in the 1st and 850th cycle; (**c**) rate performance of NTP and NTP/VGCF@C; (**d**) discharge/charge voltage profiles of an LMO-NTP/VGCF@C aqueous sodium-ion soft pack full battery in the 1st and 480th cycle; (**e**) cycle performance of an LMO-NTP/VGCF@C aqueous sodium-ion soft pack full battery.

The initial specific capacity of the pure NTP and NTP/VGCF@C composite material is 120.6 mAh/g and 118.4 mAh/g at a current density is 200 mA/g, respectively. After

200 cycles of charge and discharge, the reversible discharge capacity retention rates of NTP and NTP/VGCF@C are 91.6% and 99.2%, respectively. Obviously, the addition of carbon material increases the capacity retention by 7.6%. After 850 cycles of charge-discharge, the discharge capacity retention rate of the NTP/VGCF@C anode composite material is 96.3% (3.565 mAh/3.702 mAh), showing a better cycle stability (see Figure 6a,b). The rate capacity of these two materials at different current densities of 200 mA/g, 1A/g and 1.5A/g is shown in Figure 6c. When charging and discharging at a small rate to a large rate, and then returning to a small rate, the capacity of the composite material can be restored to the discharge capacity of 200 mA/g (117.4 mAh/g). However, pure NTP reverts to a small rate charge and discharge, and its capacity is only 84% of the initial capacity, showing a very poor rate performance. This suggests that the combination of carbon fiber and carbon coating not only maintains a robust structure for the composite material, but also acts as a conductive connection "bridge" between the NTP particles to ensure electronic transport between them. As for the full battery performance, the initial capacity of the LMO-NTP/VGCF@C aqueous sodium-ion soft pack full battery is 31.07 mAh at a rate of 0.5C, as shown in Figure 6d,e. After 480 cycles of charge and discharge, its reversible discharge capacity is 29.77 mAh, the discharge capacity retention rate is 95.8%, and its charging and discharging efficiency has been maintained at nearly 100%. Therefore, the LMO-NTP/VGCF@C aqueous sodium-ion battery system shows a long-term cycle stability performance, demonstrating a promising technique for green large-scale energy storage.

4. Conclusions

The composite anode material $NaTi_2(PO_4)_3$/VGCF@C is prepared in two steps: pure NTP synthesis, followed by the addition of VGCF and carbon coating. The microstructure and electrochemical properties of the composite material have been analyzed. The conclusions are as follows:

(1) The synthesized anode composite material is composed of NTP nanoparticles, VGCF, and an amorphous carbon coating layer. The VGCF nanofiber has a good interfacial contact with the NTP particles in the NTP/VGCF@C composite. The combination of VGCF with carbon coating makes the composite material more robust and conductive than pure NTP material.

(2) After 200 cycles of charge-discharge, the reversible capacity retention rates of NTP and NTP/VGCF@C are 91.6% and 99.2% at a current of 200 mA/g, respectively; the addition of carbon material increases the capacity retention by 7.6%, and after 850 cycles of charge-discharge, the capacity retention rate of NTP/VGCF@C is 96.3%, which is much better than that of pure NTP.

(3) The initial capacity of an LMO-NTP/VGCF@C aqueous sodium-ion full battery is 31.07 mAh at a rate of 0.5C. The reversible discharge capacity is 29.77 mAh after 480 cycles, and the discharge capacity retention rate is 95.8%. According to the results of this study, the LMO-NTP/VGCF@C aqueous sodium-ion full battery can be applied for large-scale green and safe energy storage.

Author Contributions: B.D.: Conceptualization, Formal analysis, Writing-original draft; M.L.: Data curation, Investigation; F.Z.: Methodology, Validation; Y.M.: Writing-review & editing; G.S.: Conceptualization, Validation; X.G.: Visualization; Y.C.: Resources; C.W.: Conceptualization. All authors have read and agreed to the published version of the manuscript.

Funding: This research was funded by Anhui Province Key R&D Program (grant number: 2022a05020040), the High-Level Scientific Research and Cultivation projects of Bengbu University (2021pyxm02, 00010312), an Anhui Natural Science Foundation grant (No. 1908085ME151), an Anhui Province high-end talent grant (DT18100044), and the National Foreign Expert Introduction Plan project (G20190219004), as well as University-Enterprise projects (LS0920, LS0921, RD18200058, 2019H3-7, 2020H3-8, 00011029). All individuals included in this section have consented to the acknowledgement. And The APC was funded by Anhui Province Key R&D Program (grant number: 2022a05020040).

Institutional Review Board Statement: Not applicable for studies not involving humans or animals.

Informed Consent Statement: Not applicable for studies not involving humans or animals.

Data Availability Statement: There are no new data were created.

Conflicts of Interest: The authors declare no conflict of interest.

References

1. Deng, Q.; Zheng, F.; Zhong, W.; Pan, Q.; Liu, M. P3-type $K_{0.5}Mn_{0.72}Ni_{0.15}Co_{0.13}O_2$ Microspheres as Cathode Materials for High Performance Potassium-Ion Batteries. *Chem. Eng. J.* **2020**, *392*, 123735. [CrossRef]
2. Pan, Z.H.; Yang, J.; Yang, J.; Zhang, Q.H.; Zhang, H.; Li, X.; Kou, Z.K.; Zhang, Y.F.; Chen, H.; Yan, C.L.; et al. Stitching of $Zn_3(OH)_2V_2O_7{\cdot}2H_2O$ 2D nanosheets by 1D carbon nanotubes boosts ultrahigh rate for wearable quasi-solid-state zinc-ion batteries. *ACS Nano* **2020**, *14*, 842–853. [CrossRef] [PubMed]
3. Shen, L.; Wang, Y.; Wu, F.; Shen, L.F.; Wang, Y.; Wu, F.X.; Moudrakovski, I.; van Aken, P.A.; Maier, J.; Yu, Y. Hierarchical metal sulfide/carbon spheres: A generalized synthesis and high sodium-storage performance. *Angew. Chem. Int. Ed.* **2019**, *58*, 7238–7243. [CrossRef] [PubMed]
4. Zhou, X.; Zhang, Z.; Du, Y.; Wang, Q.C.; Shen, J. A Yolk Shell Structured $FePO_4$ Cathode for High-Rate and Long-Cycling Sodium-Ion Batteries. *Angew. Chem. Int. Ed.* **2020**, *817*, 17504–17510.
5. Li, W.; Han, C.; Wang, W.; Xia, Q.; Chou, S.; Gu, Q.; Johannessen, B.; Liu, H.K.; Dou, S. Stress distortion restraint to boost the sodium-ion storage performance of a novel binary hexacyanoferrate. *Adv. Energy Mater.* **2019**, *10*, 1903006. [CrossRef]
6. Weng, W.; Xu, J.; Lai, C.; Xu, Z.; Zhou, X. Uniform Yolk-Shell Fe_7S_8@C Nanoboxes as a General Host Material for the Efficient Storage of Alkali Metal Ions. *J. Alloys Compd.* **2019**, *817*, 152732. [CrossRef]
7. Gu, Z.Y.; Guo, J.Z.; Zhao, X.X.; Wang, X.T.; Xie, D.; Sun, Z.H.; Zhao, C.D.; Liang, H.J.; Li, W.H.; Wu, X.L. High-Ionicity Fuorophosphate Lattice via Aliovalent Substitution as Advanced Cathode Materials in Sodium-Ion Batteries. *InfoMat* **2021**, *3*, 694–704. [CrossRef]
8. Deng, J.; Luo, W.B.; Chou, S.L.; Deng, J.Q.; Luo, W.B.; Chou, S.L.; Liu, H.K.; Dou, S.X. Sodium-ion batteries: From academic research to practical commercialization. *Adv. Energy Mater.* **2018**, *8*, 1701428. [CrossRef]
9. Huang, Y.Y.; Zheng, Y.H.; Li, X.; Adams, F.; Luo, W.; Huang, Y.H.; Hu, L.B. Electrode materials of sodium-ion batteries toward practical application. *ACS Energy Lett.* **2018**, *3*, 1604–1612. [CrossRef]
10. Bruce, P.G.; Scrosati, B.; Tarascon, J.M. Nanomaterials for rechargeable lithium batteries. *Angew. Chem. Int. Ed. Engl.* **2008**, *47*, 2930–2946. [CrossRef]
11. Deng, Q.; Zheng, F.H.; Zhong, W.T. Nanoscale Surface Modification of P2-Type $Na_{0.65}[Mn_{0.70}Ni_{0.16}Co_{0.14}]O_2$ Cathode Material for High-Performance Sodium-Ion Batteries. *Chem. Eng. J.* **2020**, *404*, 126446. [CrossRef]
12. Fang, G.Z.; Zhou, J.; Pan, A.Q.; Liang, S.Q. Recent advances in aqueous zinc-ion batteries. *ACS Energy Lett.* **2018**, *3*, 2480–2501. [CrossRef]
13. Zheng, G.Y.; Lin, Q.W.; Ma, J.B.; Zhang, J.; He, Y.B.; Tang, X.; Kang, F.Y.; Lv, W.; Yang, Q.H. Ultrafast Presodiation of Graphene Anodes for High-Efficiency and High-Rate Sodium-Ion Storage. *InfoMat* **2021**, *3*, 1445–1454. [CrossRef]
14. Yan, Z.H.; Yang, Q.W.; Wang, Q.H.; Ma, J.M. Nitrogen Doped Porous Carbon as Excellent Dual Anodes for Li- and Na-Ion batteries. *Chin. Chem. Lett.* **2020**, *31*, 583–588. [CrossRef]
15. Park, C.M.; Sohn, H.J. Black phosphorus and its composite for lithium rechargeable batteries. *Chem. Commun.* **2014**, *50*, 5435–5437.
16. Lian, S.T.; Lv, J.S.; Yu, Q.; Hu, G.W.; Chen, Z.; Zhou, L.; Mai, L.Q. Recent progress on TiO_2-Based anode materials for sodium-ion batteries. *J. Electrochem.* **2019**, *25*, 31–44.
17. Massaro, A.; Muñoz-García, A.B.; Maddalena, P.; Bella, F.; Meligrana, G.; Gerbaldi, C.; Pavone, M. First-principles study of Na insertion at TiO_2 anatase surfaces: New hints for Na-ion battery design. *Nanoscale Adv.* **2020**, *2*, 2745–2751. [CrossRef]
18. Longoni, G.; Cabrera, R.L.P.; Polizzi, S.; D'Arienzo, M.; Mari, C.M.; Cui, Y.; Ruffo, R. Shape controlled TiO_2 nanocrystals for Na-ion battery electrodes: The role of different exposed crystal facets on the electrochemical properties. *Nano Lett.* **2017**, *17*, 992–1000. [CrossRef]
19. Bella, F.; Muñoz-García, A.B.; Meligrana, G.; Lamberti, A.; Destro, M.; Pavone, M.; Gerbaldi, C. Unveiling the controversial mechanism of reversible Na storage in TiO_2 nanotube arrays: Amorphous versus anatase TiO_2. *Nano Res.* **2017**, *10*, 2891–2903. [CrossRef]
20. He, B.; Yin, K.B.; Gong, W.B.; Xiong, Y.W.; Zhang, Q.C.; Yang, J.; Wang, Z.X.; Wang, Z.; Chen, M.X.; Man, P.; et al. $NaTi_2(PO_4)_3$ hollow nanoparticles encapsulated in carbon nanofibers as novel anodes for flexible aqueous rechargeable sodium-ion batteries. *Nano Energy* **2021**, *82*, 105764. [CrossRef]
21. Ren, W.; Chen, X.; Zhao, C. Ultrafast aqueous potassium-ion batteries cathode for stable intermittent grid-scale energy storage. *Adv. Energy Mater.* **2018**, *8*, 1801413. [CrossRef]
22. Lee, M.H.; Kim, S.J.; Chang, D.; Kim, J.; Moon, S.; Oh, K.; Park, K.Y.; Seong, W.M.; Park, H.; Kwon, G.; et al. Toward a low-cost high-voltage sodium aqueous rechargeable battery. *Mater. Today* **2019**, *29*, 26–36. [CrossRef]
23. Fang, Y.J.; Xiao, L.F.; Qian, J.F.; Cao, Y.L.; Ai, X.P.; Huang, Y.H.; Yang, H.X. 3D graphene decorated $NaTi_2(PO_4)_3$ microspheres as a superior high-rate and ultracycle-stable anode material for sodium ion batteries. *Adv. Energy Mater.* **2016**, *6*, 1502197. [CrossRef]

24. Wu, M.G.; Ni, W.; Hu, J.; Ma, J.M. NASICON-structured $NaTi_2(PO_4)_3$ for sustainable energy storage. *Nano-Micro Lett.* **2019**, *11*, 44. [CrossRef]
25. Fu, L.; Xue, X.; Tang, Y.G.; Sun, D.; Xie, H.L.; Wang, H.Y. Size controlling and surface engineering enable $NaTi_2(PO_4)_3/C$ outstanding sodium storage properties. *Electrochim. Acta* **2018**, *289*, 21–28. [CrossRef]
26. Hsieh, Y.Y.; Fang, Y.B.; Daum, J.; Kanakaraj, S.N.; Zhang, G.Q.; Mishra, S.; Gbordzoe, S.; Shanov, V. Bio-inspired, nitrogen doped CNT-graphene hybrid with amphiphilic properties as a porous current collector for lithium-ion batteries. *Carbon* **2019**, *145*, 677–689. [CrossRef]
27. Bakierska, M.; Lis, M.; Pacek, J.; Świętosławski, M.; Gajewska, M.; Tąta, A.; Proniewicz, E.; Molenda, M. Bio-derived carbon nanostructures for high-performance lithium-ion batteries. *Carbon* **2019**, *145*, 426–432. [CrossRef]
28. Ji, L.W.; Meduri, P.; Agubra, V.; Xiao, X.C.; Alcoutlabi, M. Graphene-based nanocomposites for energy storage. *Adv. Energy Mater.* **2016**, *6*, 1502159. [CrossRef]
29. Massaro, A.; Pecoraro, A.; Muñoz-García, A.B.; Pavone, M. First-Principles study of Na intercalation and diffusion mechanisms at 2D MoS_2/graphene interfaces. *J. Phys. Chem. C* **2021**, *125*, 2276–2286. [CrossRef]
30. Wu, X.W.; Li, Y.H.; Zhao, S.Y.; Zeng, F.H.; Peng, X.C.; Xiang, Y.H.; Ruan, Q.Y.; Gao, F.; He, T.; Wu, J.H. Fabrication of F-doped, C-coated $NiCo_2O_4$ nanocomposites and its electrochemical performances for lithium-ion batteries. *Solid State Ion.* **2019**, *334*, 48–55. [CrossRef]
31. Jiang, Y.Q.; Zhu, J.; Wang, H.Y.; Li, Y.H.; Zhou, H.Z.; Meng, W.; Dai, L.; Wang, L.; He, Z.X. N-doped carbon coated $LiTi_2(PO_4)_3$ as superior anode using PANi as carbon and nitrogen bi-sources for aqueous lithium ion battery. *Electrochim. Acta* **2018**, *279*, 279–288.
32. Li, L.L.; Wen, Y.H.; Zhang, H.; Ming, H.; Pang, J.; Zhao, P.C.; Cao, G.P. Core-Shell structured $LiTi_2(PO_4)_3/C$ anode for aqueous lithium-ion batteries. *ChemElectroChem* **2019**, *6*, 1908–1914. [CrossRef]
33. Wang, D.X.; Liu, Q.; Chen, C.J.; Li, M.L.; Meng, X.; Bie, X.F.; Wei, Y.J.; Huang, Y.H.; Du, F.; Wang, C.Z.; et al. NASICON-structured $NaTi_2(PO_4)_3$@C nanocomposite as the low operation-voltage anode material for high-performance sodium-ion batteries. *ACS Appl. Mater. Interfaces* **2016**, *8*, 2238–2246. [CrossRef]
34. Deng, Q.; Cheng, Q.; Liu, X.Z. 3D porous Fluorine-Doped $NaTi_2(PO_4)_3$@C as High-Performance Sodium-Ion battery anode with broad temperature adaptability. *Chem. Eng. J.* **2022**, *430*, 132710. [CrossRef]
35. Pang, G.; Yuan, C.Z.; Nie, P.; Ding, B.; Zhu, J.J.; Zhang, X.G. Synthesis of NASICON-type $NaTi_2(PO_4)_3$-graphene nanocomposite as an anode for aqueous rechargeable Na-ion batteries. *Nanoscale* **2014**, *6*, 6328–6334. [CrossRef]
36. Wise, A.; Wu, W.; Yan, J. Using intimate carbon to enhance the performance of $NaTi_2(PO_4)_3$ anode materials: Carbon nanotubes vs. graphite. *J. Electrochem. Soc.* **2014**, *161*, A561–A567.
37. Jiang, Y.; Zeng, L.C.; Wang, J.Q.; Li, W.H.; Pan, F.S.; Yu, Y. A carbon coated NASICON structure material embedded in porous carbon enabling superior sodium storage performance: $NaTi_2(PO_4)_3$ as an example. *Nanoscale* **2015**, *7*, 14723–14729. [CrossRef]
38. Jung, Y.H.; Hong, S.T.; Kim, D.K. Electrochemical sodium-ion intercalation properties of $Na_{2.7}$ Ru_4O_9 in nonaqueous and aqueous electrolytes. *J. Electrochem. Soc.* **2013**, *160*, A897–A900. [CrossRef]
39. Beck, F.; Rüetschi, P. Rechargeable batteries with aqueous electrolytes. *Electrochim. Acta* **2000**, *45*, 2467–2482. [CrossRef]
40. Zhang, Q.; Liao, C.Y.; Zhai, T.Y.; Li, H.Q. A high rate 1.2 V aqueous sodium-ion battery based on all NASICON structured $NaTi_2(PO_4)_3$ and $Na_3V_2(PO_4)_3$. *Electrochim. Acta* **2016**, *196*, 470–478. [CrossRef]
41. Wang, H.B.; Zhang, T.R.; Chen, C.; Ling, M.; Lin, Z.; Zhang, S.Q.; Pan, F.; Liang, C.D. High-performance aqueous symmetric sodium-ion battery using NASICON-structured $Na_2VTi(PO_4)_3$. *Nano Res.* **2018**, *11*, 490–498. [CrossRef]
42. Shanbhag, S.; Bootwala, Y.; Whitacre, J.F.; Mauter, M.S. Ion transport and competition effects on $NaTi_2(PO_4)_3$ and $Na_4Mn_9O_{18}$ selective insertion electrode performance. *Langmuir ACS J. Surf. Colloids* **2017**, *33*, 12580–12591. [CrossRef] [PubMed]
43. Duan, B.; Wang, F.; Tamirat, A.G.; Suo, L.M.; Wang, Y.G.; Wang, C.S.; Xia, Y.Y. Progress in aqueous rechargeable sodium-Ion batteries. *Adv. Energy Mater.* **2018**, *8*, 1703008. [CrossRef]
44. Wu, X.Y.; Sun, M.Y.; Shen, Y.F.; Qian, J.F.; Cao, Y.L.; Ai, X.P.; Yang, H.X. Energetic aqueous rechargeable sodium-ion battery based on $Na_2CuFe(CN)_6$-$NaTi_2(PO_4)_3$ intercalation chemistry. *ChemSusChem* **2014**, *7*, 407–411. [CrossRef] [PubMed]
45. Whitacre, J.F.; Shanbhag, S.; Mohamed, A.; Polonsky, A.; Carlisle, K.; Gulakowski, J.; Wu, W.; Smith, C.; Cooney, L.; Blackwood, D.; et al. A polyionic, large-format energy storage sevice using an aqueous electrolyte and thick-format composite $NaTi_2(PO_4)_3$/activated carbon negative electrodes. *Energy Technol.* **2015**, *3*, 20–31. [CrossRef]
46. Li, Z.; Young, D.; Xiang, K.; Carter, W.C.; Chiang, Y.M. Towards high power high energy aqueous sodium-ion batteries: The $NaTi_2(PO_4)_3/Na_{0.44}MnO_2$ system. *Adv. Energy Mater.* **2013**, *3*, 290–294. [CrossRef]
47. Wang, L.; Huang, Z.N.; Wang, B.; Liu, G.J.; Cheng, M.; Yuan, Y.F.; Luo, H.; Gao, T.T.; Wang, D.L.; Yassar, R.S. Purifying the phase of $NaTi_2(PO_4)_3$ for enhanced Na^+ storage properties. *ACS Appl. Mater. Interfaces* **2019**, *11*, 10663–10671. [CrossRef]
48. Vincent, M.; Avvaru, V.S.; Haranczyk, M.; Etacheri, V. High-performance Mg-Li hybrid batteries based on pseudocapacitive anatase $Ti_{1-x}Co_xO_{2-y}$ nanosheet cathodes. *ChemSusChem* **2022**, *15*, e202102562. [CrossRef]

49. Vincent, M.; Avvaru, V.S.; Rodríguez, M.C.; Haranczyk, M.; Etacheri, V. High-rate and ultralong-life Mg-Li hybrid batteries based on highly pseudocapacitive dual-phase TiO_2 nanosheet cathodes. *J. Power Sources* **2021**, *506*, 230118. [CrossRef]
50. Ding, B.; Huang, X.N.; Cai, Z.F.; Ma, Y.Z.; Song, G.S.; Yang, W.D.; Wen, C. Effects of binders on electrochemical properties of high capacity silicon composite anodes. *Inorg. Chem. Commun.* **2020**, *113*, 107771. [CrossRef]

batteries

MDPI

Article

SOC Estimation Methods for Lithium-Ion Batteries without Current Monitoring

Zhaowei Zhang [1], Junya Shao [1], Junfu Li [1,*], Yaxuan Wang [2] and Zhenbo Wang [2]

[1] School of Automotive Engineering, Harbin Institute of Technology, Weihai 264209, China; 2200180129@stu.hit.edu.cn (Z.Z.); 2200180201@stu.hit.edu.cn (J.S.)
[2] School of Chemical Engineering and Chemistry, Harbin Institute of Technology, Harbin 150001, China; 22b925038@stu.hit.edu.cn (Y.W.); wangzhb@hit.edu.cn (Z.W.)
* Correspondence: lijunfu@hit.edu.cn

Abstract: State of charge (SOC) estimation is an important part of a battery management system (BMS). As for small portable devices powered by lithium-ion batteries, no current sensor will be configured in BMS, which presents a challenge to traditional current-based SOC estimation algorithms. In this work, an electrochemical model is developed for lithium batteries, and three methods, including the incremental seeking method, dichotomous method, and extended Kalman filter algorithm (EKF), are separately developed to establish the framework of current and SOC estimation simultaneously. The results show that the EKF algorithm performs better than the other two methods in terms of estimation accuracy and convergence speed. In addition, the estimation error of the EKF algorithm is within ±2%, which demonstrates its feasibility.

Keywords: Li-ion; SOC; current sensorless; extended Kalman filtering

1. Introduction

In recent years, the sustainable development of the environmental ecosystem has become a pressing issue due to the increasingly prevalent problems of environmental pollution. In this case, it is essential to focus on the efficient use of energy and resources. Lithium-ion batteries are a popular choice as energy storage components for small electronics and large electric vehicles due to their high energy density, long cycle life, no memory effect, and low self-discharge [1]. However, overcharging or over-discharging a lithium battery can lead to capacity degradation, shortened battery life, and even explosion.

In order to guarantee the safe and secure use of lithium-ion batteries and to extend their cycle life, a battery management system (BMS) is critical, which can effectively manage the performance of lithium-ion batteries in a comprehensive, efficient, and refined manner [2]. It is necessary to ensure optimal battery performance and longevity in various applications.

The reliability of the BMS depends on the precision of the state of charge (SOC) estimation of lithium-ion batteries. SOC indicates the remaining capacity and provides an indication of whether the battery needs to be charged or discharged. The variation characteristics of SOC are a critical performance indicator for assessing the status of lithium-ion batteries. Therefore, accurate SOC estimation is a core function of BMS [3], and it is a prerequisite to achieve additional functions such as safety control, battery equalization, and troubleshooting.

However, SOC is an internal state of the battery and cannot be directly measured. It can only be estimated based on the relationship between voltage, current, temperature, and the aging of the battery [4]. Therefore, it is crucial to develop reliable SOC estimation algorithms that consider all relevant factors to ensure optimal performance and safety of lithium-ion batteries in various applications.

In automobiles, robots, and energy storage systems, current measurement is usually achieved using shunt resistors or Hall-effect current sensors. While shunt resistors have inherent power losses and require isolation circuitry, Hall-effect sensors are typically expensive. In low-cost portable applications, such as microphones and hearing aids, the current sensor is often not equipped, considering the size and cost of the device. Therefore, it is practical to design a current sensorless SOC estimation method for low-cost portable applications.

Common SOC estimation algorithms include the open-circuit voltage method, ampere-hour integral method, electrochemical impedance spectroscopy, adaptive filtering method, and data-driven neural network method [5]. The open circuit voltage method takes a long time to collect data and is difficult to apply in practice [6], while the ampere-hour integral method is affected by the initial charge state and current stability. The electrochemical impedance spectroscopy method is only used for laboratory research [7]. In addition, the adaptive filtering method has a complex algorithm and long calculation cycle, which includes nonlinear Kalman filter, particle filter, specifically including extended Kalman filter, traceless Kalman filter, and other methods [8–12]. In neural network methods [13–15] and support vector machine methods [16–19], the SOC estimation of a battery is viewed as a regression problem, using multiple inputs (e.g., voltage, current, and environmental variables) to predict the SOC. These methods usually require a large quantity of experimental data to train the neural network and use various optimization techniques to improve precision and robustness. Ignoring the internal mechanism of the battery, the model accuracy depends on the quality of the sample data. In addition, in the field of fast charging, Xuejiao Xu et al. [20] created their own three-electrode system so as to measure the anode/cathode potential and estimate the SOC.

A battery model, which includes equivalent circuit models (ECM) [21] and electrochemical models, provides high simulation accuracy and reflects the external characteristics of the battery. The electrochemical model parameters have physical meanings and act as a bridge between the external characteristics of the battery and the internal electrochemical reaction mechanism of the battery [22–24]. In order to achieve accurate SOC estimation, the BMS requires current measurements as input to the estimator.

For the current sensorless SOC estimation method, Cambron and Cramer [25] estimated the current by an unknown input observer, and Putra et al. [26] created a new method to accomplish current estimation on the basis of Thevenin ECM. Chun et al. [27] obtained the open circuit voltage (OCV) and current information from the terminal voltages and then calculated the SOC using the ampere-hour integral method. However, these methods either used a linear relationship between OCV and SOC or utilized an overly simple battery model that led to a decrease in the accuracy of the model.

To avoid the above problems, Jing Hou et al. [28] used the variational Bayesian extended Kalman filter method to achieve simultaneous estimation of SOC and current. Experimental results showed that the mean absolute errors (MAEs) and the root mean square errors (RMSEs) of the SOC estimations of the proposed variational Bayes-based unscented Kalman filter (VB-UKF) were less than $\pm 3\%$.

In this work, three methods are proposed to estimate SOC in the absence of current sensors based on the electrochemical model of lithium-ion batteries, which include the incremental seeking method, the dichotomous method, and the improved extended Kalman filter algorithm. The remainder of this work is organized as follows: (1) In Section 2, the electrochemical model of lithium-ion batteries is established to mathematically express the mechanisms in the charging and discharging process, and the relationship function between voltage and SOC is obtained. The principles of the three methods are described; (2) In Section 3, the data test of LCO lithium-ion batteries is conducted to obtain the voltage and current data under different working conditions; (3) In Section 4, the estimation of SOC under no current monitoring by three methods is completed compared with the reference value. In addition, the effects of inaccurate initial values of SOC and different levels of voltage noise on the accuracy of the estimation results are also explored.

2. Lithium-Ion Battery Model and Algorithm Principle

2.1. Simplified Electrochemical Model

The structure of a lithium-ion battery is divided into three areas: the positive electrode, the negative electrode, and the diaphragm. The cathode material typically uses lithium ferrous phosphate (LFP), lithium cobalt oxide (LCO), lithium-nickel-cobalt-manganese oxide (LNCM), etc., while the anode material is usually graphite, and the electrolyte is mostly LiPF6 alkyl carbonate with polymer materials. During the process of charging and discharging, lithium ions are de-embedded and transferred between the positive and negative electrodes [29].

The classical electrochemical model of lithium-ion batteries is based on ten control equations in the form of partial differential equations. These equations describe the solid-phase diffusion within the particles, liquid-phase diffusion in the electrolyte, solid-phase potential equilibrium, and liquid-phase potential equilibrium in the positive, diaphragm, and negative regions. The model structure is shown in Figure 1.

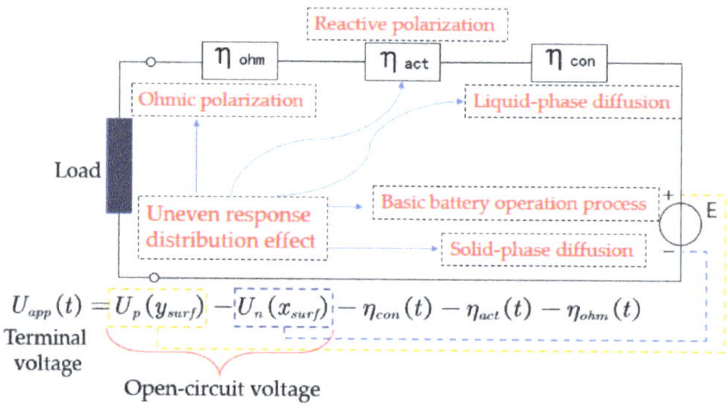

Figure 1. Lithium battery electrochemical model structure.

In this work, the internal processes of the cell were described using algebraic equations. These equations encompassed various aspects of the cell's functioning, such as the basic working process, solid-phase diffusion process, concentration polarization effect, reaction polarization effect, ohmic polarization effect, and the calculation of the terminal voltage [30]. The equations involved in the model are presented in Table 1. The meanings of the parameters involved in the above model are shown in Table 2.

2.2. Identification of Model Parameters

The intrinsic characteristics of the battery parameters remain constant and can be obtained by consulting the manufacturer or references. However, the variables to be identified in the electrochemical model include the initial lithium intercalation in the cathode, cathode capacity, offset of the lithium intercalation, and more [31]. The parameters involved can be obtained by analyzing the intrinsic connection between the cell terminal voltage change, the open-circuit voltage, and each part of the overpotential under the designed identification conditions with different forms of current excitation.

The parameter identification method referred to an excitation response analysis developed in [30]. Based on the simplified electrochemical model, using the input data of current and the corresponding response data of voltage, the relationship between the internal mechanisms and external behaviors was established quantitatively, and the parameters in the model were fitted during this process.

The parameters obtained by identification are shown in Table 3.

Table 1. Electrochemical model equations for lithium batteries.

Correlation Mechanism	Equations
Terminal voltage	$U_{app}(k) = E_{ocv}(k) - \eta_{con}(k) - \eta_{act}(k) - \eta_{ohm}(k)$
Open circuit potential correction	$E_{ocv}(k) = E_{ocv}^{ref}(k) + (T(k) - T_{ref})\frac{\mathrm{d}E_{ocv}}{\mathrm{d}T}$
Basic working principle	$x_{avg}(k) = x_0 - \int_{t_1}^{t_k} I\mathrm{d}t/Q_n, x_{surf}(k) = x_{avg}(k) - \Delta x(k), \Delta x(k) = \Delta x_1(k) + \frac{2}{7}\tau_n I(k)/Q_n,$ $y_{avg}(k) = y_0 + \int_{t_1}^{t_k} I\mathrm{d}t/Q_p, y_{surf}(k) = y_{avg}(k) + \Delta y(k), \Delta y(k) = \Delta y_1(k) + \frac{2}{7}\tau_p I(k)/Q_p,$ $E_{ocv}(k) = U_p\left[y_{surf}(k)\right] - U_n\left[x_{surf}(k)\right]$
Solid diffusion	$\Delta x_1(k+1) = \Delta x_1(k) + \frac{1}{\tau_n}[\frac{12\tau_n I(k)}{7Q_n} - \Delta x_1(k)]\Delta t, \Delta y_1(k+1) = \Delta y_1(k) + \frac{1}{\tau_p}[\frac{12\tau_p I(k)}{7Q_p} - \Delta y_1(k)]\Delta t$
Liquid-phase diffusion	$\Delta c_n(k+1) = \Delta c_n(k) + \frac{1}{\tau_e}[P_{conn} I(k) - \Delta c_n(k)]\Delta t, \Delta c_p(k+1) = \Delta c_p(k) + \frac{1}{\tau_e}[P_{comp} - \Delta c_p(k)]\Delta t,$ $\eta_{con}(k) = \frac{2RT(k)}{F}(1 - t_+)\ln\left[\frac{c_0 + \Delta c_p(k)}{c_0 - \Delta c_n(k)}\right]$
Reactive polarization	$\eta_{act}(k) = \frac{2RT(k)}{F}[\ln(\sqrt{m_n^2(k)+1} + m_n(k)) - \ln(\sqrt{m_p^2(k)+1} + m_p(k))]$ $m_p(k) = \frac{1}{6Q_p c_0^{0.5}}\frac{1}{(1 - y_{surf}(k))^{0.5}(y_{surf}(k))^{0.5}}P_{act}I(k), m_n = \frac{1}{6Q_n c_0^{0.5}}\frac{1}{(1 - x_{surf}(k))^{0.5}(x_{surf}(k))^{0.5}}P_{act}I(k)$
Ohmic polarization	$\eta_{ohm} = R_{ohm}I(k)$

Table 2. Meaning of related parameters.

Parameters	Physical Meaning
U_{app}	Terminal voltage (V)
E_{ocv}	Electromotive force (V)
η_{con}	Concentration polarization overpotential (V)
η_{act}	Reaction polarization overpotential (V)
η_{ohm}	Ohmic polarization overpotential (V)
y_{avg}, x_{avg}	Solid-phase average stoichiometric number of positive and negative electrodes (-)
y_{surf}, x_{surf}	Solid-phase surface stoichiometric number of positive and negative electrodes (-)
Δy	Deviations between y_{surf} and y_{avg} (-)
Δx	Deviations between x_{surf} and x_{avg} (-)
Δy_1	Intermediate variable of Δy (-)
Δx_1	Intermediate variable of Δx (-)
T	Battery internal temperature (K)
R	Ideal gas constant (J mol^{-1} K^{-1})
y_0, x_0	Initial lithium intercalation concentration fraction of positive and negative electrodes (-)
c_0	Initial lithium ion concentration in electrolyte (mol m^{-3})
Q_p, Q_n	Total capacity of positive and negative electrodes (A s)
Q_{all}	Total capacity of the battery (A s)
τ_p, τ_n	Solid diffusion time constants of positive and negative electrodes (s)
D_x, D_y	Lithium embedding rate of positive and negative electrodes (-)
P_{act}	Reaction polarization coefficient (m$^{-1.5}$ mol$^{0.5}$ s)
P_{con}	Proportional coefficient of liquid phase diffusion (mol m^{-3} A^{-1})
R_{ohm}	Ohm internal resistance (Ω)
y_{ofs}	Embedded lithium offset (-)
τ_e	Liquid phase diffusion time constant (s)

Table 3. Battery parameters.

Parameters	Value
y_0, x_0	0.7941, 0.4538
c_0	1000
Q_p, Q_n	19,859.04, 11,888.64
τ_p, τ_n	184.7533, 2.5501
D_x, D_y	0.7922, 0.4743
P_{act}	271,780
P_{con}	955.1863
R_{ohm}	0.0686
y_{ofs}	0.0708
τ_e	62.2686

2.3. *Principle of SOC Estimation Algorithm under No Current Monitoring*

The three SOC estimation algorithms utilized in this study were based on the simplified electrochemical model. However, the model function inputs did not take into account the temperature of the battery. The inputs solely consisted of the SOC and current, while the output was limited to the cell terminal voltage.

2.3.1. Incremental Seeking Method

This solution was based on the idea of enumeration to achieve current estimation and SOC prediction. The approach involves gradually increasing the current within fixed boundary conditions and then inputting the current into the model to calculate the voltage to identify the current values that meet the necessary requirements. Once the appropriate current value is determined, the SOC can be estimated using the ampere-hour integral method. The corresponding flowchart illustrating this method is depicted in Figure 2.

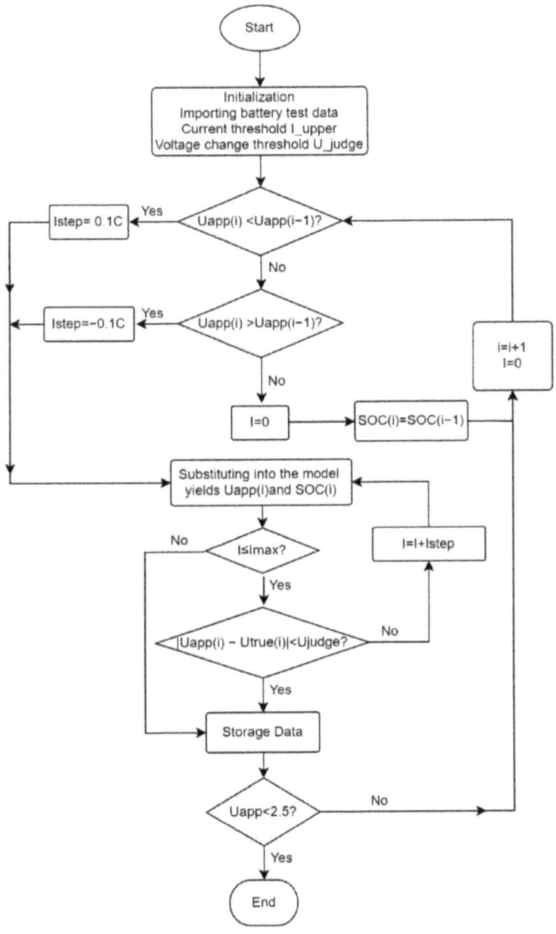

Figure 2. Flow chart of incremental seeking method.

The specific scheme was as follows: Firstly, the measured voltage was compared with the previous second's voltage to determine whether the battery was being charged or discharged. Then, the current sign was initialized, and upper and lower search boundaries were set. An initial current value of 0 was assumed. A voltage simulation was carried out using the assumed current, and the simulated voltage was compared to the measured voltage. If the error between the measured voltage and simulated voltage exceeded an acceptable range, the current value was increased based on the error until the error was within an acceptable range. The current corresponding to the simulated voltage was considered the current, and the SOC was predicted using the ampere-hour integration method.

2.3.2. Dichotomous Method

As shown in Figure 3, this scheme used the dichotomy method for current estimation and SOC prediction. The basic principle of the dichotomy method involved continuously dividing the search range into two parts within fixed boundary conditions to find the number that meets certain requirements.

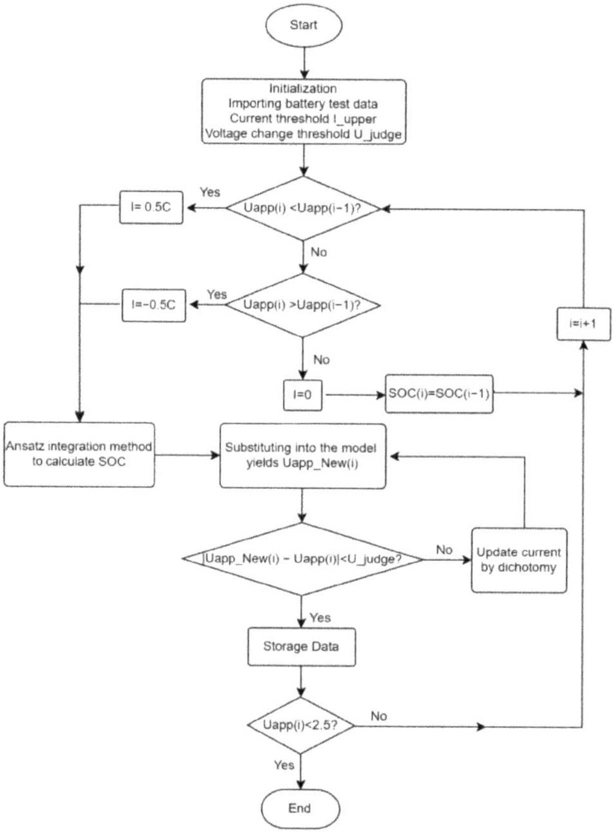

Figure 3. Dichotomous search flow chart.

The specific scheme is as follows: firstly, the measured voltage was compared with that of the previous second to determine the charging and discharging state at this time. Then, the current was initialized, and the upper and lower search boundaries were assumed. The voltage simulation was carried out based on the assumed current, and the corresponding simulation voltage was compared with the measured voltage. If the error is greater than the acceptable range, the current value is adjusted according to the error until the error between the measured voltage and the simulated voltage is in the acceptable range. At this point, the current corresponding to the simulated voltage was considered as the current at that moment. Finally, the SOC prediction was carried out by the ampere-hour integration method.

2.3.3. Extended Kalman Filter

The solution utilized a modified extended Kalman filter to achieve SOC estimation. Based on the simplified electrochemical model, the load current was taken as the unknown input, and the system state equation and the measurement equation were established. In addition, the current estimation was achieved using the modified extended Kalman filter method. Finally, the SOC estimation was completed by the ampere-hour integral method. The flow chart is shown in Figure 4.

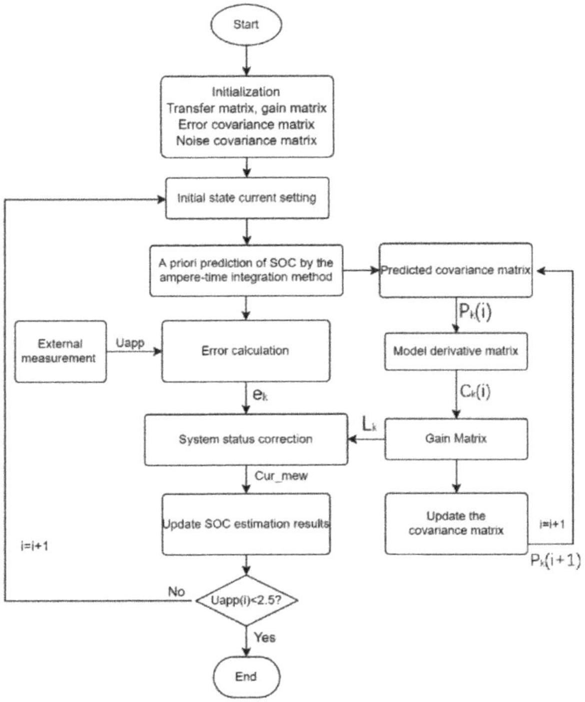

Figure 4. Extended Kalman filter estimation flow chart.

To implement this solution, an update of state variables and observation updates was required, which meant that the priori SOC estimation, succession of current values, and covariance matrix calculations were carried out. The gain matrix was then calculated, and the currents were corrected according to voltage errors. Finally, the accurate SOC estimation for a single cell was obtained by the ampere-hour integration method.

3. Battery Testing Process and Results

Different lithium cobaltate battery monoblocks from the same batch with a standard capacity of 800 mAh were selected. Charging and discharging tests were performed using Neware's battery test system to simulate the battery operating conditions at different charge and discharge multipliers. The data sampling frequency was 1 s. The specific current configurations were as follows:

The specific current configuration for the DST operating conditions test is described as follows:

(1) The lithium-ion battery was fully charged by constant current and constant voltage charging.
(2) The battery was rested in a constant temperature chamber for 1 h.
(3) The battery was discharged at a constant discharge rate of 0.25 C for 30 s, discharged at 0.5 C for 12 s, and charged at 0.25 C for 10 s.
(4) Step (3) was repeated three times.
(5) The battery was charged successively at a constant charge rate of 0.25 C for 35 s, 2 C for 10 s, and 1.25 C for 25 s.
(6) The battery was charged at a constant charge rate of 0.5 C for 10 s, discharged at 0.5 C for 30 s, and charged at 1 C for 10 s.
(7) The battery was rested for 50 s.

(8) Steps (3)–(7) were repeated until the voltage decreased to 3.7 V. Then, the battery was discharged at a constant discharge rate of 0.5 C until the voltage was 2.5 V to stop the experiment.

The HPPC condition test with the specific current configuration was conducted as follows:

(1) The lithium-ion battery was fully charged by constant current and constant voltage charging.
(2) The battery was rested in a constant temperature chamber for 1 h.
(3) The battery was discharged at 1 C for 8 min and was rested for 30 min.
(4) The battery was discharged at 3 C for 10 s and rested for 3 min.
(5) The battery was charged at 1.5 C for 10 s and rested for 2 min.
(6) Steps (3)–(5) were repeated once.
(7) The current in (4) and (5) was modified to 1.875 C. Steps (3)–(5) were repeated until the resting voltage dropped to 3.6 V.
(8) The battery was discharged at 0.9 C until the voltage was 2.5 V to stop the experiment.

4. Results and Discussion

It is worth noting that all three methods of SOC estimation without current monitoring include the ampere-hour integration method, and the estimation accuracy is greatly influenced by the initial SOC accuracy and noise. The effects of different initial SOCs and random voltage noise on the algorithm estimation results were shown and discussed as follows, respectively.

4.1. Effect of Different Initial SOC on Estimation Results

The initial SOC of the test data was set as 100%, 70%, and 90%, respectively. The estimated results were as follows.

4.1.1. Incremental Seeking Method

The estimation results using the incremental seeking method for different initial SOC states are shown in Figures 5 and 6.

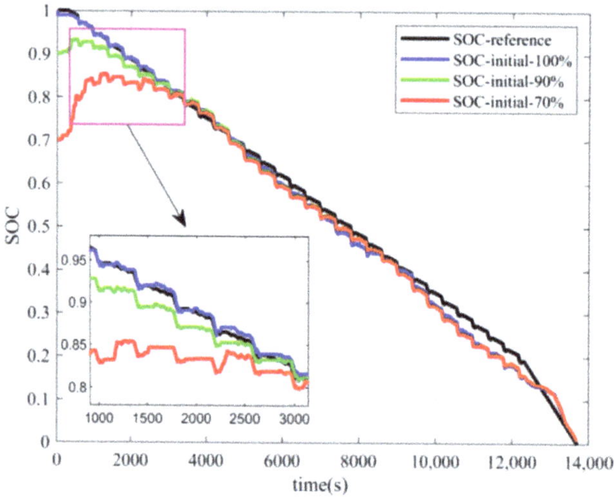

Figure 5. Estimation results of incremental seeking method under DST conditions and different initial SOC.

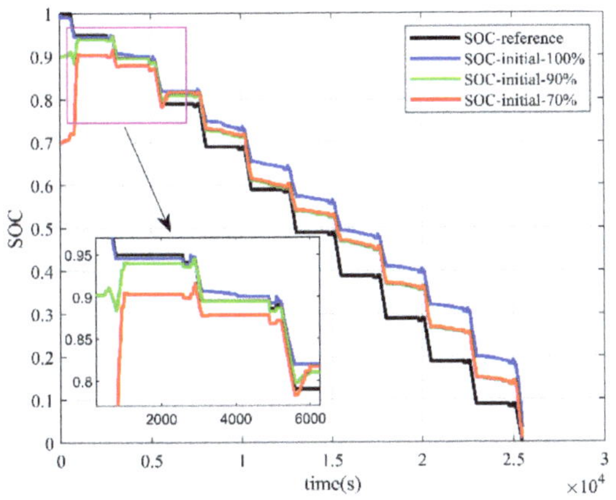

Figure 6. Estimation results of incremental seeking method under HPPC conditions and different initial SOC.

The estimated trend obtained by the incremental seeking method was consistent with the actual reference value, and the estimated results varied with the battery charge and discharge state. For example, under DST conditions, SOC decreased along a wave line until 12,500 s and then decreased along a straight line. Under HPPC conditions, SOC decreased along a step.

However, the difference between the estimation results with initial SOC values of 100%, 90%, and 70% was significant under the HPPC condition. The error was largest when the initial SOC was accurate, illustrating the instability of the algorithm. When the initial SOC state was inaccurate, the incremental seeking method converged when the SOC dropped to nearly 80%, and the convergence process was often accompanied by a large abrupt change. After convergence was completed, the initial SOC value had a large impact on the estimation results.

4.1.2. Dichotomous Method

The estimation results using the dichotomous method for different initial SOC states are shown in Figures 7 and 8.

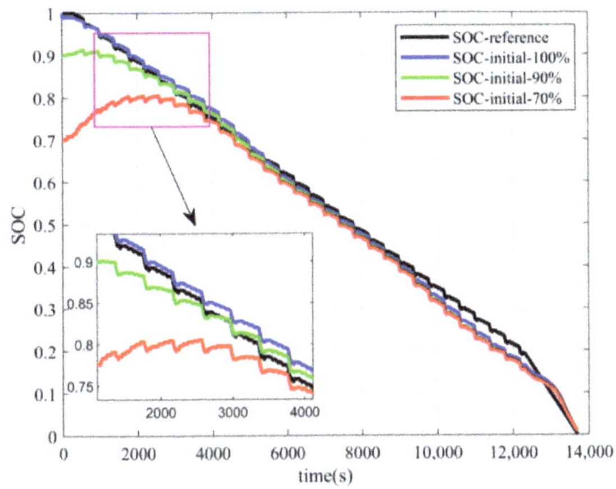

Figure 7. Estimation results of dichotomous method under DST conditions and different initial SOC.

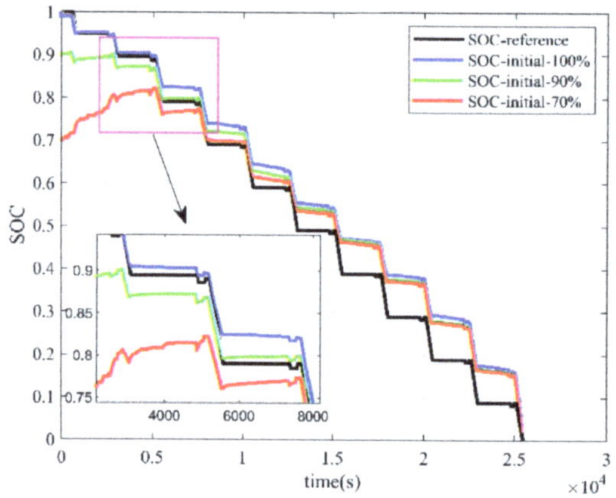

Figure 8. Estimation results of dichotomous method under HPPC conditions and different initial SOC.

When applying the dichotomy method, the overall estimation trend was consistent with the actual reference value, and the algorithmic estimation results captured the same change phenomenon as the charging and discharging state of the battery varied.

From the overall viewpoint, the dichotomous SOC estimation error was large, especially under HPPC conditions, and significant estimation bias could be seen in Figure 8. When the initial SOC state was inaccurate, if the initial SOC was 90%, it completed convergence when the SOC decreased to around 85%. In addition, if the initial SOC was 70%, it completed convergence when the SOC decreased to around 75%. The convergence process had a great degree of abrupt change, but it was less than the incremental seeking method. Inaccurate initial SOC had a small impact on the estimation results after the convergence of the algorithm.

4.1.3. EKF

The estimation results using the EKF algorithm for different initial SOC states are shown in Figures 9 and 10.

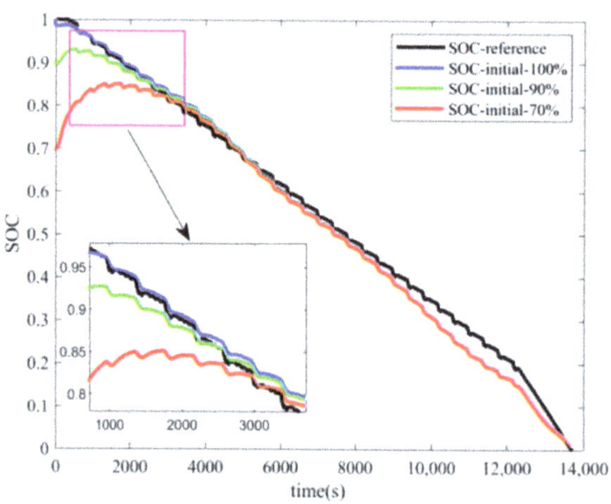

Figure 9. Estimation results of EKF algorithm for DST conditions and different initial SOC.

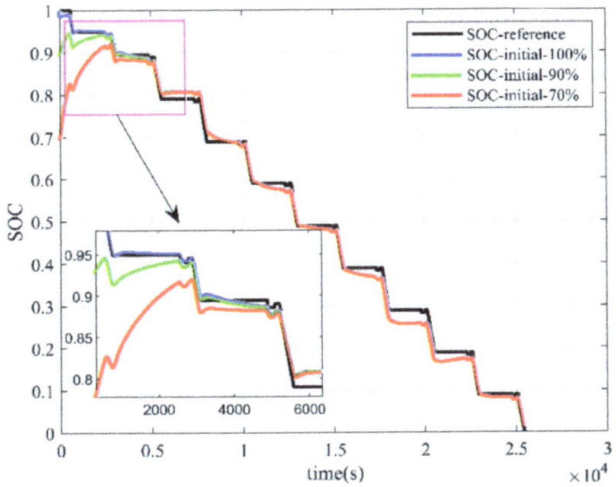

Figure 10. Estimation results of EKF algorithm for HPPC conditions and different initial SOC.

The overall estimation trend obtained using the EKF algorithm was consistent with the actual reference value, and the algorithm estimation results reflected the same variation phenomenon as the battery charge and discharge state change.

As a whole, the SOC estimation error of the EKF algorithm was small, and the average errors for the two operating conditions when the initial SOC was accurate were 1.94% and 1.13%, respectively. When the initial SOC was inaccurate, the algorithm converged quickly except for the DST condition where the initial SOC is 70%. Convergence was achieved when the SOC decreased to about 85~90%. The convergence process was smooth, and the convergence estimates were largely consistent with the reference values and almost independent of the initial SOC.

The estimation errors of the three methods for different initial SOC states are shown in Table 4.

Table 4. Estimation error at different initial SOC states.

		Accurate Initial SOC		Initial SOC 0.9		Initial SOC 0.7	
Incremental seeking method		soc	Voltage	soc	Voltage	soc	Voltage
DST	Average value	3.58%	0.042 V	3.60%	0.035 V	5.08%	0.043 V
	Maximum value	12.76%	0.300 V	11.42%	0.450 V	30%	0.659 V
HPPC	Average value	7.62%	0.025 V	7.80%	0.028 V	9.10%	0.035 V
	Maximum value	15.13%	0.373 V	15.20%	0.372 V	30%	0.669 V
Dichotomous method		soc	Voltage	soc	Voltage	soc	Voltage
DST	Average value	1.87%	0.014 V	2.58%	0.016 V	4.17%	0.084 V
	Maximum value	5.02%	0.416 V	10.00%	0.416 V	30%	0.495 V
HPPC	Average value	4.51%	0.028 V	4.43%	0.030 V	6.37%	0.044 V
	Maximum value	8.82%	0.352 V	10.00%	0.296 V	30%	0.288 V
EKF		soc	Voltage	soc	Voltage	soc	Voltage
DST	Average value	1.94%	0.034 V	2.36%	0.035 V	3.82%	0.036 V
	Maximum value	5.27%	0.414 V	10.21%	0.414 V	30.15%	0.414 V
HPPC	Average value	1.13%	0.0098 V	1.41%	0.010 V	2.33%	0.011 V
	Maximum value	3.69%	0.3872 V	10.20%	0.387 V	30.14%	0.387 V

In terms of overall SOC estimation performance, the EKF algorithm performed the best and met the actual demand. The dichotomy method was the second-best and could accurately estimate SOC when the initial state was reliable. However, the incremental seeking method had the worst performance, with the highest error, failing to meet the actual demand and requiring further improvement.

Regarding the impact of different initial SOC states on algorithm estimation, the EKF algorithm was highly resistant to interference and almost unaffected. The dichotomy method was less affected, while the incremental seeking method was the most unstable. Inaccurate initial states had a significant impact on the results of the incremental seeking method.

Concerning the convergence of algorithms under different initial SOC states, the EKF converged fast and seamlessly, while the dichotomy method converged slightly slower than the EKF. In contrast, the incremental seeking method had the slowest convergence rate, with a sudden change in the convergence process caused by the algorithm's principle.

If the initial SOC error was too large, there was a significant increase in the error for all three methods. This indicates that the error correction capability of the three methods is limited due to the lack of current input. However, it is worth noting that the initial SOC error is usually not so large in portable devices. Furthermore, the SOC error can be calibrated gradually during the rest mode.

Jing Hou et al. [24] conducted pulse discharge experiments to validate their proposed VB-UKF method. The average absolute errors of the two methods, VB-UKF and the unscented recursive three-step filter (URTSF), were 1.52% and 2.65% when the initial SOC was accurate. When the initial SOC was 80%, the errors of the two methods were 1.67% and 2.28%, respectively. When the initial SOC was 60%, the errors of the two methods were 2.16% and 2.58%, respectively. It can be seen that the estimation effect of the EKF algorithm is perfectly acceptable.

4.2. *Effect of Different Voltage Noise on Algorithm Estimation Results*

The test data was free of voltage noise, and random Gaussian noise with the mean of 0 and 3σ of 10 mV, 50 mV, and 100 mV was superimposed, respectively. The estimated results are shown below.

4.2.1. Incremental Seeking Method

The SOC estimation results under different voltage noise using the incremental seeking method are shown in Figures 11 and 12.

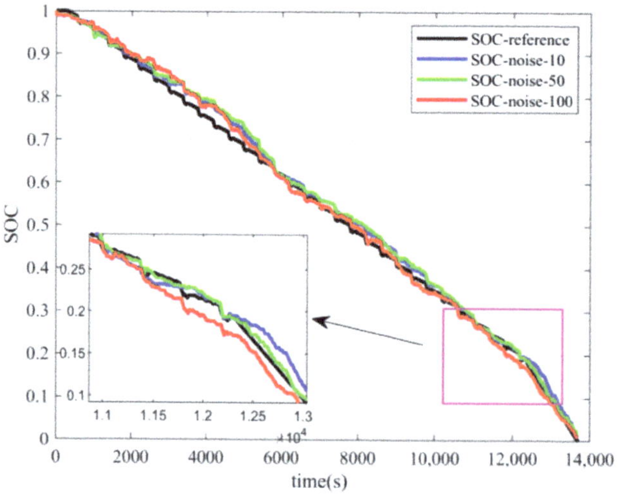

Figure 11. Estimation results of incremental seeking method under DST conditions and different voltage noise.

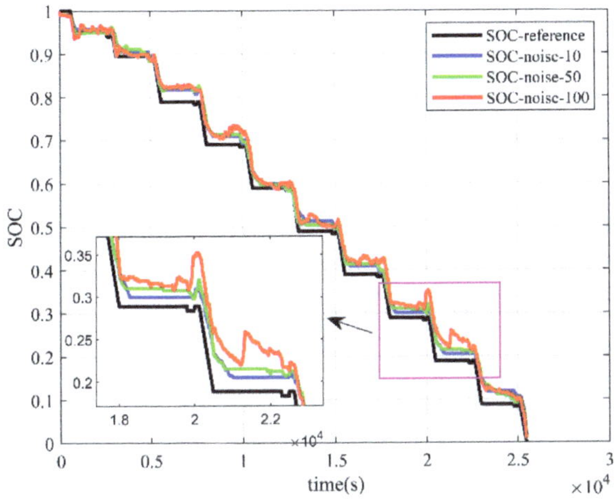

Figure 12. Estimation results of incremental seeking method under HPPC conditions and different voltage noise.

When applying the incremental seeking method under voltage noise, the estimation trend was consistent with the actual reference value, and the estimation results reflected the same changes as the battery charging and discharging states.

However, the presence of voltage noise caused more irregular fluctuations in the algorithm's estimation accuracy, resulting in large errors. This suggests that the incremental seeking method is susceptible to noise.

4.2.2. Dichotomous Method

The SOC estimation results using the dichotomous method with different voltage noises are shown in Figures 13 and 14.

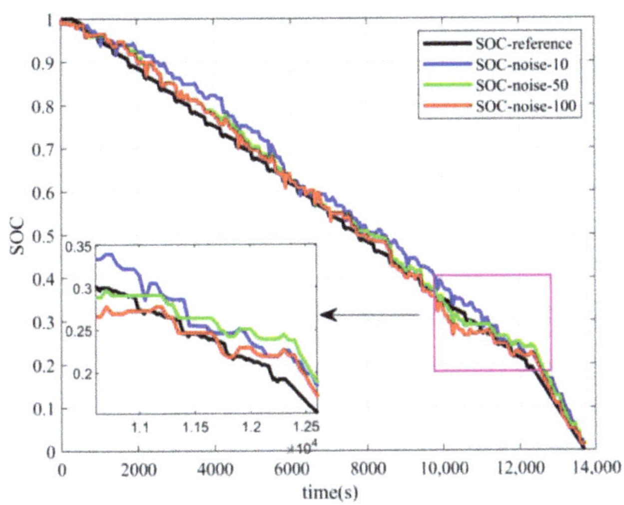

Figure 13. Estimation results of dichotomous method under DST conditions and different voltage noise.

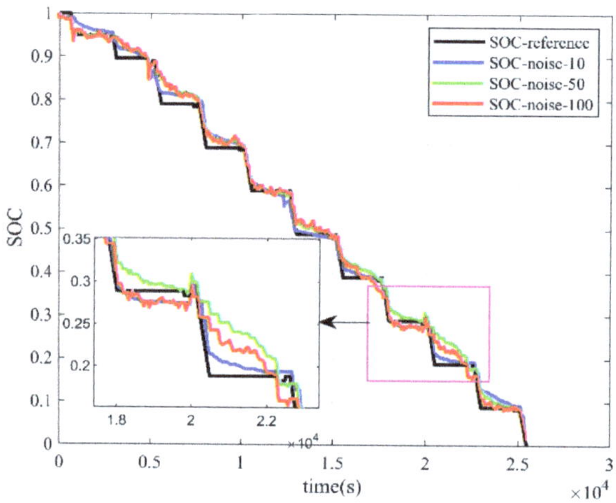

Figure 14. Estimation results of dichotomous method under HPPC conditions and different voltage noise.

When using the dichotomous method under voltage noise, the estimation trend was consistent with the actual reference value, and the estimation result reflected the same changes as the battery charging and discharging states.

The mean error of SOC estimation is small when using the dichotomous method. However, appropriate voltage noise could improve the estimation accuracy of the algorithm, while excessive noise led to significant deviations in the results under the DST condition. This shows the poor stability of the dichotomous method.

4.2.3. EKF

The SOC estimation results using the EKF algorithm with different voltage noise are shown in Figures 15 and 16.

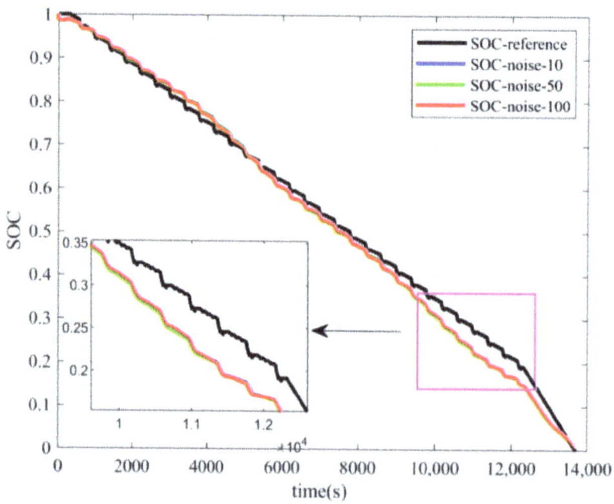

Figure 15. Estimation results of the EKF algorithm for DST conditions and different voltage noise.

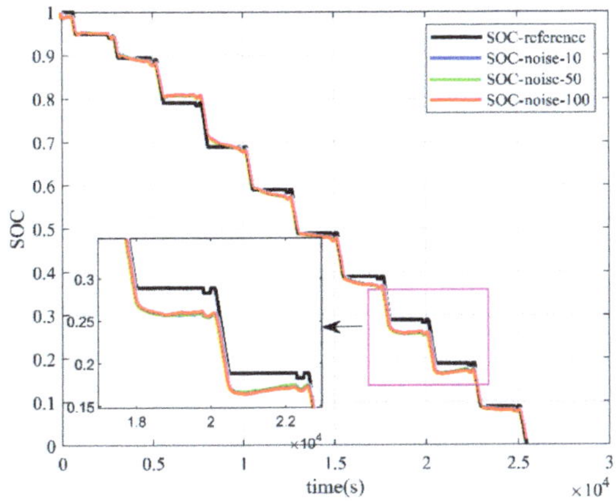

Figure 16. Estimation results of the EKF algorithm for HPPC conditions and different voltage noise.

When using the EKF method under voltage noise, the estimation trend remained consistent with the actual reference value, and the estimation results could keep the same change when the charge and discharge state of the battery changed.

Overall, the EKF algorithm demonstrated a high estimation accuracy even under voltage noise, and the SOC estimation results were consistently reliable despite the presence of different levels of voltage noise.

The estimation errors of the three methods under different voltage noise are shown in Table 5.

Table 5. Estimation error under different voltage noise.

		No Noise		10 mV Noise		50 mV Noise		100 mV Noise	
Incremental seeking method		soc	Voltage	soc	Voltage	soc	Voltage	soc	Voltage
DST	Average value	3.58%	0.042 V	5.92%	0.044 V	5.81%	0.051 V	5.01%	0.060 V
	Maximum value	12.76%	0.300 V	11.99%	0.422 V	11.31%	0.342 V	11.14%	0.418 V
HPPC	Average value	7.62%	0.025 V	6.63%	0.026 V	6.63%	0.033 V	7.27%	0.056 V
	Maximum value	15.13%	0.373 V	14.78%	0.411 V	14.86%	0.374 V	15.58%	0.406 V
Dichotomous method		soc	Voltage	soc	Voltage	soc	Voltage	soc	Voltage
DST	Average value	1.87%	0.014 V	3.66%	0.077 V	1.83%	0.074 V	1.56%	0.082 V
	Maximum value	5.02%	0.416 V	7.34%	1.020 V	5.09%	0.685 V	7.19%	0.732 V
HPPC	Average value	4.51%	0.028 V	1.38%	0.024 V	1.54%	0.040 V	1.58%	0.057 V
	Maximum value	8.82%	0.352 V	5.74%	0.937 V	7.18%	0.520 V	6.96%	0.539 V
EKF		soc	Voltage	soc	Voltage	soc	Voltage	soc	Voltage
DST	Average value	1.94%	0.034 V	1.94%	0.034 V	1.96%	0.037 V	1.90%	0.045 V
	Maximum value	5.27%	0.414 V	5.30%	0.420 V	5.42%	0.448 V	4.98%	0.415 V
HPPC	Average value	1.13%	0.0098 V	1.13%	0.011 V	1.11%	0.020 V	1.11%	0.032 V
	Maximum value	3.69%	0.3872 V	3.67%	0.384 V	3.69%	0.375 V	3.76%	0.359 V

From the experimental results, we can see that the error of the EKF algorithm under different noises was within 2%, which would meet the practical requirements. The dichotomous method was the second-best and had a certain ability to fight against noise. However, the incremental seeking method exhibited the poorest performance, with the highest error, failing to meet practical requirements and requiring further improvements.

With regard to the impact of voltage noise on the algorithm's estimation performance, the EKF algorithm effectively filtered out the noise and was almost unaffected. In contrast, both the incremental seeking and dichotomous methods were more susceptible to voltage noise, resulting in lower stability and accuracy.

4.3. Algorithm Computational Efficiency

Because the process of completing the overall battery SOC estimation was long, the computational efficiency of the three algorithms was considered by comparing the time required to perform a single SOC estimation among them.

The algorithms were run in the following environment: under PC conditions using processor: Intel (R) Core (TM) i5-10200H CPU @ 2.40 GHz, RAM configuration:16.00 GB, and MATLAB software.

The time consumption of all algorithms is shown in Table 6. The incremental pathfinding method took 0.024 s and 0.025 s to complete a single SOC estimation without a current sensor under the two operating conditions. The dichotomous method took 0.018 s and 0.02 s. The EKF algorithm took 0.012 s and 0.007 s. The ratio of the calculated efficiencies of the three methods at DST conditions is about 2:3:4.

Table 6. Algorithm operation schedule.

Work Conditions \ Algorithm		Incremental Seeking Method	Dichotomous Method	EKF
DST	Total time	331 s	250 s	171 s
	Average data estimation time per frame	0.024 s	0.018 s	0.012 s
HPPC	Total time	636 s	498 s	191.0 s
	Average data estimation time per frame	0.025 s	0.020 s	0.007 s

4.4. Repeatability Verification

In order to avoid the chance of single battery data, more test data of LCO lithium-ion batteries were used to verify the reliability of the algorithm. The HPPC operation data of more batches of batteries were used to verify the algorithm. The average error is shown in Table 7.

Table 7. Estimation error at different initial SOC states in HPPC conditions.

	Accurate Initial SOC		Initial SOC 0.9		Initial SOC 0.7	
Incremental seeking method	soc	Voltage	soc	Voltage	soc	Voltage
Average value	7.59%	0.026 V	7.74%	0.030 V	9.16%	0.041 V
Maximum value	15.72%	0.336 V	15.21%	0.339 V	30%	0.697 V
Dichotomous method	soc	Voltage	soc	Voltage	soc	Voltage
Average value	3.23%	0.026 V	3.41%	0.029 V	5.42%	0.037 V
Maximum value	7.73%	0.336 V	10.00%	0.287 V	30%	0.276 V
EKF	soc	Voltage	soc	Voltage	soc	Voltage
Average value	3.06%	0.010 V	3.67%	0.010 V	4.73%	0.011 V
Maximum value	9.07%	0.308 V	11.22%	0.4003 V	30.16%	0.4003 V

In order to ensure the reliability of the algorithm, the simulation results of constant current charge-discharge test data of the LFP lithium-ion battery are also presented. When the initial SOC is accurate, the average error results of the EKF algorithm are shown in Table 8.

Table 8. Average error of EKF algorithm for eight batteries.

Battery Number	A	B	C	D	E	F	G	H
SOC	1.99%	1.67%	1.08%	1.76%	1.73%	1.52%	1.46%	1.56%
Voltage	0.0185 V	0.0168 V	0.0203 V	0.0189 V	0.0177 V	0.173 V	0.0180 V	0.0180 V

The SOC estimation results of the battery numbered A are shown in Figure 17.

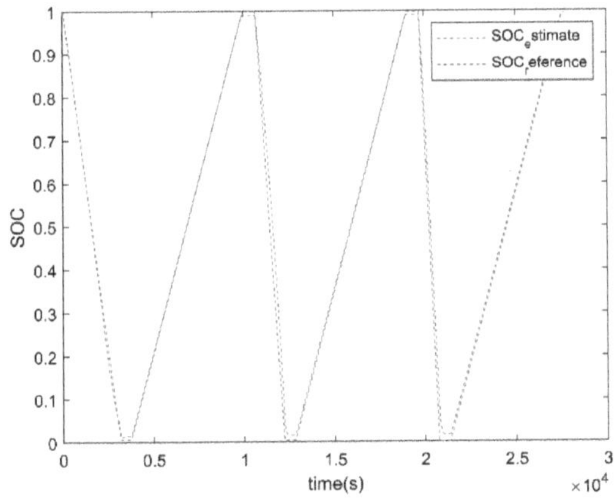

Figure 17. Estimation results of the EKF algorithm for HPPC conditions.

When using different battery test data, the EKF algorithm can still maintain a good estimation effect.

5. Conclusions

It is important to obtain current information for accurate SOC estimation of lithium-ion batteries. However, due to concerns about cost, size, and power consumption, current sensors are sometimes not equipped in portable devices.

In this work, three methods were developed for lithium battery SOC estimation that provide new solutions for SOC estimation for small portable devices in the absence of current monitoring and fill the gap in existing studies. The dichotomous and incremental seeking methods were found to be sensitive to initial SOC accuracy, and estimation errors exceeded the allowed range. For instance, using the dichotomous method led to a 6.37% error at an initial SOC of 70% under HPPC conditions. The incremental seeking method was also more susceptible to voltage noise than the other two methods, with a 10 mV noise causing an estimation error of 5.92% in DST conditions.

The dichotomous method and the extended Kalman filter method demonstrated superior accuracy in the presence of certain measurement noises, with estimation errors under 100 mV noise controlled within ±2%. The extended Kalman filter algorithm was particularly effective in filtering out the noise, with error variation following noise addition not exceeding 0.04%. Additionally, the extended Kalman filter method outperformed the other two methods in terms of estimation speed, with an average time of 0.01 s per frame of data estimated.

In addition, it is worth noting that the estimation error of SOC will be larger when the initial SOC error is large due to the absence of current input. However, measures can be taken to avoid excessive errors. For example, the battery is left for a period of time, and the battery is calibrated by OCV after it is fully charged. In that case, this weakness does not affect the usefulness and sophistication of the algorithms used.

Considering the aging process of the battery, the battery model parameters will change, leading to an increase in SOC estimation error. In order to solve this problem, a possible approach is to use an improved extended Kalman filter (IEKF) algorithm [32] or a model adaptive EKF (MAEKF) [33] algorithm. The general idea is to select the parameters of the electrical model with high sensitivity. In view of the fact that the battery voltage derivative changes abruptly twice with time when the discharge current is constant, the parameters are updated using IEKF or MAEKF. Thus, the excellent accuracy of the model is maintained during the aging process of the battery.

Author Contributions: Conceptualization, J.L.; methodology, J.L.; software, J.L.; validation, Z.Z.; formal analysis, Z.Z.; investigation, Z.Z.; resources, J.L.; data curation, Z.Z.; writing—original draft preparation, Z.Z.; writing—review and editing, J.S., J.L., Y.W.; visualization, Z.Z.; supervision, Z.W.; project administration, J.L. All authors have read and agreed to the published version of the manuscript.

Funding: This research received no external funding.

Data Availability Statement: No new data were created or analyzed in this study. Data sharing is not applicable to this article.

Conflicts of Interest: The authors declare no conflict of interest.

References

1. Kulova, T.L.; Fateev, V.N.; Seregina, E.A.; Grigoriev, A.S. A Brief Review of Post-Lithium-Ion Batteries. *Int. J. Electrochem. Sci.* **2020**, *15*, 7242–7259. [CrossRef]
2. Garcia Elvira, D.; Machado, R.; Plett, G.L.; Trimboli, M.S.; Valderrama Blavi, H.; Cid Pastor, A.; Martinez Salamero, L. Simplified Li Ion Cell Model for BMS Coupling an Equivalent Circuit Dynamic Model with a Zero Dimensional Physics Based SEI Model. *J. Electrochem. Soc.* **2021**, *168*, 110526. [CrossRef]
3. Miron-Alexe, V. Portable lithium-ion battery ups with bms function for raspberry pi and other iot embedded systems. *J. Sci. Arts* **2022**, *22*, 763–780. [CrossRef]
4. Wang, C.; Liu, Z.; Sun, Y.; Gao, Y.; Yan, P. Aging Behavior of Lithium Titanate Battery under High-Rate Discharging Cycle. *Energies* **2021**, *14*, 5482. [CrossRef]
5. Wu, L.; Liu, K.; Pang, H.; Jin, J. Online SOC Estimation Based on Simplified Electrochemical Model for Lithium-Ion Batteries Considering Current Bias. *Energies* **2021**, *14*, 5482. [CrossRef]
6. Li, J.; Zhao, M.; Dai, C.; Wang, Z.; Pecht, M. A mathematical method for open-circuit potential curve acquisition for lithium-ion batteries. *J. Electroanal. Chem.* **2021**, *895*, 115488. [CrossRef]
7. Orazem, M.E. Electrochemical impedance spectroscopy: The journey to physical understanding. *J. Solid State Electrochem.* **2020**, *24*, 2151–2153. [CrossRef]
8. Xing, L.; Zhan, M.; Guo, M.; Ling, L. Parameter identification and SOC estimation for power battery based on multi-timescale double Kalman filter algorithm. *Int. J. Comput. Sci. Eng.* **2022**, *25*, 619–628. [CrossRef]
9. Fang, C.; Jin, Z.; Wu, J.; Liu, C. Estimation of Lithium-Ion Battery SOC Model Based on AGA-FOUKF Algorithm. *Front. Energy Res.* **2021**, *9*, 769818. [CrossRef]
10. Huang, Z.; Fang, Y.; Xu, J. Soc estimation of li-ion battery based on improved ekf algorithm. *Int. J. Automot. Technol.* **2021**, *22*, 335–340. [CrossRef]
11. Xing, L.; Wu, X.; Ling, L.; Lu, L.; Qi, L. Lithium Battery SOC Estimation Based on Multi-Innovation Unscented and Fractional Order Square Root Cubature Kalman Filter. *Appl. Sci.* **2022**, *12*, 9524. [CrossRef]
12. Chen, L.; Wu, X.; Tenreiro Machado, J.A.; Lopes, A.M.; Li, P.; Dong, X. State-of-Charge Estimation of Lithium-Ion Batteries Based on Fractional-Order Square-Root Unscented Kalman Filter. *Fractal Fract.* **2022**, *6*, 52. [CrossRef]
13. Wang, Q.; Ye, M.; Wei, M.; Lian, G.; Li, Y. Deep convolutional neural network based closed-loop SOC estimation for lithium-ion batteries in hierarchical scenarios. *Energy* **2023**, *263*, 125718. [CrossRef]
14. Cui, Z.; Kang, L.; Li, L.; Wang, L.; Wang, K. A hybrid neural network model with improved input for state of charge estimation of lithium-ion battery at low temperatures. *Renew. Energy* **2022**, *198*, 1328–1340. [CrossRef]
15. Chen, N.; Zhao, X.; Chen, J.; Xu, X.; Zhang, P.; Gui, W. Design of a Non-Linear Observer for SOC of Lithium-Ion Battery Based on Neural Network. *Energies* **2022**, *15*, 3835. [CrossRef]
16. Zhou, Y.; Wang, S.; Xie, Y.; Zhu, T.; Fernandez, C. An improved particle swarm optimization-least squares support vector machine-unscented Kalman filtering algorithm on SOC estimation of lithium-ion battery. *Int. J. Green Energy* **2023**. [CrossRef]
17. Manoharan, A.; Begam, K.M.; Aparow, V.R.; Sooriamoorthy, D. Artificial Neural Networks, Gradient Boosting and Support Vector Machines for electric vehicle battery state estimation: A review. *J. Energy Storage* **2022**, *55*, 105384. [CrossRef]
18. Wu, X.; Mi, L.; Tan, W.; Qin, J.; Zhao, M. State of Charge (SOC) Estimation of Ni-MH Battery Based on Least Square Support Vector Machines. In Proceedings of the International Conference on Mechatronics and Intelligent Materials, Lijiang, China, 21–22 May 2011; pp. 1204–1209.

19. Xuan, L.; Qian, L.; Chen, J.; Bai, X.; Wu, B. State-of-Charge Prediction of Battery Management System Based on Principal Component Analysis and Improved Support Vector Machine for Regression. *IEEE Access* **2020**, *8*, 164693–164704. [CrossRef]
20. Xu, X.; Yue, X.; Chen, Y.; Liang, Z. Li Plating Regulation on Fast-Charging Graphite Anodes by a Triglyme-$LiNO_3$ Synergistic Electrolyte Additive. *Angew. Chem. Int. Ed.* **2023**, *62*, e202306963. [CrossRef] [PubMed]
21. He, J.; Meng, S.; Yan, F. A Comparative Study of SOC Estimation Based on Equivalent Circuit Models. *Front. Energy Res.* **2022**, *10*, 914291. [CrossRef]
22. Ruan, H.; Sun, B.; Zhang, W.; Su, X.; He, X. Quantitative Analysis of Performance Decrease and Fast-Charging Limitation for Lithium-Ion Batteries at Low Temperature Based on the Electrochemical Model. *IEEE Trans. Intell. Transp. Syst.* **2021**, *22*, 640–650. [CrossRef]
23. Wu, S.; Pan, W.; Zhu, M. A Collaborative Estimation Scheme for Lithium-Ion Battery State of Charge and State of Health Based on Electrochemical Model. *J. Electrochem. Soc.* **2022**, *169*, 090516. [CrossRef]
24. Wang, J.; Zhang, L.; Xu, D.; Zhang, P.; Zhang, G. A Simplified Fractional Order Equivalent Circuit Model and Adaptive Online Parameter Identification Method for Lithium-Ion Batteries. *Math. Probl. Eng.* **2019**, *2019*, 6019236. [CrossRef]
25. Cambron, D.C.; Cramer, A.M. A Lithium-Ion Battery Current Estimation Technique Using an Unknown Input Observer. *IEEE Trans. Veh. Technol.* **2017**, *66*, 6707–6714. [CrossRef]
26. Putra, W.S.; Dewangga, B.R.; Cahyadi, A.; Wahyunggoro, O.; IEEE. Current Estimation Using Thevenin Battery Model. In Proceedings of the Joint International Conference On Electric Vehicular Technology and Industrial, Mechanical, Electrical, and Chemical Engineering (ICEVT & IMECE) 2015, Surakarta, Indonesia, 4–5 November 2015; pp. 5–9.
27. Chun, C.Y.; Baek, J.; Seo, G.S.; Cho, B.H.; Kim, J.; Chang, I.K.; Lee, S. Current sensor-less state-of-charge estimation algorithm for lithium-ion batteries utilizing filtered terminal voltage. *J. Power Sources* **2015**, *273*, 255–263. [CrossRef]
28. Hou, J.; Yang, Y.; Gao, T. A Variational Bayes Based State-of-Charge Estimation for Lithium-Ion Batteries without Sensing Current. *IEEE Access* **2021**, *9*, 84651–84665. [CrossRef]
29. Pei, M.; Shi, H.; Yao, F.; Liang, S.; Xu, Z.; Pei, X.; Wang, S.; Hu, Y. 3D printing of advanced lithium batteries: A designing strategy of electrode/electrolyte architectures. *J. Mater. Chem. A* **2021**, *9*, 25237–25257. [CrossRef]
30. Li, J.; Wang, L.; Lyu, C.; Liu, E.; Xing, Y.; Pecht, M. A parameter estimation method for a simplified electrochemical model for Li-ion batteries. *Electrochim. Acta* **2018**, *275*, 50–58. [CrossRef]
31. Liu, Y.; Tang, S.; Li, L.; Liu, F.; Jiang, L.; Jia, M.; Ai, Y.; Yao, C.; Gu, H. Simulation and parameter identification based on electrochemical-thermal coupling model of power lithium ion-battery. *J. Alloys Compd.* **2020**, *844*, 156003. [CrossRef]
32. Sepasi, S.; Ghorbani, R.; Liaw, B.Y. Improved extended Kalman filter for state of charge estimation of battery pack. *J. Power Sources* **2014**, *255*, 368–376. [CrossRef]
33. Sepasi, S.; Ghorbani, R.; Liaw, B.Y. A novel on-board state-of-charge estimation method for aged Li-ion batteries based on model adaptive extended Kalman filter. *J. Power Sources* **2014**, *245*, 337–344. [CrossRef]

batteries

MDPI

Article

Multi-Layer TiO_{2-x}-PEDOT-Decorated Industrial Fe_2O_3 Composites as Anode Materials for Cycle-Performance-Enhanced Lithium-Ion Batteries

Yangzhou Ma [1,2], Qi Li [1], Haoduo Li [1], Zhenfei Cai [1], Shuai Wang [1], Li Zhang [1], Jian Li [1], Guangsheng Song [1,*], Youlong Xu [2,*] and Tingfeng Yi [3]

1 Key Laboratory of Green Fabrication and Surface Technology of Advanced Metal Materials, Ministry of Education, School of Materials Science and Engineering, Anhui University of Technology, Maanshan 243002, China; yangzhou.ma@outlook.com (Y.M.); liqi855@163.com (Q.L.); m18691685169@163.com (H.L.); zhenfei_cai@163.com (Z.C.); ws_ahut@163.com (S.W.); zhangli1@etcbattery.com (L.Z.); jianli13625593904@outlook.com (J.L.)

2 Electronic Materials Research Laboratory, Key Laboratory of the Ministry of Education, Xi'an Jiaotong University, Xi'an 710049, China

3 School of Resources and Materials, Northeastern University at Qinhuangdao, Qinhuangdao 066004, China; yitingfeng@neuq.edu.cn

* Correspondence: song_ahut@163.com (G.S.); ylxu@mail.xjtu.edu.cn (Y.X.)

Abstract: An industrial submicron-sized Fe_2O_3 with no special shape was decorated by a multi-layer coating of oxygen-deficient TiO_{2-x} and conducting polymer PEDOT (poly 3,4-ethylenedioxythiophene). A facile sol–gel method followed by an EDOT polymerization process was adopted to synthesize the hierarchical coating composite. The microstructure and phase composition were characterized using an X-ray diffraction analysis (XRD), scanning electron microscopy (SEM), and transmission electron microscopy (TEM). In particular, the existence state of PEDOT was determined using Fourier transform infrared (FT-IR) and a thermogravimetric (TG) analysis. The characterization results indicated the dual phase was well-coated on the Fe_2O_3 and its thickness was nano scale. Electrochemical characterization indicated that the multi-layer coating was helpful for significantly enhancing the cycle stability of the Fe_2O_3, and its electrochemical performance was even better than that of the single-layer coating samples. The synergistic effects of the ceramic phase and conducting polymer were demonstrated to be useful for improving electrochemical properties. The obtained FTP-24 sample exhibited a specific discharge capacity of 588.9 mAh/g after 360 cycles at a current density of 100 mA/g, which effectively improved the intrinsic cycling performance of the Fe_2O_3, with a corresponding discharge capacity of 50 mAh/g after 30 cycles.

Keywords: Fe_2O_3; TiO_{2-x}-PEDOT; multi-layer coating; anode material; high stability; Li-ion batteries

1. Introduction

Rapid development in the field of energy storage has posed higher demands on new materials, leading to the development of multiple energy storage devices and materials [1–5], such as lithium-, sodium-, and potassium-ion batteries [6–9]. Lithium-ion batteries have been widely applied on power grids, electric vehicles, and portable devices in recent years, which is due to their high energy density, cycle stability, and environmentally friendly features compared to some traditional energy storage devices [10–13]. To meet the demand of the high energy storage of Li-ion batteries, a large number of anode materials, such as silicon [14,15] and transition metal oxides [16–18] etc., have been developed to substitute graphite. Iron oxide (Fe_2O_3) has been demonstrated to be one of the most promising candidates due to its low cost, considerable specific capacity (~1000 mAh/g), and environmental friendliness [19–22]. However, the anode faces the problem of rapid capacity decay and a low Coulombic efficiency, which is due to its large volume expansion during cycling, particle agglomeration, and the inherently low electrical conductivity of the metal oxide.

Coating is the most direct and effective strategy for overcoming these problems [23–26]. Buffered and high-ionic/electronic-conductivity materials can be introduced on the surfaces of Fe_2O_3 particles to prolong their cycling life, inhibit particle agglomeration, and ensure fast Li-ion diffusion [27–30]. Carbon, metal oxides, and conductive polymers, as the most frequent coating materials, have been introduced to Fe_2O_3-based composites. All coating materials should be modified for better application as anode materials, for example, Fe_2O_3 nanoparticles anchored on N-doped carbon remained at 496 mAh g^{-1} after 1000 cycles at 500 mA g^{-1}, which effectively suppressed the electrode pulverization [31]. Multi-layer, yolk-shell structured Fe_2O_3@carbon was designed to prolong the cycle life, which possessed a high charge capacity of 800.6 mAh g^{-1} after 300 cycles at 4 A g^{-1}. Meanwhile, the phase and structural evolution was uncovered by in situ TEM measurements, and the improved structural properties of multiple layers contributed to the enhancement of battery performance [32]. Oxides are more likely to be used as coating materials due to their better Li-ion conductivity. TiO_2 has attracted much attention among various oxide candidates due to it having a higher ionic conductivity than amorphous carbon and an excellent electrochemical stability. The Fe_2O_3@TiO_2 core–shell structure has been demonstrated to be beneficial to electrochemical performance. Zeng et al. reported core–shell nanospheres with a high reversible specific capacity of 497.3 mAh g^{-1} up to 100 cycles [33]. We introduced oxygen defects into a TiO_2 coating layer to further improve its conductivity, and obtained TiO_2-coated α-Fe_2O_3 composites with a high cycling performance in our previous work [34]. Some flexible materials were adopted to be added into the Fe_2O_3-TiO_2 composites to inhibit the pulverization issue in the charge/discharge process. Fu et al. chose CNTs as the core supporting material, and coated with Fe_2O_3-TiO_2. The composites remained at a specific capacity of 770 mAh g^{-1} after 200 cycles under a current density of 200 mA g^{-1} [35]. Conductive polymers, such as polypyrrole (PPY) [36,37], polyaniline (PANI), and poly 3,4-ethylenedioxythiophene (PEDOT) [38,39] etc., were applied in energy storage devices for a long time for their long cycling performance. Jeong et al. [40] designed a hierarchical hollow spheres structure with PANI as a coating layer, and the composites exhibited an excellent rate performance and cycling stability during 100 cycles, remaining at 732 mAh g^{-1}. The electrical conductivity of the entire electrodes was increased significantly by the introduction of conductive polymers.

PEDOT possesses a better stability of ion-doped states compared to other conductive polymers [41]. It has been demonstrated that the incorporation of PEDOT can enhance the electrochemical performance of Fe_2O_3 [42]. Based on our previous work of oxygen-deficient TiO_2-coated Fe_2O_3 composites, we adopted a PEDOT coating on the composites to further improve their electrical conductivity. The hierarchical structure was helpful for improving the α-Fe_2O_3 anode cycling stability.

2. Materials and Methods

2.1. Synthesis of Fe_2O_3/TiO_{2-x}/PEDOT Composites

The composites were synthesized via a simple polymerization method for Fe_2O_3/TiO_{2-x} composites reported in our previous study [34]. A certain amount of precursor particles, Fe_2O_3/TiO_2 with oxygen defects, were stirred and suspended in deionized water. The thickness of the PEDOT coated on the particles was regulated by the polymerization time. The EDOT monomer at 100 μL was dropped in the suspension liquid and then 0.1 g of camphorsulfonic acid (HCSA) was introduced into the mixture as a dopant. The stirring times were set as 8 h, 24 h, 40 h, and 56 h after 0.5 g of ammonium persulfate was added into the liquid. The final products were obtained with the washing and drying process, labeled as FTP-8, FTP-24, FTP-40, and FTP-56. The full synthesis process is briefly illustrated in Figure 1.

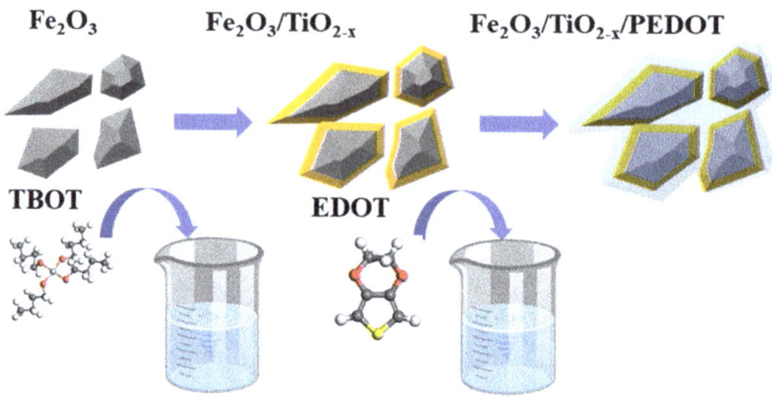

Figure 1. Schematic illustration of the preparation process of Fe_2O_3/TiO_{2-x}/PEDOT anode materials.

2.2. Characterization

The compounds' surface morphologies and distributions of elements were examined using a combination of FESEM (Field Emission Scanning Electron Microscopy; NANO SEM 430) and EDS (Energy Dispersive X-ray Spectroscopy). The composites' coating layers and lattice structures were characterized using TEM (Transmission Electron Microscopy; JEM-2100, JOEL). The powder's structural information was detected using an XRD (X-ray Diffraction) analyzer (Bruke D8), utilizing Cu-Kα radiation (λ = 1.5406 Å) within the 2θ range of 10−80°. The compounds' compositions and oxidation states were determined through XPS (X-ray Photoelectron Spectroscopy, Thermo K-Alpha+, Al Kα 1486.6 eV). The characteristic peaks of PEDOT were identified using FT-IR (Fourier Transform Infrared Spectrum) performed on a Nicolet 6700 spectrophotometer. The content of PEDOT in the mixture was quantified via a TG (Thermogravimetric) analysis conducted under air, ranging from room temperature to 900 °C, employing a DTG-60H model from Shimadzu, Kyoto, Japan.

2.3. Electrochemical Measurement

The working electrodes consisted of 80 wt% active materials, 10 wt% acetylene black, and 10 wt% CMC + citric acid mixture in deionized water, ensuring a homogeneous slurry. This slurry was then evenly applied onto a copper foil and dried in a vacuum oven at 60 °C for 24 h. Afterward, the coated electrodes were cut into disks with a diameter of 14 mm. The average mass loading of the active material in each electrode was 1.6 mg. The specific capacity calculation was based on the total mass of all the active materials present on the electrodes. To assemble the CR2016-type half-cell, a thin Li plate was utilized as the anode electrode, while Celgard 2400 served as the separator. The entire assembly process was carried out in an argon glove box (MBRAUN Unilab), with the water and oxygen levels being maintained below 0.1 ppm. Galvanostatic charge/discharge characterization was performed at room temperature using a Neware battery test system (Neware Technology Co., Ltd., Shenzhen, China). The cut-off voltage was set at 0.01−2.5 V. In order to assess the coating effect, electrochemical impedance spectroscopy (EIS) was conducted prior to the cycling test. This analysis evaluated the electronic resistance and ion diffusion rate within a frequency range from 10 mHz to 10 MHz, with a 10 mV bias. The measurements were conducted using a CS310H electrochemical workstation system (Corrtest, Co. Ltd., Wuhan, China). Furthermore, cyclic voltammetry (CV) measurements were performed with a scan rate of 0.1 $mV \cdot s^{-1}$ between 0.01 and 3.0 V in order to gather additional information.

3. Results

3.1. Structural and Physicochemical Analysis of the FTP Powders

The XRD results of the Fe_2O_3/TiO_{2-x}/PEDOT composites with various EDOT polymerization durations are shown in Figure 2a. For each sample, most of the high-intensity diffraction peaks indicated a Hematite α-Fe_2O_3 phase (ICDD#01-073-2234), which meant that the main component of the composites was the commercial Fe_2O_3. It was noted that some tiny peaks appeared at 30.2° and 57.3°, manifesting a Maghemite γ-Fe_2O_3 phase (ICDD#00-039-1346). This phase, which benefited the cycling performance, was formed along with the reduction process of the TiO_2 coating. The characteristic peaks of TiO_2 and PEDOT cannot be detected clearly in the curves; thus, the TIO_2 and PEDOT coating layer may exist in the composites as an amorphous phase, and the PEDOT-coating process had no effect on the phase structure of the Fe_2O_3/TiO_2 composites.

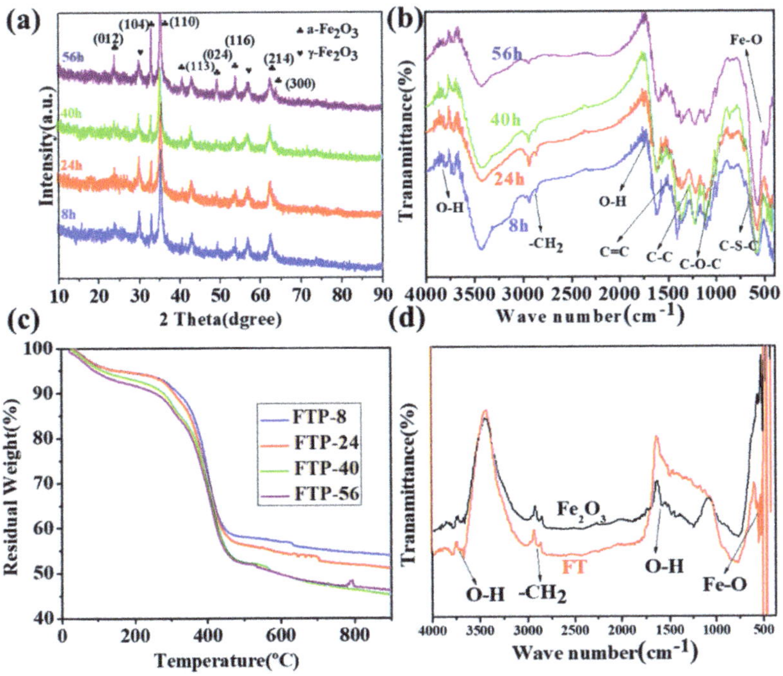

Figure 2. (**a**) XRD patterns, (**b**) FT-IR results, and (**c**) TG curves of Fe_2O_3/TiO_{2-x}/PEDOT composites with multiple polymerization times, and (**d**) FT-IR results of Fe_2O_3 and Fe_2O_3/TiO_{2-x} as control groups.

In order to confirm the existence of PEDOT in the final composites, FT-IR was performed for all the composites with multiple polymerization durations. It can be seen that the characteristic peaks of Fe_2O_3 were shown at 470 cm^{-1} (Fe-O), 2871 cm^{-1} (C-H), 3400 cm^{-1}, and 1640 cm^{-1} (O-H). Meanwhile, the characteristic peaks corresponding to PEDOT were detected at 1315 cm^{-1} (C-C), 1515 cm^{-1} (C=C), 1048 cm^{-1} (C-O-C), and 674 cm^{-1} (C-S-C). It can be demonstrated that the EDOT was well polymerized and deposited in the composites. In addition, the FTP-56 and FTP-40 samples had higher-intensity peaks at 674 cm^{-1} due to longer polymerization durations, leading to more polymer being produced. Further evidence to verify the existence of PEDOT is the FT-IR results of the pure Fe_2O_3 and Fe_2O_3 coated with TiO_2, which are illustrated in Figure 2d. The characteristic peaks corresponding to PEDOT cannot be found in the spectrum; instead, some detected peaks only corresponded to the Fe_2O_3. Thus, the PEDOT was well-added into the composites.

Figure 2c shows the TG results of the Fe_2O_3/TiO_{2-x}/PEDOT composites. The weight loss below 100 °C was due to the moisture and some oligomer in the composites. When the temperature was increased to above 350 °C, the PEDOT started combustion and the weight loss became quite significant. The weight residuals of FTP-8 to FTP-56 were 53.2%, 50.3%, 44.2%, and 45.3%, respectively. The thickness of the polymer coating layer had a positive correlation with the polymerization duration. As the polymerization duration increased, the polymerization rate slowed down. When the duration exceeded 40 h, the EDOT in the solution was used up. The content of PEDOT in FTP-40 and FTP-56 was almost the same, which led to similar electrochemical performances.

The morphology and structure information is illustrated in Figure 3. The SEM image of raw commercial Fe_2O_3 is shown in Figure 3a. The mean particle size was 2–4 μm and possessed an irregular shape. The TiO_{2-x}-coating process had slight effect on the particle size in our previous investigation [34]. However, with the PEDOT layer added into the composites, the process had a significant influence on the particle size, as shown in Figure 3b. For each PEDOT coating, the composites had a similar morphology. FTP-24 was chosen as a representative of all the FTP samples. It can be seen that the particle size increased to 10–15 μm and some secondary small particles were distributed randomly around the main particle. This can be attributed to the Fe_2O_3 particle aggregation in the EDOT polymerization process. This intricate process often begins with monomer adsorption onto the surface of iron oxide particles, followed by subsequent chemical reactions that lead to the formation of polymer chains. As a result, small particles are interconnected, giving rise to larger particles, and, consequently, an increase in particle size. Thus, the PEDOT intertwined on the Fe_2O_3 particles had a confinement effect on the particles to inhibit the expansion in the Li insertion process, which helped to improve the cycle performance. The corresponding EDS mapping results of FTP-24 are shown in Figure 3c. The main observable elements of Fe, O, and Ti were uniformly distributed, which meant the composite had no local segregation.

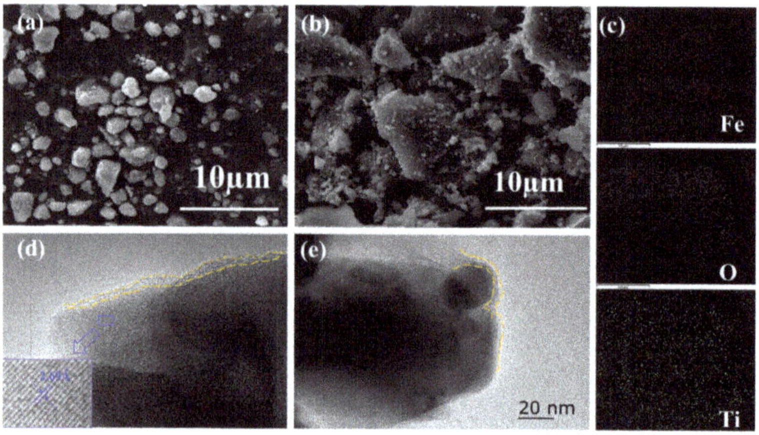

Figure 3. The morphology and structure characterization of the composites. (**a**) SEM image of raw commercial Fe_2O_3. (**b**) SEM image of FTP-24 as a representative of the PEDOT-coated composites. (**c**) The corresponding EDS mapping analysis of FTP-24. (**d**,**e**) TEM images of FTP-8 and FTP-24 with the lattice fringes in the insert.

Figure 3d,e illustrates the TEM images of FTP-8 and FTP-24. It can be seen that the amorphous layer was uniformly covered on the crystalline phase. The interplanar spacing of intra particles was 2.69 Å, corresponding to the (112) plane of Hematite Fe_2O_3. The thickness of the amorphous coating layer of FTP-8 was about 1–3 nm; however, the thickness increased to 5–8 nm in the FTP-24 sample, which was due to its longer polymerization duration. Moreover, a double-layer coating structure is observed in Figure 3e, which may be attributed to the TiO_{2-x} and PEDOT layer.

3.2. Electrochemical Performance of the Compounds

The cyclic voltammogram results are shown in Figure 4. For the FTP samples, the first distinct cathodic peak with a high current density appeared at about 0 to 0.4 V, corresponding to the SEI film formation and irreversible lithiation reaction [43]. The reversible conversion reaction from Fe_2O_3 to the cubic $Li_2(Fe_2O_3)$ and further to Fe^0 took place as well in the first cathodic scanning process. The anodic board peaks, appearing at 1.06 and 1.53 V, can be ascribed to the Fe^0 two-step oxidation to Fe^{2+} and to Fe^{3+}. All the first-cycle curves are illustrated in Figure 4e. As the polymerization duration increased, especially for the FTP-56 sample, the cathodic peaks at 0.4 V became more distinct, indicating a greater irreversible lithiation reaction and more lithium loss occurring in the first cycle. After the first cycle, as shown in Figure 4a–d, the CV curves became stable. The cathodic peak shifted to 0.75 V and the current density was slightly weaker, indicating the reversible lithiation process. The CV curves of bare Fe_2O_3 are shown in Figure 4f. It can be seen that the irreversible reaction peak at 0.61 V was much sharper, signifying more lithium loss in the first cycles for the SEI formation. After the first cycle, the capacity loss was significant, as shown by observing the current density decrease in the first three cycles. Thus, in comparison to the PEDOT-coated samples, the bare Fe_2O_3 exhibited a quite unstable state in the electrochemical process.

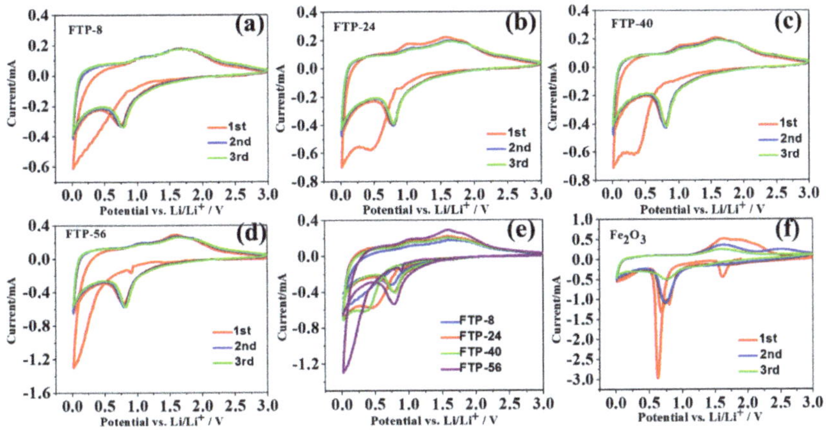

Figure 4. First three cycles of cyclic voltammograms of (**a**) FTP-8, (**b**) FTP-24, (**c**) FTP-40, (**d**) FTP-56, (**e**) first cycle of CV results of all the FTP samples, and (**f**) first three cycles of CV results of bare Fe_2O_3.

The constant current charge–discharge curves (current density, 100 mA/g) of the first three cycles of the obtained samples are illustrated in Figure 5. The optimal one of the FTP samples, FTP-24, was chosen to study the PEDOT coating's effects on the electrochemical performance. The first specific discharge and charge capacities were 1543.4 and 1062.1 mAh/g, respectively, and the corresponding initial coulombic efficiency (ICE) was 68.81%. The results were well-matched to the CV analysis, in which the low ICE was caused by the irreversible lithiation process and the formation of SEI film. In the following cycles, the capacity decay became less noticeable and the electrode exhibited a quite stable charge–discharge plateau. In order to further understand the PEDOT and TiO_2 coatings' synergistic effects on the electrochemical performance, samples without PEDOT (FT), without TiO_2 (FP-24), and without both (Fe_2O_3) were prepared, and their corresponding results are shown in Figure 5b–d, respectively. In the first three cycles of the three samples, the capacity decay was significant, rather than the FTP-24 sample only being significant in the first cycle. In the initial discharge process, there were two typical plateaus at about 1.7 and 0.75 V in the Fe_2O_3@PEDOT-24h and Fe_2O_3 samples, conforming to the CV results. Meanwhile, the plateau at 1.7 V disappeared in the Fe_2O_3@TiO_2 sample. It is worth noting that the PEDOT coating effectively inhibited the rapid capacity decay, and the stable plateau at 0.75 V remained in the discharge process as well. The initial specific discharge, charge capacity, and Coulombic efficiency of all the FTP samples and control groups are listed in Table 1. It can be seen that the TiO_2 coating was the key factor leading to the decrease in ICE, and PEDOT had a slight positive effect on enhancing the ICE. All the FTP samples possessed a similar ICE, however, FTP-24 showed the highest initial charge–discharge specific capacity. In comparison to the other FTP samples, FTP-24 exhibited a higher capacity retention as well.

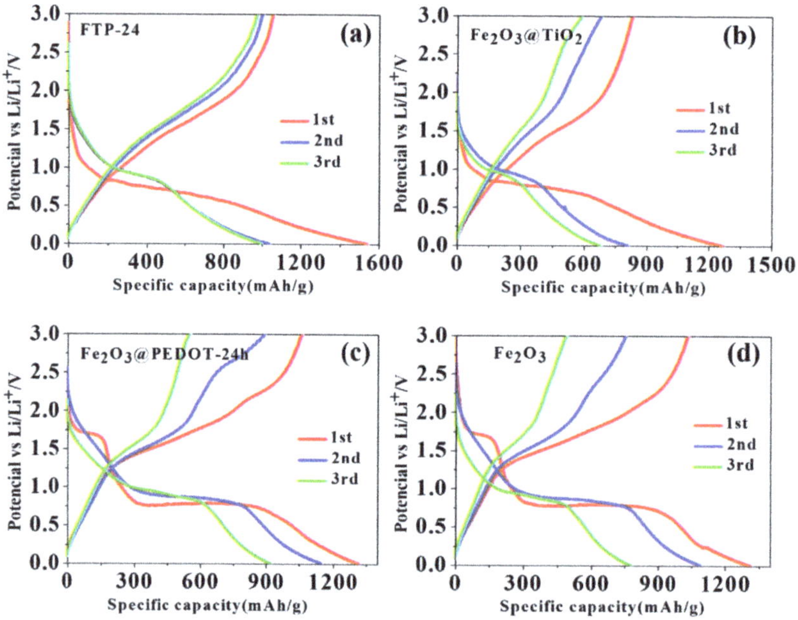

Figure 5. The constant current charge–discharge curves (current density, 100 mA/g) of first three cycles of FTP-24 (**a**), Fe_2O_3@TiO_2 (**b**), Fe_2O_3@PEDOT-24h (**c**), and Fe_2O_3 (**d**).

Table 1. The charge–discharge specific capacity of the initial cycle and the initial coulombic efficiency of all the FTP samples and the control groups at a current of 100 mA/g.

Sample	Initial Charge Specific Capacity (mAh/g)	Initial Discharge Specific Capacity (mAh/g)	Initial Coulombic Efficiency (%)	Charge Capacity after 150 Cycles (mAh/g)	Discharge Capacity after 150 Cycles (mAh/g)
FTP-8	731.1	1107.0	66.04	513.6	518.4
FTP-24	1062.1	1543.4	68.81	623.4	628.2
FTP-40	761.4	1180.1	64.52	433.3	437.6
FTP-56	659.7	986.7	66.86	358.6	356.9
FT	833.6	1269.3	65.67	230.0	228.6
FP-24	1062.5	1318.2	80.60	97.1	97.5
Fe_2O_3	1033.3	1313.8	78.65	75.87	76.0

The cycle performances and impedance spectra are illustrated in Figure 6. The FTP samples, especially FTP-24, exhibited a better cycle performance at 100 mA/g than those without dual-hierarchical decoration, as shown in Figure 6a. This phenomenon was due to the synergistic effect of the TiO_2 and PEDOT coating on inhibiting the Fe_2O_3 volume change in the charge–discharge process. As a good Li-ion conductor, TiO_2 isolated the direct contact between the Fe_2O_3 and the electrolyte, meanwhile, PEDOT relieved the particle pulverization issue, owing to its flexibility. It was noticed that the FT sample exhibited a lower specific capacity than in our previous report, which was due to the change in the binder. The sodium alginate was proven ineffective when we introduced PEDOT into the samples.

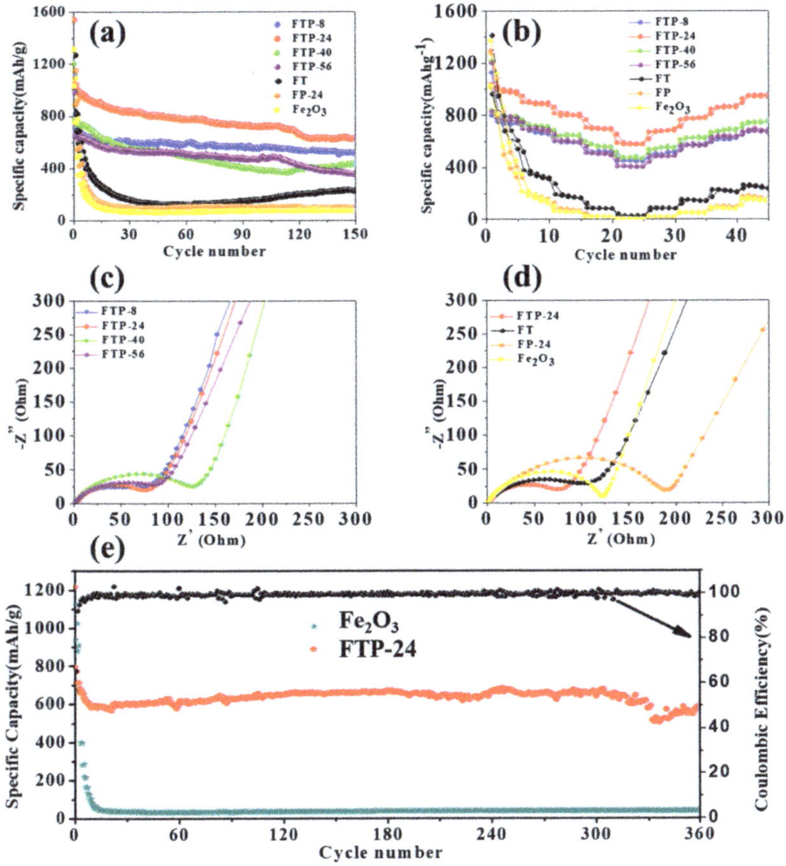

Figure 6. The electrochemical performance of the FTP, FT, FP, and bare Fe_2O_3 samples. (**a**) Cycle performance at a current density of 100 mA/g, (**b**) rate performance, (**c**,**d**) impedance spectra, and (**e**) long cycle performance comparison of FTP-24 and Fe_2O_3, 360 cycles at 100 mA/g.

The rate performance is shown in Figure 6b and current densities of 25, 50, 100, 200, 400, 200, 100, 50, and 25 mA/g were selected. It can be seen that the specific capacity decreased more slightly with an increase in the current density of the FTP samples, especially FTP-24, in comparison to that of the non-FTP samples. The specific capacity recovered to the initial value when the current density was back. The non-FTP samples exhibited fast decay with increasing the current density, even to zero under 400 mA/g. This inferred the dual hierarchical decoration availably improved the ionic and electronic conductivity of the commercial Fe_2O_3 materials.

To further understand the enhancement effect on the ionic and electronic conductivity of the TiO_2 and PEDOT decoration, impedance spectra were performed, as shown in

Figure 6c,d. By comparing the Nyquist plots at a high frequency in Figure 6c, the FTP samples exhibited a low charge transfer resistance (R_{ct}, ~80 Ω), except FTP-40 (~140 Ω), which inferred that more PEDOT coating did not determinately lead to a decrease in transfer resistance. This phenomenon was confirmed by the FP-24 sample, which possessed the highest Rct value (shown in Figure 6d). The results explained that the FTP-40 sample showed a poor cycle performance out of all the FTP samples. In the low-frequency region, the slopes of the spectra were similar to each other, which meant the double-layer coating had slight inhibiting effect on the Li-ion diffusion. In Figure 6d, the FTP sample showed great advantage in reducing the R_{ct}, compared to the single-layer or bare Fe_2O_3, and even the resistance had a significant increase in the single PEDOT-coated sample, which demonstrated that the synergistic effect of the dual coating was quite important. The primary factor contributing to the higher Rct was the interface formed between the PEDOT and Fe_2O_3, imposing constraints on the electron transport. However, by introducing TiO_2 with oxygen defects as an intermediary, we managed to significantly enhance the interfacial contact.

The long cycle performance of the FTP-24 sample is illustrated in Figure 6e. It can be seen that the specific capacity and cycle stability were enhanced significantly by the dual hierarchical decoration, compared to the bare Fe_2O_3.The discharge specific capacity remained at 588.9 mAh/g after 360 cycles at a current density of 100 mA/g. To emphasize the significance of our Fe_2O_3 composites, we include a comparison with other outstanding reported anodes in Table 2. As evident from the table and its long cycle performance, the modified industrial Fe_2O_3 exhibited substantial promise as an anode material for Li-ion batteries.

Table 2. A comparison between our study and previously reported data for reference.

Materials	Current Density (mA/h)	Initial Discharge Specific Capacity (mAh/g)	Reversible Capacity (mAh/g)	Capacity Retention (%)	Ref.
0.2rGO/Fe_2O_3 −175 °C	100	1372	435 (50 Cycles)	31.7	[44]
Fe_3O_4@graphene	100	1625	849 (100 Cycles)	52.2	[45]
Silicon-Carbon	100	1090	200 (100 Cycles)	18.3	[46]
10%-SC	100	1227	800 (100 Cycles)	65.2	[47]
CuO-NiO/rGO	100	990	680 (50 Cycles)	68.7	[48]
Carbon@SnS_2 core–shell microspheres	100	1611	500 (50 Cycles)	31.0	[49]
FTP-24	100	1543	588.9 (360 Cycles)	40.7	This work

4. Conclusions

A facile polymerization method was applied to decorate industrial Fe_2O_3 with a TiO_2 and PEDOT dual coating for application as an anode material in Li-ion batteries. The synergistic effect of the hierarchical layer on the enhancement of cycle stability was demonstrated by the structural and electrochemical characterization. The ceramic-phase TiO_2 isolated the direct contact between the Fe_2O_3 and the electrolyte, meanwhile, PEDOT relieved the particle pulverization issue, owing to its flexibility. The optimum FTP-24 sample, corresponding to a polymerization duration of 24 h, exhibited the most excellent specific capacity and cycle performance. Too long of a polymerization duration would affect the electronic conductivity and Li-ion transportation in the electrode. Due to the dual hierarchical decoration, the composites based on the industrial Fe_2O_3 possessed 588.9 mAh/g after 360 cycles at a current density of 100 mA/g, which could make them excellent candidates as commercial anode materials.

Author Contributions: Conceptualization, Y.M. and G.S.; methodology, Y.X. and T.Y.; software, Q.L.; validation, L.Z., Q.L. and H.L.; formal analysis, Z.C. and Y.M.; investigation, S.W.; resources, G.S.; data curation, J.L.; writing—original draft preparation, Y.M.; writing—review and editing, Y.X., T.Y. and G.S.; visualization, Q.L.; supervision, G.S. and Y.X.; project administration, G.S.; funding acquisition, Y.M. and G.S. All authors have read and agreed to the published version of the manuscript.

Funding: This work was supported by Natural Science Foundation of China (No. 52207246), Key projects of natural science research in universities of Anhui Province (KJ2020A0263), Anhui Natural Science Foundation (No. 1908085ME151), and China Postdoctoral Science Foundation (No. 2020M673404). We are also grateful to the open project (Key Lab Marine Mat & Related Technol, CAS, 2019K07; State Key Lab of Advanced Special Steel, Shanghai Key Lab of Adv. Ferrometallurgy SKLASS 2019-23) for providing some characterization support.

Data Availability Statement: The data presented in this study are available on request from the corresponding author. The data are not publicly available due to privacy.

Conflicts of Interest: The authors declare no conflict of interest.

References

1. Ding, B.; Ding, M.; Ma, Y.; Song, G.; Li, Z.; Ge, J.; Yang, W.; Wen, C. Effect of MgO on electrochemical properties of silicon-based anode composite material. *Solid State Sci.* **2022**, *131*, 106940. [CrossRef]
2. Wei, T.; Zhang, N.; Ji, Y.; Zhang, J.; Zhu, Y.; Yi, T. Nanosized zinc oxides-based materials for electrochemical energy storage and conversion: Batteries and supercapacitors. *Chin. Chem. Lett.* **2022**, *33*, 714–729. [CrossRef]
3. Xu, L.; Chon, M.J.; Mills, B.; Thompson, C.V. Mechanical stress and morphology evolution in RuO_2 thin film electrodes during lithiation and delithiation. *J. Power Sources* **2022**, *552*, 232260. [CrossRef]
4. Xu, L.; Thompson, C.V. Mechanisms of the cyclic (de)lithiation of RuO_2. *J. Mater. Chem. A* **2020**, *8*, 21872–21881. [CrossRef]
5. Huang, Y.; Li, Q.; Wang, S.; Cai, Z.; Ma, Y.; Song, G.; Yang, W.; Wen, C.; Xie, Y. Enhanced electrochemical performance of silicon anode materials with titanium hydride treatment. *J. Electroanal. Chem.* **2023**, *933*, 117292. [CrossRef]
6. Gu, Z.Y.; Heng, Y.L.; Guo, J.Z.; Cao, J.M.; Wang, X.T.; Zhao, X.X.; Sun, Z.H.; Zheng, S.H.; Liang, H.J.; Li, B.; et al. Nano self-assembly of fluorophosphate cathode induced by surface energy evolution towards high-rate and stable sodium-ion batteries. *Nano Res.* **2023**, *16*, 439–448. [CrossRef]
7. Du, K.D.; Meng, Y.F.; Zhao, X.X.; Wang, X.T.; Luo, X.X.; Zhang, W.; Wu, X.L. A unique co-recovery strategy of cathode and anode from spent $LiFePO_4$ battery. *Sci. China Mater.* **2022**, *65*, 637–645. [CrossRef]
8. Zhang, K.Y.; Gu, Z.Y.; Ang, E.H.; Guo, J.Z.; Wang, X.T.; Wang, Y.; Wu, X.L. Advanced polyanionic electrode materials for potassium-ion batteries: Progresses, challenges and application prospects. *Mater. Today* **2022**, *54*, 189–201. [CrossRef]
9. Xu, X.; Li, F.; Zhang, D.; Ji, S.; Huo, Y.; Liu, J. Facile construction of $CoSn/Co_3Sn_2$@C nanocages as anode for superior lithium-/sodium-ion storage. *Carbon Neutralization* **2023**, *2*, 54–62. [CrossRef]
10. Bai, Y.; Muralidharan, N.; Sun, Y.K.; Passerini, S.; Stanley Whittingham, M.; Belharouak, I. Energy and environmental aspects in recycling lithium-ion batteries: Concept of Battery Identity Global Passport. *Mater. Today* **2020**, *41*, 304–315. [CrossRef]
11. Qu, X.; Huang, H.; Wan, T.; Hu, L.; Yu, Z.; Liu, Y.; Dou, A.; Zhou, Y.; Su, M.; Peng, X.; et al. An integrated surface coating strategy to enhance the electrochemical performance of nickel-rich layered cathodes. *Nano Energy* **2022**, *91*, 106665. [CrossRef]
12. Liu, D.; Fan, X.; Li, Z.; Liu, T.; Sun, M.; Qian, C.; Ling, M.; Liu, Y.; Liang, C. A cation/anion co-doped $Li_{1.12}Na_{0.08}Ni_{0.2}Mn_{0.6}O_{1.95}F_{0.05}$ cathode for lithium ion batteries. *Nano Energy* **2019**, *58*, 786–796. [CrossRef]
13. Cai, Z.; Wang, S.; Zhu, H.; Tang, X.; Ma, Y.; Yu, D.Y.W.; Zhang, S.; Song, G.; Yang, W.; Xu, Y.; et al. Improvement of stability and capacity of Co-free, Li-rich layered oxide $Li_{1.2}Ni_{0.2}Mn_{0.6}O_2$ cathode material through defect control. *J. Colloid Interface Sci.* **2023**, *630*, 281–289. [CrossRef] [PubMed]
14. Sun, M.; Chu, X.; Wang, Z.; Yang, H.; Yang, Z.; Ma, J.; Zhou, B.; Yang, T.; Chen, L. Rosa roxburghii-like hierarchical hollow sandwich-structure C@Fe_2O_3@C microspheres as second nanomaterialsfor superior lithium storage. *J. Alloys Compd.* **2021**, *855*, 157518. [CrossRef]
15. Ding, B.; Cai, Z.; Ahsan, Z.; Ma, Y.; Zhang, S.; Song, G.; Yuan, C.; Yang, W.; Wen, C. A Review of Metal Silicides for Lithium-Ion Battery Anode Application. *Acta Metall. Sin.* **2021**, *34*, 291–308. [CrossRef]
16. Li, F.; Wang, Z.; Liu, W.; Yan, T.; Zhai, C.; Wu, P.; Zhou, Y. Double-Network Gel-Enabled Uniform Incorporation of Metallic Matrices with Silicon Anodes Realizing Enhanced Lithium Storage. *ACS Appl. Energy Mater.* **2019**, *2*, 2268–2275. [CrossRef]
17. Fang, R.; Xiao, W.; Miao, C.; Mei, P.; Zhang, Y.; Yan, X.; Jiang, Y. Enhanced lithium storage performance of core-shell structural Si@TiO_2/NC composite anode via facile sol-gel and in situ N-doped carbon coating processes. *Electrochim. Acta* **2019**, *317*, 575–582. [CrossRef]
18. Zhong, H.; Huang, W.; Wei, Y.; Yang, X.; Jiang, C.; Liu, H.; Zhang, W.; Liang, C.; Dai, L.; Xu, X. Facile Constructing Hierarchical Fe_3O_4@C Nanocomposites as Anode for Superior Lithium-Ion Storage. *Batteries* **2023**, *9*, 403. [CrossRef]
19. Reddy, M.V.; Yu, T.; Sow, C.H.; Shen, Z.X.; Lim, C.T.; Rao, G.V.S.; Chowdari, B.V.R. α-Fe_2O_3 nanoflakes as an anode material for li-ion batteries. *Adv. Funct. Mater.* **2007**, *17*, 2792–2799. [CrossRef]

20. Liu, H.; Zhu, J.; Li, Z.; Shi, Z.; Zhu, J.; Mei, H. Fe_2O_3/N doped rGO anode hybridized with NiCo LDH/$Co(OH)_2$ cathode for battery-like supercapacitor. *Chem. Eng. J.* **2021**, *403*, 126325. [CrossRef]
21. Hwang, J.; Yadav, D.; Yang, H.; Jeon, I.; Yang, D.; Seo, J.W.; Kang, M.; Jeong, S.Y.; Cho, C.R. In Situ Electrochemical Impedance Measurements of α-Fe_2O_3 Nanofibers: Unravelling the Li-Ion Conduction Mechanism in Li-Ion Batteries. *Batteries* **2022**, *8*, 44. [CrossRef]
22. Liu, Y.; Lei, J.; Chen, Y.; Liang, C.; Ni, J. Hierarchical-Structured Fe_2O_3 Anode with Exposed (001) Facet for Enhanced Lithium Storage Performance. *Nanomaterials* **2023**, *13*, 2025. [CrossRef] [PubMed]
23. Lai, H.; Li, J.-Q.; Lin, Y.-Y.; Sun, Y.-H. Flower-like multi-shells MO/Fe_2O_3 (M = Cu, Mn and Ni) hollow microspheres for lithium ion battery with super cycle stability. *Inorg. Chem. Commun.* **2023**, *154*, 110947. [CrossRef]
24. Li, J.; Yan, L. Synthesis of Fe_2O_3/rGO Based Composites as Anode material for Lithium ion batteries using sewage sludge as source of Iron (III) oxide. *Int. J. Electrochem. Sci.* **2022**, *17*, 221173. [CrossRef]
25. Pan, Y.; Luo, C.; Yang, D.; Sun, P.; Chen, J.; Sui, Z.; Tian, Q. Ultrathin porous Fe_2O_3@C nanosheets: Novel preparation strategy and high lithium storage. *Appl. Surf. Sci.* **2023**, *635*, 157763. [CrossRef]
26. Ge, Y.; Gao, W.; Yuan, Z.; Tao, S.; Kong, F.; Qian, B. Synergistic effect on the improved lithium ion storage performance in the porous Fe_2O_3@Fe_3C@C composite. *Mater. Res. Bull.* **2023**, *164*, 112287. [CrossRef]
27. Xu, Z.; Sun, M.; Wu, S.; Chen, Y.; Li, L.; Zou, X.; Chen, L.; Yang, H.; Pang, H. Interfacial engineering of graphene aerogel encapsulated $FeSe_2$-Fe_2O_3 heterojunction nanotubes for enhanced lithium storage. *J. Alloys Compd.* **2023**, *934*, 167939. [CrossRef]
28. Jin, Y.; Zhang, W.; Qing, T.; Chen, G.; Zeng, S.; Huang, J. Constructing 3D sandwich-like carbon coated Fe_2O_3/helical carbon nanofibers composite as a superior lithium-ion batteries anode. *J. Electroanal. Chem.* **2023**, *929*, 117098. [CrossRef]
29. Chen, L.; Zhao, C.; Lu, Y.; Wan, L.; Yan, K.; Bai, Y.; Liu, Z.; Yang, X.; Su, Y.; Wu, F. Facile Synthesizing Yolk-Shelled Fe_3O_4@Carbon Nanocavities with Balanced Physiochemical Synergism as Efficient Hosts for High-Performance Lithium–Sulfur Batteries. *Batteries* **2023**, *9*, 295. [CrossRef]
30. Shi, W.C.; Shao, Y.R.; Shen, Y.; Li, Z.F.; Hu, T.L. Facile synthesis of porous iron oxides heterostructural nanosheets as an enhanced anode material for lithium ion battery. *Mater. Lett.* **2022**, *321*, 132454. [CrossRef]
31. Zhu, J.; Wei, L.; Hu, J.; Xue, C. Anchoring iron oxide nanoparticles on polypyrrole/rGO derived nitrogen-doped carbon as lithium-ion battery anode. *J. Alloys Compd.* **2017**, *723*, 729–735. [CrossRef]
32. Zheng, Z.; Li, P.; Huang, J.; Liu, H.; Zao, Y.; Hu, Z.; Zhang, L.; Chen, H.; Wang, M.S.; Peng, D.L.; et al. High performance columnar-like Fe_2O_3@carbon composite anode via yolk@shell structural design. *J. Energy Chem.* **2020**, *41*, 126–134. [CrossRef]
33. Qin, G.; Zeng, M.; Wu, X.; Wen, J.; Li, J. Fabrication of Fe_2O_3@TiO_2 core–shell nanospheres as anode materials for lithium-ion batteries. *J. Mater. Sci. Mater. Electron.* **2018**, *29*, 12944–12950. [CrossRef]
34. Ma, Y.; Zhang, L.; Cai, Z.; Huang, X.; Ding, B.; Ahsan, Z.; Song, G.; Xu, Y.; Yang, W.; Wen, C. Study of TiO_2 -Coated α-Fe_2O_3 Composites and the Oxygen-Defects Effect on the Application as the Anode Materials of High-Performance Li-Ion Batteries. *ACS Appl. Energy Mater.* **2020**, *3*, 11666–11673. [CrossRef]
35. Fu, Y.; Wei, Q.; Lu, B.; Wang, X.; Sun, S. Stem-like nano-heterostructural MWCNTs/α-Fe_2O_3@TiO_2 composite with high lithium storage capability. *J. Alloys Compd.* **2016**, *684*, 419–427. [CrossRef]
36. Zhu, J.; Xu, Y.; Wang, J.; Wang, J.; Bai, Y.; Du, X. Morphology controllable nano-sheet polypyrrole-graphene composites for high-rate supercapacitor. *Phys. Chem. Chem. Phys.* **2015**, *17*, 19885–19894. [CrossRef] [PubMed]
37. Zeng, Y.; Xie, D.; Ai, S.; Huang, H.; Zheng, Z.; Xie, S.; Liu, P.; Wang, S.; Zhang, M.; Cheng, F. Facile fabrication of core-shell α-Fe_2O_3@PPy imbedded into porous biomass-derived carbon for enhanced lithium storage. *J. Energy Storage* **2023**, *67*, 107625. [CrossRef]
38. Rasool, M.; Chiu, H.C.; Zank, B.; Zeng, Y.; Zhou, J.; Zaghib, K.; Perepichka, D.F.; Demopoulos, G.P. PEDOT Encapsulated and Mechanochemically Engineered Silicate Nanocrystals for High Energy Density Cathodes. *Adv. Mater. Interfaces* **2020**, *7*, 202000226. [CrossRef]
39. da Silva, R.J.; Lima, R.M.A.P.; de Oliveira, M.C.A.; Alcaraz-Espinoza, J.J.; de Melo, C.P.; de Oliveira, H.P. Supercapacitors based on (carbon nanostructure)/PEDOT/(eggshell membrane) electrodes. *J. Electroanal. Chem.* **2020**, *856*, 113658. [CrossRef]
40. Jeong, J.M.; Choi, B.G.; Lee, S.C.; Lee, K.G.; Chang, S.J.; Han, Y.K.; Lee, Y.B.; Lee, H.U.; Kwon, S.; Lee, G.; et al. Hierarchical hollow spheres of Fe_2O_3@polyaniline for lithium ion battery anodes. *Adv. Mater.* **2013**, *25*, 6250–6255. [CrossRef]
41. Chen, H.W.; Li, C. PEDOT: Fundamentals and Its Nanocomposites for Energy Storage. *Chin. J. Polym. Sci.* **2020**, *38*, 435–448. [CrossRef]
42. Wu, C.; Chen, J.; Liang, L.; Li, N. N-doped/oxygen-deficient Fe_2O_{3-x} @PEDOT cruciform nanosheet arrays: Boosting the performance of all-solid-state flexible supercapacitors. *Chem. Phys. Lett.* **2023**, *823*, 2–6. [CrossRef]
43. Chaudhari, S.; Srinivasan, M. 1D hollow α-Fe_2O_3 electrospun nanofibers as high performance anode material for lithium ion batteries. *J. Mater. Chem.* **2012**, *22*, 23049–23056. [CrossRef]
44. Zhu, G.; Gao, B.; Tu, G.; Liu, H.; Wang, M. Improving Effect of Graphene on Electrochemical Properties of Fe_2O_3 Anode Materials. *Metals* **2022**, *12*, 593. [CrossRef]
45. Suo, Y.; Zhao, Q.Q.; Meng, J.K.; Li, J.; Zheng, X.C.; Guan, X.X.; Liu, Y.S.; Zhang, J.M. Fabrication and electrochemical properties of 3D Fe_3O_4@graphene aerogel composites as lithium-ion battery anodes. *Mater. Lett.* **2016**, *174*, 36–39. [CrossRef]
46. Shen, L.; Wang, D.; Ali, K.; Li, M.; Shi, Z. Electrochemical Preparation of Nano-Sized Silicon as a Lithium-Ion Battery Anode Material. *J. Electrochem. Soc.* **2021**, *168*, 120509. [CrossRef]

47. Jiang, J.; Ou, Y.; Jiang, Y.; Hu, X.; Xing, C.; Liu, S.; Liu, X.; Li, W.; Zhao, B. Preparation of SiOx–TiO_2/Si/CNTs composite microspheres as novel anodes for lithium-ion battery with good cycle stability. *J. Mater. Sci. Mater. Electron.* **2022**, *33*, 11025–11037. [CrossRef]
48. Ranjbar-Azad, M.; Behpour, M. Facile in situ co-precipitation synthesis of CuO–NiO/rGO nanocomposite for lithium-ion battery anodes. *J. Mater. Sci. Mater. Electron.* **2021**, *32*, 18043–18056. [CrossRef]
49. Yu, G.; Chen, X.; Wang, A.; Wang, Y. Carbon@SnS_2 core-shell microspheres for lithium-ion battery anode materials. *Ionics* **2018**, *24*, 2915–2923. [CrossRef]

batteries

MDPI

Article

Li-Ion Batteries with a Binder-Free Cathode of Carbon Nanotubes-$LiFePO_4$-Al Foam

Ying Jin [1,†], Shaoxin Wei [1,†], Zhoufei Yang [1,†], Chaojie Cui [1,2,*], Jin Wang [1], Dongliang Li [2] and Weizhong Qian [1,2,*]

[1] Department of Chemical Engineering, Tsinghua University, Beijing 100084, China
[2] Ordos Laboratory, Ordos 017000, China
* Correspondence: cuicj06@tsinghua.edu.cn (C.C.); qianwz@tsinghua.edu.cn (W.Q.)
† These authors contributed equally to this work.

Abstract: With the increasing demand for Li resources worldwide, the easy recycling of Li-ion batteries materials becomes essential. We report a binder-free cathode consisting of carbon nanotubes (CNTs) and $LiFePO_4$ (LFP) nanoparticles embedded in a 3D Al network. The electrode stability depends on the CNT ratio, where 3% CNT-wrapping LFPs provide a stable structure free of detachment from Al foam, as observed on Al foil. The binder-free cathode sheet exhibited excellent performance for high-rate discharge and long-term cycle life. Materials on the cathode can be easily detached with ultrasonic treatment when immersed in organic solvent, which is advantageous for a green and high-efficiency strategy of recycling all valuable materials compared to the binder-used electrode.

Keywords: Li-ion battery recycling; binder-free cathode; carbon nanotubes; Al foam; 3D porous current collector

1. Introduction

The development of the secondary Li-ion battery (LIB) industry directly leads to an application burst of electrical vehicles worldwide, contributing to clean technology for carbon neutrality [1–4]. However, the huge demand of Li-based cathode materials increases the Li supply pressure, considering that the reserve of the Li element on Earth is actually low compared to other elements (Al, Si, Fe, etc.) [5,6]. The shortage of Li compounds brings very unstable Li prices, which is unhealthy to the development of the industry. In turn, this stresses the importance of recycling Li resources from spent LIBs [7–10].

In general, for LIB fabrication, binders are used in large amounts to adhere Li cathode materials tightly onto the Al foil (current collector). This tight interaction ensures the close contact of the cathode materials with the current collector, for electronic transfer and for long-term stability [11,12]. But such structure result in a high difficulty in recycling Li resources from spent batteries. Up to now, although various methods (wet chemistry and high-temperature melting) have been proposed for recycling Li inside LIBs [13–16], most of them are destructive to the battery structure. And the separation of various complicated and even toxic components of LIBs from each other is still costly and sometimes full of risk.

In the present work, we propose a new concept for fabricating a binder-free cathode with 3D Al foam (current collector) to fix the cathode materials. In detail, in using $LiFePO_4$ (LFP) nanoparticles to model the cathode material, they were firstly mixed with carbon nanotubes (CNTs) to form a composite slurry, and then used to fill the pores of the Al foam, following the compression procedure. LIBs with such configuration work well for charging and discharging to deliver a high performance. The contact between the LFP nanoparticles with the CNTs was good, where 1D CNTs wrapped the LFP nanoparticles tightly owing to the strong interaction of the van der Waals forces between the two kinds of nanoparticles [17–20]. And the contact between the composite powder and numerous Al wires, which composed the uniform 3D network, was also excellent based on the mechanical compression [21–23]. In considering the recycling of LFP materials, the cathode sheets of spent LIBs were immersed in organic solutions. The assisted external ultrasonic operation enabled the easy detachment of CNTs and LFP powders from the Al foam network. Most of the CNTs and LFP could be extracted from the cathode and transferred to solution under ultrasonic conditions. The present strategies may offer new options for the field.

2. Materials and Methods

2.1. Material Preparation

CNTs prepared with the chemical vapor deposition (CVD) method had a diameter range of 7–11 nm and length of several micrometers [24–27]. After purification with acidic washing and high-temperature vacuum treatment, their purity met the requirements of the LIB industry. LFP nanoparticles (purchased from Pulead Technology Industry Co., Ltd., Beijing, China) had a particle size of 100 nm, which had a thin carbon layer on the external surface. Al foam was prepared with magnetically controlled spouting Al particles on porous polyurethane (PU) foam from the Al target, followed with the removal of the PU template through controlled oxidation below 600 °C [21,22]. The pristine Al foam used here was 1 mm thick, 50 cm long, and 20 cm wide. Its porosity, mass density, and Al purity were 95%, 120 g m^{-2}, and 99.995%, respectively.

2.2. Cathode Sheet Fabrication

Firstly, LFP nanoparticles, CNTs, and conductive carbon black (Super P) were mixed in an *N*-methyl-2-pyrrolidone (NMP) solution and stirred in a homogenizer (SK-300II, Kakuhunter Co., Ltd., Shiga Prefecture, Japan) for 2 h. The viscosity of the slurry was controlled at 1500–3000 mPa·s. Secondly, 3D Al foam was immersed into the slurry to sufficiently absorb the LFP and CNT mixture inside. Then, the cathode sheet was dried at 70 °C for 2 h to remove all NMP, following roll compression by a rolling machine (MSK-2150, MTI Corporation, Richmond, CA, USA).

Three binder-free cathodes with different compositions (1% CNTs + 94% LFP, 3% CNTs + 92% LFP, and 5% CNTs + 90% LFP, where the mass fraction of Super P was controlled at 5%) were prepared and compared, which were marked as 1% CNTs, 3% CNTs and 5% CNTs, respectively. These cathode sheets were die-cut to a size of 5 cm wide, 7.6 cm long, and 190 μm thick. The mass loading and compaction density of active materials in the cathode sheets were 22 mg cm^{-2} and 1.16 g cm^{-3}.

2.3. Anode Sheet and Pouch Cell Fabrication

An anode sheet was directly purchased from Guangdong Canrd New Energy Technology Co., Ltd. (Guangzhou, China). It was a typically fast-charging graphite anode adhered onto a Cu foil (current collector). The N/P ratio was controlled at 1.10.

To fabricate a pouch cell, 5 cathode sheets and 6 anode sheets were stacked with polypropylene (PP) membrane (20 μm thick) one by one and sealed with an Al plastic film. The space inside the electrodes and pouch cell was filled with $LiPF_6$-based electrolyte, which was purchased from Zhangjiagang Guotai Huarong New Chemical Materials Co., Ltd. (Suzhou, China) (LB-4927BY). The areal capacity and total capacity of the pouch cell were about 3.05 mAh cm^{-2} and 580 mAh.

2.4. Electrochemical Test

Charge and discharge tests on the pouch cell were performed using a battery testing system (CT6002A, Wuhan Land Electronics Co., Ltd., Wuhan, China) at 25 °C. During the initial formation process, the fresh pouch cell was firstly charged at 0.05 C for 2 h and then charged at 0.1 C for 5 h, followed by being charged to 3.8 V at 0.2 C and discharged to 2.4 V at 0.2 C to finish the first charge–discharge cycle. It was then charged and discharged at 0.2 C for 3 cycles in a voltage range of 2.4–3.8 V, including a constant-voltage charge process at 3.8 V until a cut-off current rate of 0.02 C. During the rate tests, the pouch cell was charged at 1 C (including the constant-voltage charge process) and discharged at rates from 1 C to 20 C (constant-current mode). A cyclic test was performed by charging and discharging the pouch cell at 1 C. The specific capacity (mAh g^{-1}) was calculated using the discharge capacity and the total mass of the LFP in the cathodes. The energy density (E, Wh kg^{-1}) and average power density (P, W kg^{-1}) were calculated using the following equations:

$$E = \frac{I \times \int U dt}{3600 \times m} \quad (1)$$

$$P = \frac{3600 \times E}{t} \quad (2)$$

Here, I is the discharge current (A), U is voltage (V), t is discharge time (s), and m (kg) is the total mass of active materials (LFP and graphite) in the cathode and anode.

3. Results and Discussion

Figure S1 shows scanning electron microscope (SEM) images of the LFP nanoparticles and transmission electron microscope (TEM) images of the CNT powders. The particle size of the LFP nanoparticles was about 100 nm, and the diameter of the CNTs was about 10 nm with walls of about 15 layers. When these materials were mixed into binder-free slurry and coated on conventional Al foil, a severe delamination of the coating occurred after drying due to inadequate adhesion. In contrast, the 3D skeleton structure of the Al foam can provide excellent support for the active materials, forming a stable binder-free cathode (Figure S2). Further, we placed three Al foam-based binder-free cathodes in the middle of a folded white paper, ensuring tight contact, and then observed the delamination of the coatings. As shown in Figure 1a, some delamination of coating materials occurred in the 1% CNT cathode, whereas no such phenomenon was seen in the other two cathodes, proving the important adhesive effect of CNTs. The micrograph in Figure 1b shows that pristine Al foam was constructed from lots of interconnected and interlaced Al wires, exhibiting a 3D porous structure with pore diameters in the range of several hundred micrometers. When the slurry was coated on Al foam without compression, cathode materials were primarily filled within the pores of the Al foam under a low-loading condition. While increasing the loading, a small portion of the coatings can also be supported on the surface of Al wires. After roll compression, most of the active materials are squeezed into the internal pores of the cathode, resulting in an extremely flat surface (Figure 1c). In addition, the Al wires interconnected to form a 3D conductive network and support structure, in tight contact with the active materials. As shown in Figure 1d, the CNTs were wrapped around the LFP nanoparticles uniformly on a microscopic scale. This interaction through van der Waals forces is tight to ensure that the external layer interacts with the inner layer tightly. In summary, Al wires form a continuous network to fix LFP and CNT particles on a macroscopic scale, and the CNTs are also wrapped around the LFP through the strong interaction at a microscopic scale, which are the key effects of avoiding the detachment of LFP nanoparticles from the matrix.

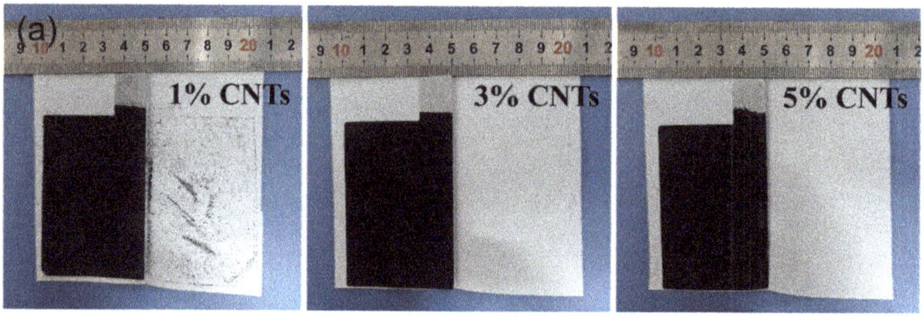

Figure 1. *Cont.*

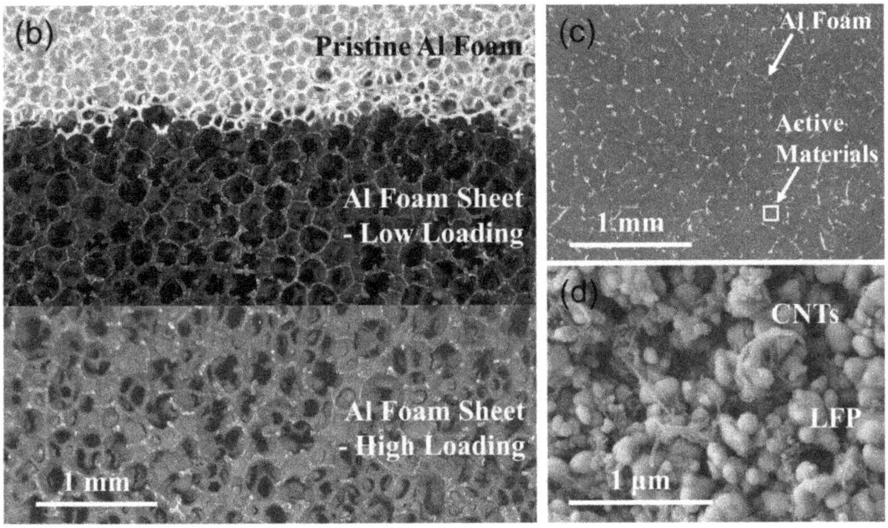

Figure 1. (**a**) Optical images of the binder-free cathodes of CNTs-$LiFePO_4$-Al foam with three compositions. The coating materials in the 1% CNT cathode exhibited delamination after compression within a folded white paper, which was not observed in the other two cathodes. (**b**) Optical micrograph of pristine Al foam and Al foam-based binder-free cathode with low and high loading of active materials (3% CNTs, uncompressed). (**c**) Optical micrograph of Al foam-based binder-free cathode (3% CNTs, compressed); the golden wires are Al foam, and the black regions are active materials. (**d**) Enlarged SEM image of the white region in (**c**).

The electrochemical performances of the pouch cells with three binder-free cathodes of CNTs-$LiFePO_4$-Al foam were tested to verify the availability and superiority of the strategy proposed in this work. The specific capacities at different discharge rates in Figure 2a show that the three cathodes had similar performances at 0.2 C–6 C, and the 3% CNT cathode had the most superior performance at a high rate of 10 C–20 C. To explain the optimal 3% CNT content, we measured the tap density of the LFP and CNTs powders used in this work, which were about 0.9 g cm^{-3} and 0.035 g cm^{-3}. Accordingly, the volume ratio of the LFP to CNTs in the cathodes of 1% CNTs (94% LFP + 1% CNTs), 3% CNTs (92% LFP + 3% CNTs), and 5% CNTs (90% LFP + 5% CNTs) could be determined as 3.66, 1.19, and 0.70, respectively. The volume fraction of CNTs increased with the rise in their content, and the volume ratio in the 3% CNT cathode was closest to 1. For 1% CNTs, insufficient conductivity and adhesion (Figure 1a) may be the reasons for the relatively lower specific capacity at high discharge rates. For 5% CNTs, an excessive content of CNTs would lead to poor dispersion and severe spontaneous aggregation (Figure S3), which diminished the improvement effect on the conductivity of LFP. It may even hinder the Li-ion diffusion of LFP particles, thereby reducing the specific capacity at high discharge rates. Overall, the 3% CNT cathode exhibited the best rate performance among these binder-free cathodes. It can withstand a discharge rate as high as 20 C, with specific capacities at 4 C/10 C/20 C being 127 mAh g^{-1}, 109 mAh g^{-1}, and 71 mAh g^{-1}, and the corresponding capacity retention (compared to 0.2 C) reaching 91%, 78%, and 51%, where the values always remained at a high level (Figure 2b). These results indicate that the CNTs-Al foam-assisted binder-free cathode showed extremely excellent electrochemical performance.

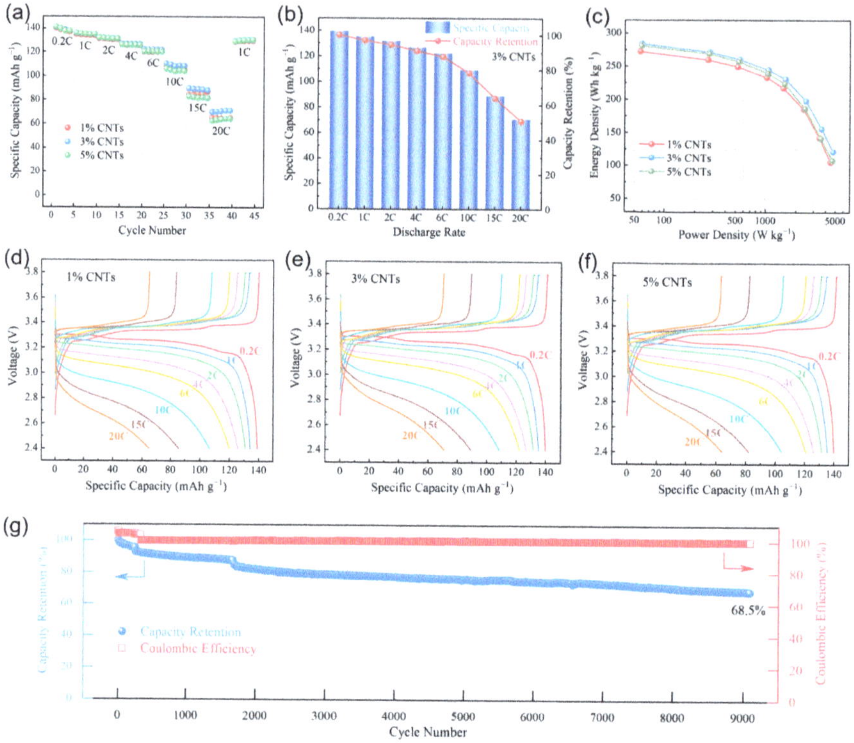

Figure 2. (**a**) Specific capacity of the binder-free cathodes of CNTs-$LiFePO_4$-Al foam at different discharge rates. (**b**) Capacity retention of the binder-free cathode with 3% CNTs at different discharge rates compared to 0.2 C. (**c**) Energy density and power density (Ragone plot) of the pouch cells with binder-free cathodes. (**d**) Voltage–capacity plot of the binder-free cathode with 1% CNTs. (**e**) Voltage–capacity plot of the binder-free cathode with 3% CNTs. (**f**) Voltage–capacity plot of the binder-free cathode with 5% CNTs. (**g**) Cycle performance of the binder-free cathode with 3% CNTs.

Based on the total mass of active materials (LFP and graphite) in the cathode and anode, we calculated the energy density and average power density of the three pouch cells with binder-free cathodes (Figure 2c). Similarly, 3% CNTs exhibited the most outstanding performance. The energy density of 3% CNTs could reach 284 Wh kg^{-1} at a power density of 59 W kg^{-1}, and it remained at 245 Wh kg^{-1} at 1064 W kg^{-1} and 121 Wh kg^{-1} at 4739 W kg^{-1}. As shown in Figures 2d–f and S4, the three pouch cells exhibited well charge–discharge characteristics at all rates. Within the discharge rates of 0.2 C–6 C, the voltage polarization did not significantly increase, demonstrating the typical characteristic of LFP materials with a voltage plateau. Even at high discharge rates of 10 C–20 C, all three binder-free cathodes still possessed a high specific capacity and high discharge voltage. Specifically, the median voltages of 3% CNTs while discharging at 0.2 C, 2 C, 6 C, 10 C, and 20 C were 3.259 V, 3.180 V, 3.048 V, 2.922 V, and 2.725 V, respectively, always higher than that of 1% CNTs (3.258 V, 3.173 V, 3.023 V, 2.891 V, and 2.671 V) and 5% CNTs (3.258 V, 3.172 V, 3.024 V, 2.885 V, and 2.704 V), again demonstrating the best discharge performance of 3% CNTs among these binder-free cathodes. Furthermore, we also performed the cyclic test at 1 C charge–discharge using the pouch cell with the binder-free cathode of 3% CNTs. Figure 2g shows that the coulombic efficiency was stable at near 100%, with the discharge capacity slowly decreasing as the cycle number increased. Specifically, the capacity retention after 2000, 5000, and 9000 cycles was 82%, 76%, and 69%, respectively. These electrochemical test results confirm that the Li-ion batteries with a binder-free cathode of CNTs-$LiFePO_4$-Al foam possessed remarkable rate performance and cycle performance, which we attribute to three key factors. Firstly, the strong interaction between CNTs and LFP, along with an appropriate volume ratio and the supporting effect of Al foam with a 3D skeleton, confer good stability to the cathode, enabling it to withstand high-current impacts and maintaining a stable electrode structure over long cycles. Secondly, the CNTs and Al foam effectively enhanced the electronic conductivity within the cathode coatings and between the coatings and current collector at microscopic and macroscopic scales, allowing for small polarization and high capacity even at very high discharge rates. Furthermore, the porous structure of the CNTs and Al foam facilitate electrolyte seepage, which is beneficial for improving Li-ion diffusion performance and ensuring sufficient ion supply over extended cycling periods [21,23]. These performances of the present binder-free cathode of CNTs-$LiFePO_4$-Al foam are comparable or superior to those of other binder-free electrodes [28–33], proving that it is a simple but advanced strategy for fabricating a binder-free cathode using Al foam and CNTs. As for the fast-charging aspect, we believe that it is more closely related to the anode performance (such as anionic activity regulation, solid electrolyte interphase modification, anodic kinetic enhancement, etc.) [34–36], which was not deeply investigated in this study.

After the discharge test, the pouch cell with the binder-free cathode was treated to recycle Li compounds. Figure 3a shows disassembled binder-free cathode sheets with 3% CNTs. There was little black cathode material detachment on the PP membrane, indicating that the supporting effect of the CNTs and Al foam remained stable even after high-rate discharge. Further, the binder-free cathode was immersed into NMP solvent and subjected to ultrasonic treatment. It can be clearly seen in Figure 3b that the black powders were detached from the Al foam and some regions of silver-white Al foam occurred after ultrasonication for 40 min. When increasing the ultrasonic time to 100 min, the black areas (cathode material coating) became lighter in color, while the silver-white areas (Al foam) became larger, indicating that more cathode materials could be detached as ultrasonic time increased. The micrograph in Figure 3c reveals that most of the cathode materials had been detached in the NMP solvent with only a little residual on the Al wires (as the arrows mark) after ultrasonic treatment for 100 min. Meanwhile, the NMP solvent in the beaker bottle became very black. The same procedure of ultrasonic treatment was conducted for the binder-free cathode of 5% CNTs (Figure S5). However, there was still lots of black powder residual on the Al foam cathode after ultrasonic for 100 min. This difference indicates that excessive CNTs may lead to excessive adhesion, making it difficult for active materials in spent LIBs to detach from Al foam, thus complicating the recycling process of Li compounds.

In order to validate the feasibility of recycling LFP nanoparticles from the binder-free cathode of CNTs-$LiFePO_4$-Al foam, we transferred a portion of NMP suspension after 100 min of ultrasonication into a centrifuge tube, which at this point was uniformly dark black (Figure 3d). After standing the tube vertically for 12 h, it was observed that almost all the solids in the solution naturally settled at the bottom, leaving a supernatant with a very low solid concentration at the top (Figure 3e). This phenomenon indicates that a solid–liquid separation process of the suspension after ultrasonic treatment can be easily achieved through natural sedimentation and filtration. Of course, the application of high-speed centrifugation and other techniques can further enhance the recycling efficiency of cathode materials. Figure 3f shows a SEM image of solid materials in the suspension with 40 min ultrasonication, where the majority were LFP nanoparticles with a minority of CNTs, suggesting that LFP with its higher density was easier to detach from the binder-free cathode during the ultrasonic process. When increasing the time to 100 min, a significant number of CNTs can also be detached (Figure 3g). In short, the strategy to recycling the valuable materials (LFP and CNTs) in a binder-free cathode of CNTs-$LiFePO_4$-Al foam through ultrasonication was proved to be feasible and efficient. In addition, the spent ultrasonic solvent can also be used for many times. Figure 4 summarizes the recycling process of LFP and CNTs nanomaterials from a spent binder-free cathode of CNTs-$LiFePO_4$-Al foam. It was estimated that the ultrasonic treatment was effective in recycling 95–97% of all materials on the cathode side. The high recycling ratio of LFP nanoparticles was due to the small size of LFP, which detached gradually from the pores of the Al foam, although the pores were somewhat narrowed after compression. Only very few CNTs adhered to the Al foam owing to their strong van der Waals interaction. Apparently, it is a green and high-efficiency strategy for recycling all valuable materials compared to the binder-used electrode.

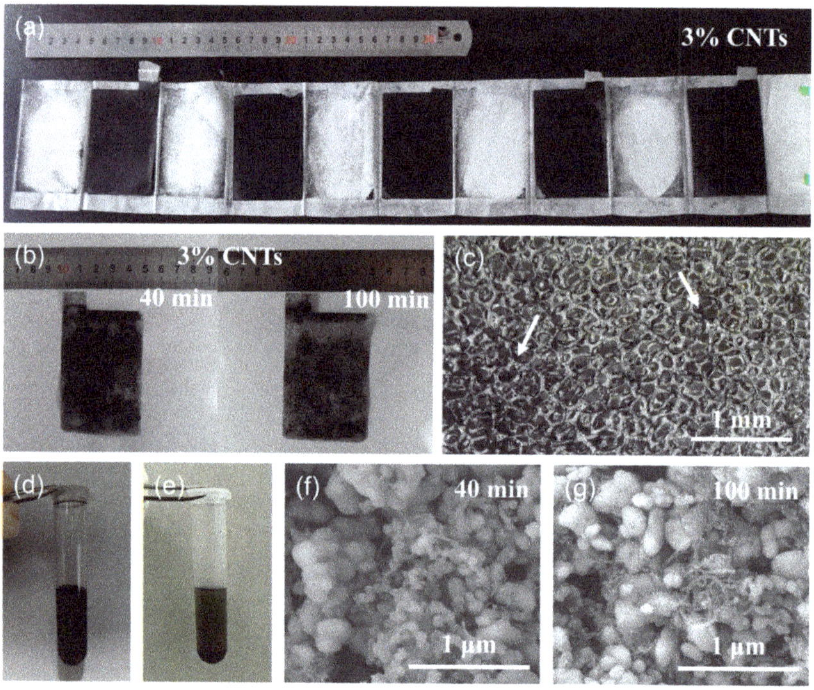

Figure 3. (**a**) Optical image of the disassembled binder-free cathode sheets with 3% CNTs. (**b**) Optical images of the binder-free cathode with 3% CNTs after ultrasonic treatment for 40 min and 100 min. (**c**) Optical micrograph of the binder-free cathode with 3% CNTs after ultrasonic treatment for 100 min. (**d**) The NMP suspension with 100 min ultrasonication. (**e**) The solution in (**d**) after standing vertically for 12 h. (**f**) SEM image of solid materials in the suspension with 40 min ultrasonication. (**g**) SEM image of solid materials in the suspension with 100 min ultrasonication.

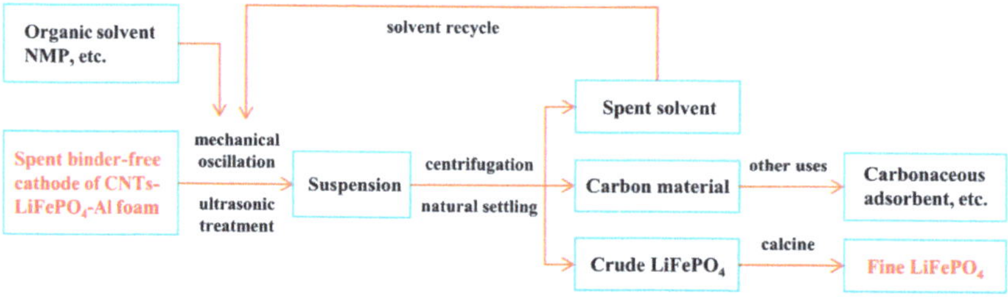

Figure 4. Recycling process of LFP and CNT nanomaterials from spent binder-free cathode of CNTs-$LiFePO_4$-Al foam. After a green and high-efficiency procedure, lots of fine LFP and CNTs could be recycled from the spent binder-free cathode due to the application of Al foam as current collector and the well-design ratio of LFP and CNTs.

4. Conclusions

We proposed a strategy to fabricate a binder-free cathode using Al foam and CNTs. Owing to the supporting effect and adhesive effect of Al foam and CNTs at macroscopic and microscopic scales, the binder-free cathode of CNTs-$LiFePO_4$-Al foam exhibited good stability with no obvious material detachment before and after the electrochemical tests. Thanks to the enhanced conductivity of the 3D Al conductive network and wrapped CNTs with high electrical conductivity, it also showed excellent performance for high-rate discharge as high as 20 C and long-term life for 9000 cycles. In addition, this electrode structure is also conducive to the recycling of valuable materials in cathodes (LFP and CNTs) through simple ultrasonication, sedimentation, filtration, etc. In considering that the electrode must have sufficient adhesion during the charge–discharge process, yet not too strong to hinder ultrasonic separation, the content of CNTs in the binder-free cathode should be carefully optimized, where 3% CNTs of the mass fraction were optimum in this work.

Supplementary Materials: The following supporting information can be downloaded at https://www.mdpi.com/article/10.3390/batteries10080261/s1: Figure S1: (a,b) Scanning electron microscope (SEM) images of LFP nanoparticles. (c,d) Transmission electron microscope (TEM) images of CNT powders; Figure S2: Optical image of the binder-free cathodes with 3% CNTs: (a) Al foil-based and (b) Al foam-based; Figure S3: SEM image of 5% CNTs cathode, showing that the excessive content of CNTs lead to poor dispersion and severe aggregation; Figure S4: The differential capacity curves (dQ/dV) of the pouch cells with binder-free cathodes: (a) 1% CNTs, (b) 3% CNTs, and (c) 5% CNTs; Figure S5: (a) Optical images of the binder-free cathode with 5% CNTs after ultrasonic treatment for 40 min and 100 min. (b) Optical micrograph of the binder-free cathode with 5% CNTs after ultrasonic treatment for 100 min.

Author Contributions: Conceptualization, C.C. and W.Q.; methodology, Y.J. and Z.Y.; validation, S.W., J.W. and D.L.; investigation, Y.J., J.W. and D.L.; resources, Y.J.; data curation, S.W. and Z.Y.; writing—original draft preparation, Y.J., S.W. and Z.Y.; writing—review and editing, C.C. and W.Q.; visualization, S.W. and Z.Y.; supervision, W.Q; funding acquisition, C.C. and W.Q. All authors have read and agreed to the published version of the manuscript.

Funding: This research was funded by the National Natural Science Foundation of China (No. 22109085 and No. 21975142).

Data Availability Statement: The data that support the findings of this study are available from the corresponding authors upon reasonable request.

Conflicts of Interest: The authors declare no conflicts of interest.

References

1. Liu, C.; Li, F.; Ma, L.-P.; Cheng, H.-M. Advanced Materials for Energy Storage. *Adv. Mater.* **2010**, *22*, E28–E62. [CrossRef] [PubMed]
2. Harper, G.; Sommerville, R.; Kendrick, E.; Driscoll, L.; Slater, P.; Stolkin, R.; Walton, A.; Christensen, P.; Heidrich, O.; Lambert, S.; et al. Recycling lithium-ion batteries from electric vehicles. *Nature* **2019**, *575*, 75–86. [CrossRef] [PubMed]
3. Xu, J.; Cai, X.; Cai, S.; Shao, Y.; Hu, C.; Lu, S.; Ding, S. High-Energy Lithium-Ion Batteries: Recent Progress and a Promising Future in Applications. *Energy Environ. Mater.* **2023**, *6*, e12450. [CrossRef]
4. Zhang, M.; Wang, L.; Wang, S.; Ma, T.; Jia, F.; Zhan, C. A Critical Review on the Recycling Strategy of Lithium Iron Phosphate from Electric Vehicles. *Small Methods* **2023**, *7*, 2300125. [CrossRef] [PubMed]
5. Zeng, X.; Li, J. Implications for the carrying capacity of lithium reserve in China. *Resour. Conserv. Recycl.* **2013**, *80*, 58–63. [CrossRef]
6. Wu, J.; Mackenzie, A.; Sharma, N. Recycling lithium-ion batteries: Adding value with multiple lives. *Green Chem.* **2020**, *22*, 2244–2254. [CrossRef]
7. Mancini, M.; Hoffmann, M.F.; Martin, J.; Weirather-Koestner, D.; Axmann, P.; Wohlfahrt-Mehrens, M. A proof-of-concept of direct recycling of anode and cathode active materials: From spent batteries to performance in new Li-ion cells. *J. Power Sources* **2024**, *595*, 233997. [CrossRef]
8. Lv, W.; Wang, Z.; Cao, H.; Sun, Y.; Zhang, Y.; Sun, Z. A Critical Review and Analysis on the Recycling of Spent Lithium-Ion Batteries. *Acs Sustain. Chem. Eng.* **2018**, *6*, 1504–1521. [CrossRef]

9. He, Y.; Yuan, X.; Zhang, G.; Wang, H.; Zhang, T.; Xie, W.; Li, L. A critical review of current technologies for the liberation of electrode materials from foils in the recycling process of spent lithium-ion batteries. *Sci. Total Environ.* **2021**, *766*, 142382. [CrossRef] [PubMed]
10. Du, K.; Ang, E.H.; Wu, X.; Liu, Y. Progresses in Sustainable Recycling Technology of Spent Lithium-Ion Batteries. *Energy Environ. Mater.* **2022**, *5*, 1012–1036. [CrossRef]
11. Su, X.; Fang, H.; Yang, H.; Zou, F.; Li, G.; Wang, L.; Liao, H.; Guan, W.; Hu, X. Cellulose sulfate lithium as a conductive binder for $LiFePO_4$ cathode with long cycle life. *Carbohydr. Polym.* **2023**, *313*, 120848. [CrossRef] [PubMed]
12. Chou, S.-L.; Pan, Y.; Wang, J.-Z.; Liu, H.-K.; Dou, S.-X. Small things make a big difference: Binder effects on the performance of Li and Na batteries. *Phys. Chem. Chem. Phys.* **2014**, *16*, 20347–20359. [CrossRef] [PubMed]
13. Liu, P.; Xiao, L.; Chen, Y.; Tang, Y.; Wu, J.; Chen, H. Recovering valuable metals from $LiNi_xCo_yMn_{1-x-y}O_2$ cathode materials of spent lithium ion batteries via a combination of reduction roasting and stepwise leaching. *J. Alloys Compd.* **2019**, *783*, 743–752. [CrossRef]
14. He, S.; Xiang, W.; He, W.; Yu, F.; Liu, Z. Recovery of spent $LiCoO_2$ cathode material: Thermodynamic analysis and experiments for precipitation and separation of elements. *Chem. Eng. J.* **2022**, *429*, 132371. [CrossRef]
15. Tao, R.; Xing, P.; Li, H.; Sun, Z.; Wu, Y. Recovery of spent $LiCoO_2$ lithium-ion battery via environmentally friendly pyrolysis and hydrometallurgical leaching. *Resour. Conserv. Recycl.* **2022**, *176*, 105921. [CrossRef]
16. Zhou, M.; Li, B.; Li, J.; Xu, Z. Pyrometallurgical Technology in the Recycling of a Spent Lithium Ion Battery: Evolution and the Challenge. *Acs EsT Eng.* **2021**, *1*, 1369–1382. [CrossRef]
17. Landi, B.J.; Ganter, M.J.; Cress, C.D.; DiLeo, R.A.; Raffaelle, R.P. Carbon nanotubes for lithium ion batteries. *Energ. Environ. Sci.* **2009**, *2*, 638–654. [CrossRef]
18. Luo, S.; Wang, K.; Wang, J.; Jiang, K.; Li, Q.; Fan, S. Binder-Free $LiCoO_2$/Carbon Nanotube Cathodes for High-Performance Lithium Ion Batteries. *Adv. Mater.* **2012**, *24*, 2294–2298. [CrossRef] [PubMed]
19. Zhu, X.; Wu, X.; The Nam Long, D.; Tian, Y.; Zhao, H.; Chen, P. Binder-free flexible $LiMn_2O_4$/carbon nanotube network as high power cathode for rechargeable hybrid aqueous battery. *J. Power Sources* **2016**, *326*, 498–504. [CrossRef]
20. Guo, Y.; Li, X.; Wang, Z.; Wang, J.; Guo, H.; Yan, G. Free-standing ultrathick $LiMn_2O_4$@single-wall carbon nanotubes electrode with high areal capacity. *J. Energy Chem.* **2022**, *73*, 452–459. [CrossRef]
21. Yang, Z.; Tian, J.; Ye, Z.; Jin, Y.; Cui, C.; Xie, Q.; Wang, J.; Zhang, G.; Dong, Z.; Miao, Y.; et al. High energy and high power density supercapacitor with 3D Al foam-based thick graphene electrode: Fabrication and simulation. *Energy Storage Mater.* **2020**, *33*, 18–25. [CrossRef]
22. Yang, Z.; Wang, J.; Cui, C.; Jin, Y.; Zhang, G.; Zhou, H.; Qian, W. High power density & energy density Li-ion battery with aluminum foam enhanced electrode: Fabrication and simulation. *J. Power Sources* **2022**, *524*, 230977. [CrossRef]
23. Wei, S.; Cui, C.; Jin, Y.; Wang, J.; Wang, J.; Li, D.; Qian, W. Enhancement of Li intercalation kinetics of $LiFePO_4$ nanoparticles with mesoporous carbon. *Energy Mater.* **2024**, *4*, 400062. [CrossRef]
24. Qian, W.; Liu, T.; Wang, Z.; Yu, H.; Li, Z.; Wei, F.; Luo, G. Effect of adding nickel to iron-alumina catalysts on the morphology of as-grown carbon nanotubes. *Carbon* **2003**, *41*, 2487–2493. [CrossRef]
25. Qian, W.; Wei, F.; Wang, Z.; Liu, T.; Yu, H.; Luo, G.; Xiang, L.; Deng, X. Production of carbon nanotubes in a packed bed and a fluidized bed. *AIChE J.* **2003**, *49*, 619–625. [CrossRef]
26. Qian, W.; Liu, T.; Wei, F.; Wang, Z.; Luo, G.; Yu, H.; Li, Z. The evaluation of the gross defects of carbon nanotubes in a continuous CVD process. *Carbon* **2003**, *41*, 2613–2617. [CrossRef]
27. Qian, W.; Wei, F.; Liu, T.; Wang, Z. The formation mechanism of the coaxial carbon-metal nanowires in a chemical vapor deposition process. *Solid State Commun.* **2003**, *126*, 365–367. [CrossRef]
28. Delaporte, N.; Ossonon, D.B.; Zaghib, K.; Belanger, D. Fabrication of Current Collectors and Binder-Free Electrodes on Separators Used in Lithium-Ion Batteries. *Batter. Supercaps* **2020**, *3*, 638–646. [CrossRef]
29. Ezzedine, M.; Jardali, F.; Florea, I.; Cojocaru, C.-S. Nanostructured S@VACNTs Cathode with Lithium Sulfate Barrier Layer for Exceptionally Stable Cycling in Lithium-Sulfur Batteries. *J. Electrochem. Soc.* **2024**, *171*, 050531. [CrossRef]
30. Pang, X.; Lee, H.; Rong, J.; Zhu, Q.; Xu, S. Self-Thermal Management in Filtered Selenium-Terminated MXene Films for Flexible Safe Batteries. *Small* **2024**, *20*, 2309580. [CrossRef] [PubMed]
31. Guo, X.; Lan, T.; Feng, X.; Li, D.; Chen, Y. Binder-Free Thin-Film Electrode Fabricated by Spray Drying Method: A Case of $LiFePO_4$. *J. Electrochem. Energy Convers. Storage* **2021**, *18*, 011016. [CrossRef]
32. Wei, T.; Tong, J.; Gao, F.; Weng, J.; Tong, Q. High-rate performance of binder-free $LiFePO_4$ cathode prepared by using various types of conductive carbons. *J. Electroanal. Chem.* **2022**, *927*, 116966. [CrossRef]
33. Ramos-Fajardo, J.M.; Pelaez-Tirado, I.M.; Marin-Rueda, J.R.; Castro-Garcia, M.; Canales-Vazquez, J.; Perez-Flores, J.C. LFP-based binder-free electrodes produced via fused filament fabrication. *J. Phys. Energy* **2023**, *5*, 035010. [CrossRef]
34. Xu, S.; Peng, B.; Pang, X.; Huang, F. Anionic Activity in Fast-Charging Batteries: Recent Advances, Prospects, and Challenges. *ACS Mater. Lett.* **2022**, *4*, 2195–2209. [CrossRef]

35. Li, S.; Wang, K.; Zhang, G.; Li, S.; Xu, Y.; Zhang, X.; Zhang, X.; Zheng, S.; Sun, X.; Ma, Y. Fast Charging Anode Materials for Lithium-Ion Batteries: Current Status and Perspectives. *Adv. Funct. Mater.* **2022**, *32*, 2200796. [CrossRef]
36. Cai, W.; Yao, Y.-X.; Zhu, G.-L.; Yan, C.; Jiang, L.-L.; He, C.; Huang, J.-Q.; Zhang, Q. A review on energy chemistry of fast-charging anodes. *Chem. Soc. Rev.* **2020**, *49*, 3806–3833. [CrossRef] [PubMed]

batteries

MDPI

Article

Effects of Crystalline Diamond Nanoparticles on Silicon Thin Films as an Anode for a Lithium-Ion Battery

Yonhua Tzeng *, Cheng-Ying Jhan, Shi-Hong Sung and Yu-Yang Chiou

Institute of Microelectronics, National Cheng Kung University, Tainan 70101, Taiwan;
q18101505@gs.ncku.edu.tw (C.-Y.J.); q16111253@gs.ncku.edu.tw (S.-H.S.); q16121135@gs.ncku.edu.tw (Y.-Y.C.)
* Correspondence: tzengyo@mail.ncku.edu.tw

Abstract: Crystalline diamond nanoparticles which are 3.6 nm in size adhering to thin-film silicon results in a hydrophilic silicon surface for uniform wetting by electrolytes and serves as a current spreader for the prevention of a local high-lithium-ion current density. The excellent physical integrity of an anode made of diamond on silicon and the long-life and high-capacity-retention cycling performance are thus achieved for lithium-ion batteries. A specific capacity of 1860 mAh/g(si) was retained after 200 cycles of discharge/charge at an areal current density of 0.2 mA/cm^2. This is compared to 1626 mAh/g(si) for a thin-film-silicon anode without the additive of diamond nanoparticles.

Keywords: lithium-ion battery; thin film; silicon; anode; diamond; electrolyte wetting

1. Introduction

Rapid development in electric vehicles, renewable energy, high-performance computing, and mobile electronic devices needs high-performance energy storage devices [1–3]. Graphite is the predominant anode material for the most popular mobile energy storage devices of lithium-ion batteries (LIBs), although the specific capacity (372 mAh/g) of graphite is not as high as desirable. Silicon (Si) offers a high theoretical specific capacity (3579 mAh/g for crystalline $Li_{15}Si_4$) and a relatively low discharge voltage (about 0.4 V vs. Li/Li^+), making it an attractive candidate for next-generation anode materials for high-energy-density LIBs [4]. However, the low electrical conductivity of silicon and its large volume expansion due to alloying with lithium (about 400%) adversely affect the charge–discharge cycling stability and electrochemical performance of silicon-based anodes [5,6]. Many efforts were made to overcome barriers against the wide adoption of silicon-based anodes. Focus has been on the reduction of the silicon size [7], hybridization [8], and novel structures [9]. Si-based nanoparticles [10], nanowires [11], nanotubes [12], and thin films [13] represent nanoscale structures for the improved cycling stability of silicon-based anodes. Silicon thin films have garnered special attention due to their ease of fabrication, low chemical residue, and precise stoichiometry. A flexible silicon thin-film electrode is especially welcome and suitable for wearable electronics.

Qiu et al. [14] reported Si-thin-film-based LIBs fabricated by the RF magnetron sputtering of amorphous carbon/silicon/carbon/silicon multi-layer thin film structures. An electrode with an initial discharge specific capacity of 1888.74 mAh/g was reported to retained a specific capacity of 1243.56 mAh/g after 150 cycles of discharge/charge.

Liu et al. [15] inserted silver nanoparticles (AgNPs) between RF-sputtered amorphous Si thin films. An anode with the densest AgNPs insertion reached 1250 mAh/g at 10 C with 46% capacity retention. The results demonstrated the feasibility of multi-layer thin-film-silicon anodes by interfacial engineering.

In this study, we fabricated silicon thin films of about a 100 nm thickness by thermal evaporation on copper foils. Single-crystal diamond nanoparticles of a 3.6 nm size in water suspension were attached to the silicon surface by ultrasonication. These processes repeated twice to make a two-layer Si/D/Si/D anode of a total thickness of about 200 nm.

The diamond surface layer and interfacial layer resulted in a uniform lithium ion current density, which in turn reduced hot spots, where the local high-lithium-ion current density induces excessive internal stress, resulting in the fracture of the silicon thin films. The diamond nanoparticles' decoration further improves the hydrophilicity of the silicon surface for uniform wetting by water-based electrolyte to form a uniform and stable solid-electrolyte interphase (SEI) on the surface of the anode. The thin-film silicon-based anode retains excellent physical integrity with the reduced exposure of the fresh silicon surface due to the fracture of the silicon films. Thinner and more uniform SEI is formed, leaving electrolyte better preserved for the cycling operation of the battery.

2. Materials and Methods

Silicon films and Si/D/Si/D multilayer films were prepared by thermal evaporation deposition. The process utilized n-type crystalline silicon (purity 99.999%) as the evaporation source and copper foils (diameter 14 mm) as the current collectors. Copper foils were cleaned with acetone and alcohol before the deposition of silicon films.

The deposition process was carried out at a background pressure of 3×10^{-6} torr. The deposition rate for the Si films was 0.4–0.5 nm/min, and the deposition was conducted at 200 °C substrate temperature.

A 200 nm-thick silicon film was deposited on a copper foil and used as a reference anode for a lithium-ion battery. For Si/D/Si/D multilayer films, a 100 nm-thick silicon film was first deposited on a copper foil. The silicon film was immersed in a water solution containing 3.6 nm diamond particles in an ultrasonic agitator. The solution was stirred at 60 rpm for 5 min for uniform dispersion of the diamond nanoparticles on the silicon film. Excess diamond particles were removed by rinsing the sample surface with methanol. The electrode was then dried at 100 °C. The same process repeats when multi-layer structures were made.

Details of the fabrication of coin cells and the materials' characterization have been presented in prior publications. Please refer to references [16–20].

3. Results

Figure 1 shows schematic diagrams of (top) Si and (bottom) Si/D/Si/D thin-film anode structures. The thin-film silicon anode is deposited on a copper foil current collector by thermal evaporation, with a total thickness of 200 nm. In addition, the silicon–diamond composite film was deposited with an initial 100 nm silicon film, followed by immersing the electrode in a 3.6 nm nano-diamonds (NDs) suspension in water and using ultrasonic agitation to evenly distribute the nano-diamonds (NDs) on the surface of the silicon film. After drying on a heated plate, another 100 nm silicon film was deposited, and the 3.6 nm diamond film was dispersed on the silicon anode again using the same method.

We use this schematic diagram to illustrate the effect of nano-diamonds (NDs) on the uniform lithiation of the thin-film electrode. The red line represents the pathway of the lithium ions, and the blue diamond-shaped marks represent nano-diamond (ND) particles. In samples without a diamond layer, uneven lithiation during the cycling electrochemical process led to excessive reactions in some local areas, causing the continuous cracking of the thin film and regeneration of the SEI layer at those locations. The severe deformation of the local silicon layer can lead to issues such as the fracture of the active silicon material and electrolyte consumption by the formation of SEI on the newly exposed silicon surface.

Figure 2 displays the scanning electron microscope (SEM) images of anodes made of silicon thin films and Si/D/Si/D thin films. Clustered silicon grains can be seen on the surface of the silicon anode. Nevertheless, except local silicon clusters, the surface morphology appears to be smoother for the silicon anode than the one with diamond nanoparticles on the silicon surface and at the interface between the two silicon layers.

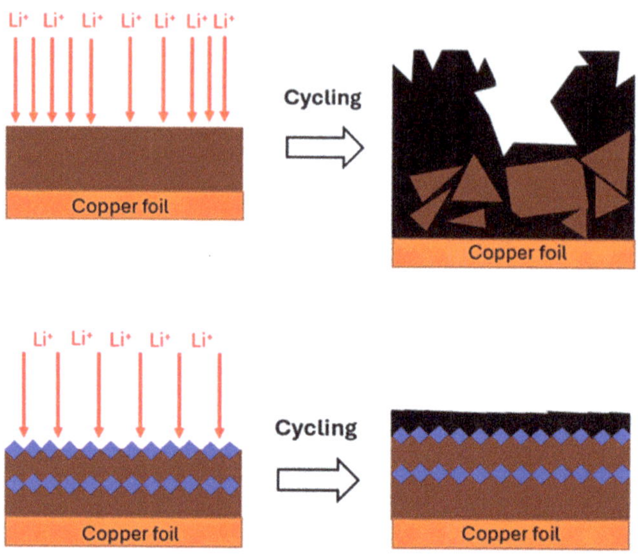

Figure 1. Schematic of the thin-film anodes showing the effects of nano-diamonds (NDs) on the uniform lithiation of the anode made of (**top**) Si and (**bottom**) Si/D/Si/D. Red arrows represent the transportation of lithium ions. Blue diamond shape represents nano-diamond (ND) particles. The black area represents SEI layers. Samples without diamond layers exhibited structural instability during cycling, leading to drastic changes in local and overall thickness of the anode.

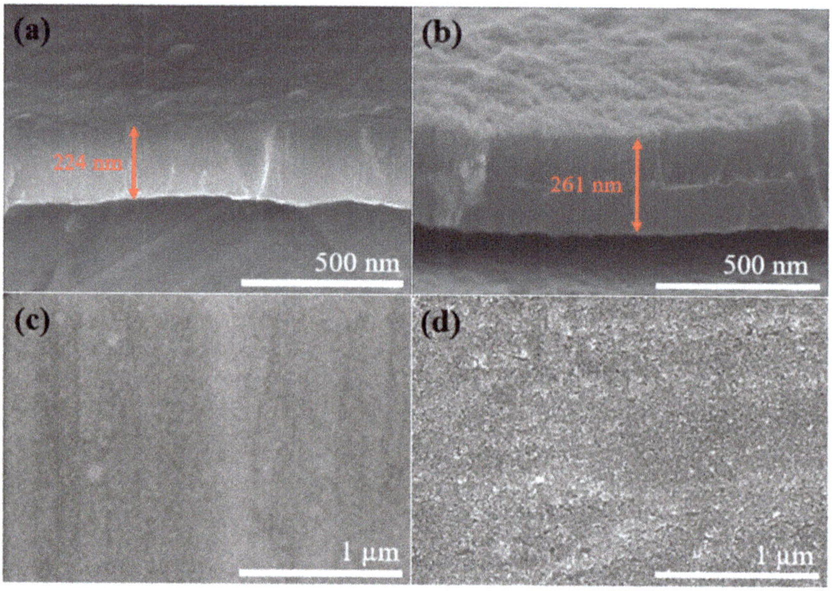

Figure 2. SEM cross-sectional images of (**a**) Si anode and (**b**) Si/D/Si/D anode; SEM top-view images of the (**c**) Si anode and (**d**) Si/D/Si/D anode.

Figure 2a shows a thermal evaporation deposited silicon anode with a thickness of approximately 224 nm, while Figure 2b illustrates the layered Si/D/Si/D thin-film electrode. The rough surface morphology of the Si/D/Si/D anode might be attributed to the thicker anode. Since the thicker the silicon anode is, the worse the cycling performance is, the adverse effect of being a slightly thicker Si/D/Si/D anode than that of the Si anode will not affect the conclusion of the performance comparison as long as the thicker Si/D/Si/D anode performs better.

Compared to the characteristic Raman peak shown in Figure 3 of the crystalline silicon at 520 cm^{-1}, the characteristic silicon peak of the deposited silicon thin film is at a lower wavenumber of 493 cm^{-1}. This is due to the lower crystallinity of the deposited silicon thin film. Additionally, the red shift of the characteristic peak is consistent with the characteristics of the small-grain-sized polycrystalline silicon. The silicon thin films are amorphous or, at best, nanocrystalline [21,22].

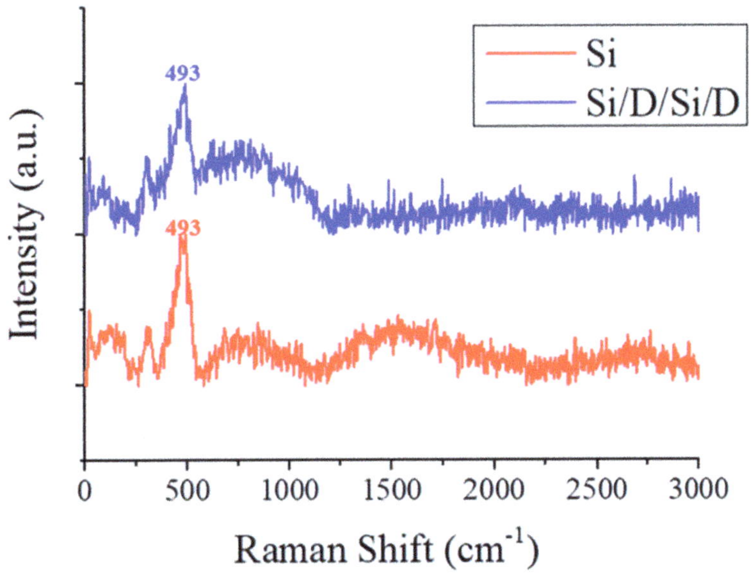

Figure 3. Raman spectra of a Si anode and a Si/D/Si/D anode.

Due to the superior cycling performance of the diamond-on-silicon anode, we conducted cyclic voltammetry (CV) tests on both anodes, i.e., silicon and Si/D/Si/D films, to understand their electrochemical performance. The results are shown in Figure 4.

The cathodic peak below 0.3 V is related to the formation of a series of amorphous phases of Li_xSi. It is worth noting that during the first charge–discharge cycle, the reduction peaks of the Si film at 0.22 V and 0.07 V are relatively unstable, as shown in Figure 4a. This phenomenon may be attributed to the uneven charging and discharging caused by different-sized amorphous silicon particles on the copper foil. Additionally, the reaction peak (0.44 V) of lithium ions migrating from the silicon anode in the first, the second, and the third scans of the Si film is inconsistent, indicating the poor reversibility of the reaction in this sample [23,24].

On the other hand, for the Si/D/Si/D anode, Figure 4b shows that the characteristic peaks of the sample do not change as much as those for the silicon anode with different scans. The almost identical CV graphs of the three scans corresponding to the first three cycles indicate good electrochemical stability and suggest uniform alloying and de-alloying processes.

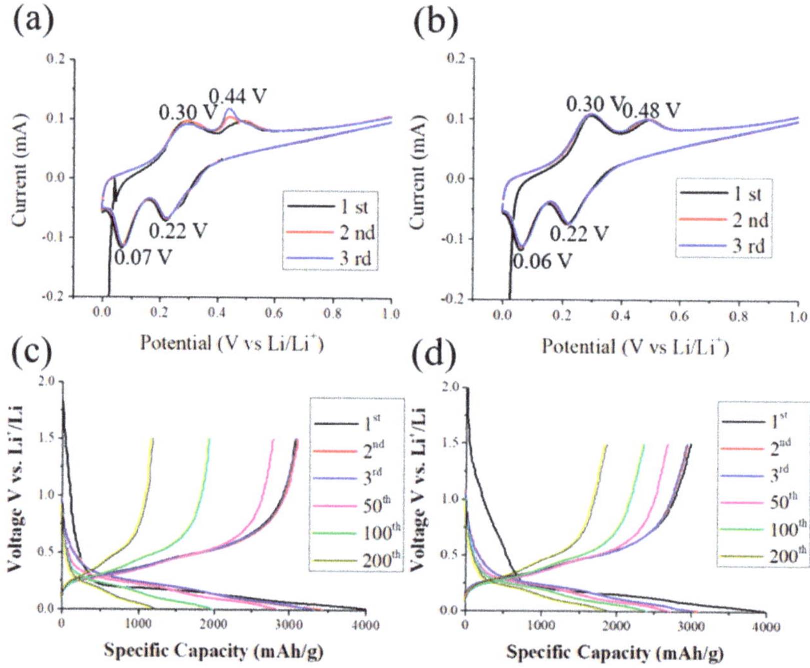

Figure 4. CV curves of (**a**) Si anode and (**b**) Si/D/Si/D anode at a scan rate of 0.05 mV/s. Charge–discharge curves of several cycle numbers for the (**c**) Si anode and (**d**) Si/D/Si/D anode.

We tested both electrodes using a battery charge–discharge system in a voltage window between 0.01 V and 1.50 V and at a current density of 0.05 mA/cm^2. The reversible capacities of the Si anode and the Si/D/Si/D anode for the first cycle were 0.143 mAh/cm^2 (3077 mAh/g (si)) and 0.140 mAh/cm^2 (2983 mAh/g (si)), respectively. In the first cycle, more SEI layers were observed in the Si/D/Si/D film during the initial electrochemical process. This is possibly due to the increase in the specific surface area of the rough electrode surface caused by diamond nanoparticles, leading to more initial SEI formation. However, the initial Coulombic efficiency of the Si/D/Si/D anode (75.1%) is comparable to that of the silicon anode (76.4%), indicating that the Si/D/Si/D structure can provide both a uniform and stable lithium ion transport path and a stable SEI film, avoiding the entrapment of silicon–lithium alloys during the cycling process. The calculation of the specific capacity is based on the weight of the silicon.

The Si/D/Si/D anode exhibits superior electrochemical reversibility and retains a reversible capacity of 0.11 mAh/cm^2 (2356 mAh/g (si)) after 100 cycles, which is compared to 0.09 mAh/cm^2 (1938 mAh/g (si)) for the silicon anode. After 200 cycles, the reversible capacity of the Si/D/Si/D anode was still retained at 0.087 mAh/cm^2 (1862 mAh/g (si)), with a Coulombic efficiency of 99%, as displayed in Figure 4d. In contrast, the silicon anode shows a reversible capacity of 0.054 mAh/cm^2 (1159 mAh/g (si)) and a Coulombic efficiency of 98.4%, as shown in Figure 4c. The multi-layer diamond-on-silicon anode has a stable SEI and a uniform ion transportation, thus exhibiting better stability in long-term cycling and a more stable Coulombic efficiency [25,26].

Figure 5a shows a poor cycling performance of the silicon anode and much better cycling performance of the Si/D/Si/D anode. Diamond nanoparticles serve to spread out the lithium ion current and improve the wettability of the silicon anode by the electrolyte, resulting in reduced internal local stress due to non-uniform volume changes by the silicon–lithium alloying and de-alloying processes. The much-improved physical integrity results in the desirable higher retention of the charge capacity. When the charge–discharge cycle is maintained for two hundred cycles, the improvement in retention is more obvious.

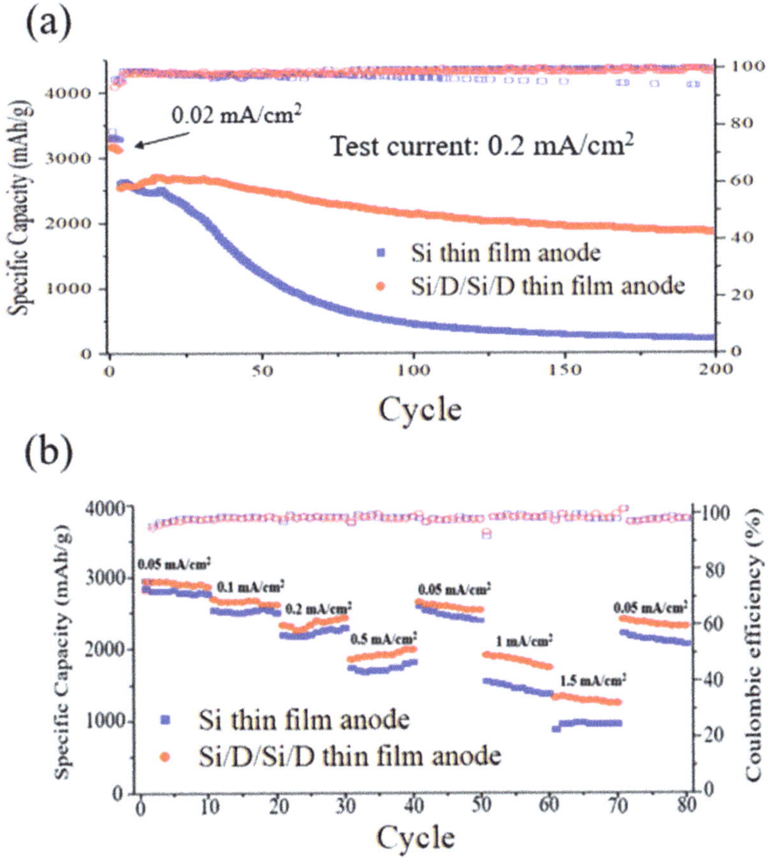

Figure 5. (**a**) Cycle performance of Si thin-film anode and Si/D/Si/D thin-film anode during the first three cycles of charge and discharge at a current density of 0.02 mA/cm^2, and for subsequent cycles at a current density of 0.2 mA/cm^2. (**b**) C-Rate performance of Si anode and Si/D/Si/D anode.

Figure 5b shows the C-Rate performance of the Si anode and Si/D/Si/D anode. We gradually increased the test current from 0.05 mA/cm^2 to 0.5 mA/cm^2. At this point, the specific capacity of the Si/D/Si/D film was approximately 2000 mAh/g(si), while the silicon anode retained about 1750 mAh/g(si). When the test current was reduced back to 0.05 mA/cm^2, the specific capacity of the Si/D/Si/D anode was restored to 2500 mAh/g(si), and the silicon anode's specific capacity was approximately 2400 mAh/g(si). We further increased the test current to 1.5 mA/cm^2 (15 C). Thanks to the uniform ion transportation of the two-layer diamond-on-silicon anode, a specific capacity of about 1350 mAh/g(si) at the equivalent of 15 C was retained. In contrast, the silicon anode only retained a specific capacity of about 1000 mAh/g(si) using the same test current. The uneven charge–discharge process led to structural damages, which degraded the anode performance in the subsequent lower-current tests.

Furthermore, the effectiveness of the surface diamond nanoparticles in uniformly dispersing the charge–discharge current while enhancing the strength of the SEI layer was demonstrated. The diamond-on-silicon anode retained a specific capacity of 1600 mAh/g(si) after 10 charge–discharge cycles at 1 mA/cm^2 and 1400 mAh/g(si) after 10 charge–discharge cycles at 1.5 mA/cm^2, both of which were much higher than that of the silicon anode [15,27].

Figure 6a,c show optical microscope surface images of a silicon anode after (a) 200 cycles and (c) 80 cycles of C-rate testing. Figure 6b,d show optical microscope surface images of the Si/D/Si/D anode after completing (b) 200 cycles and (d) 80 cycles of C-rate testing, respectively. For the silicon anode, we observed that in some areas, the silicon film detached from the copper current collector. The overall electrochemical reaction was uneven with more silicon detachment on one side than the other of the electrode. The partial detachment of the silicon film from the current collector was displayed.

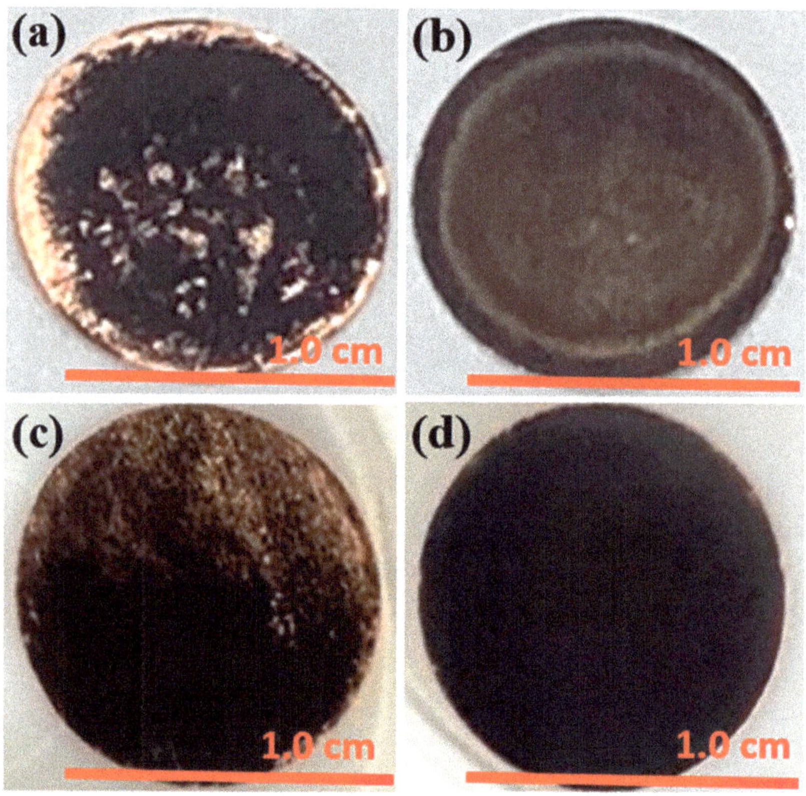

Figure 6. Optical microscope images of (**a**) Si anode and (**b**) Si/D/Si/D anode after 200 cycles. (**c**) Si anode and (**d**) Si/D/Si/D anode after 80 C-rate testing cycles.

On the other hand, the overall electrode structure of the two-layer diamond-on-silicon anode remained intact after cycling. This confirms that the diamond layers evenly disperse the charge–discharge current, avoiding local excessive reactions, which could lead to local stress and electrode polarization and fracture [27,28].

After a long-term cycling test, the anode surface was characterized using secondary electron microscopy and EDS as shown in Figure 7, to observe changes in the microstructure, elemental composition, and cross-sectional thickness. The formation of separated silicon islands was noted. They might be attributed to the plastic deformation of the copper foil current collector. During cycling, the alloying and de-alloying induced volume changes increased strain and led to the formation of islands with gaps between them. The delamination of silicon from copper was observed. The formation of copper–silicon–lithium compounds (Cu_2LiSi) during cycling might have weakened the adhesion of silicon to copper, resulting in the delamination of silicon and the loss of electrical contact.

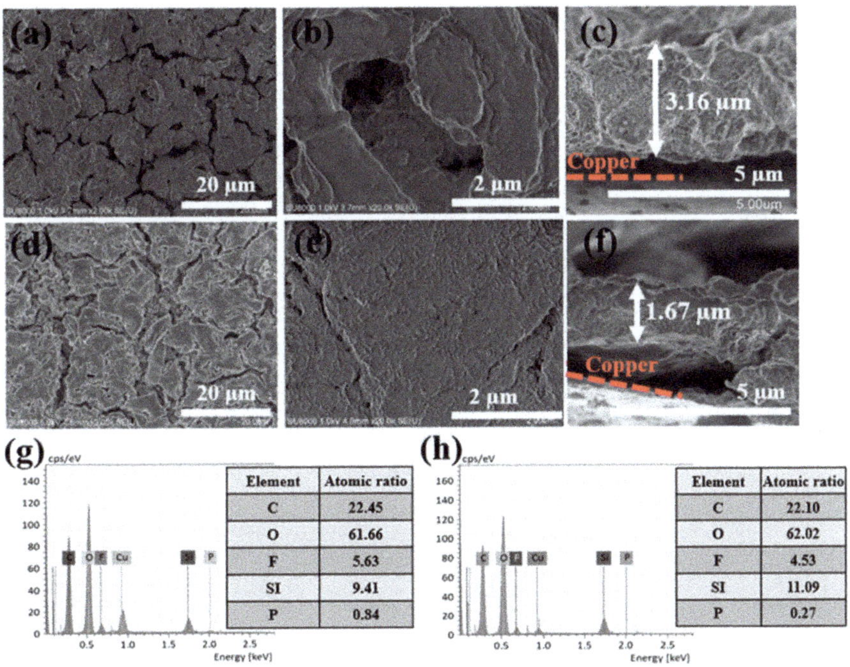

Figure 7. SEM images of (**a**,**b**) the surface of Si anode after 200 cycles at different magnification, (**c**) a cross-section of a Si anode; SEM images of (**d**,**e**) the surface of a Si/D/Si/D anode after 200 cycles at different magnification, (**f**) a cross-section of a Si/D/Si/D anode; EDS analysis of (**g**) a silicon anode and (**h**) a Si/D/Si/D anode after 200 cycles.

The surface of the silicon anode exhibited severe roughness and showed numerous protrusions and deep holes in the anode after long-term cycling, indicating uneven reactions in different areas of the anode. This unevenness could partially be attributed to the irregularity of the copper surface or the deposition of uneven silicon films. Without the diamond layer for spreading the lithium transport on the surface area, where a higher lithium ion current density would lead to more alloying and thus a larger volume expansion, non-even volume expansion causes localized stress, which results in cracks to form in the silicon films and the exposure of fresh silicon for the formation of additional SEI layers.

Volume expansion and the build-up of SEI layers caused the anode thickness to increase to nearly 3 µm in some areas of the anode after 200 cycles of the charge–discharge operation. Volume expansion is caused by continuous generation of additional solid electrolyte interphase (SEI). Stress induced by uneven volume expansion might lead to the loss of electrical contacts between the active silicon material and its electrical conductive network. Additionally, the growth of SEI and the increase in the electrical resistance of the anode hindered the rapid diffusion of Li ions and electron conduction, and even trapped some Li irreversibly in the Si anode.

The surface of the silicon anode exhibited a higher phosphorus content than that of the Si/D/Si/D anode. It might originate from phosphates (POx), which is the main component of the SEI, or arise from side reactions between the silicon and electrolyte. The remaining fluoride content likely originated from residual electrolyte trapped in the solid electrolyte interphase.

The two-layer diamond-on-silicon thin-film anode displayed a smoother surface and far less of an increase in thickness after cycling than the silicon anode. A more stable solid electrolyte interphase and lower polarization were achieved by the surface layer and the interfacial layer diamond nanoparticles in a thin-film silicon anode for the lithium ion battery [29,30].

X-ray Photoelectron Spectroscopy (XPS) analyses of the chemical composition of both anodes are shown in Figure 8. The concentrations of silicon in the Si anode and the Si/D/Si/D anode are approximately 1.7 and 1.2 atomic %, respectively. Further differences exist in the carbon-to-oxygen ratios and fluorine concentrations. The Si/D/Si/D anode has a higher carbon-to-oxygen ratio of 1.14. Without diamond, the ratio is only 0.52. The fluorine concentration in the Si/D/Si/D anode is higher at 24.7 atomic %, compared to 14.6 atomic % for the silicon anode.

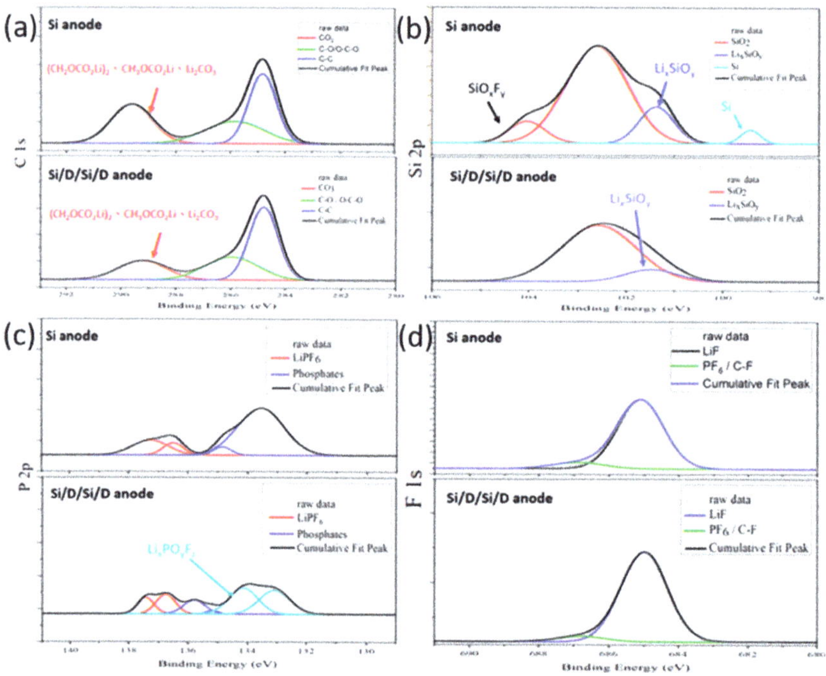

Figure 8. XPS (**a**) C 1s spectra of Si anode and Si/D/Si/D anode; (**b**) Si 2p spectra of Si anode and Si/D/Si/D anode, SiO_2 has two bands shown in red color; (**c**) P 2p spectra of Si anode and Si/D/Si/D anode; and (**d**) F 1s spectra of Si anode and Si/D/Si/D anode after 200 cycles of discharge and charge operation.

These results indicate that the application of diamond nanoparticles in the Si/D/Si/D anode affects the compositional contents of the solid electrolyte interphase (SEI), leading to higher concentrations of carbon and fluorine. This also suggests a significant impact of nano-diamonds (NDs) on the decomposition, regeneration, and build-up processes of the SEI, potentially influencing the cycling performance and the stability of lithium-ion batteries.

The carbon spectral results reveal that SEI components are located mainly near the anode surface of the silicon anode after long-term cycling. This indicates the more rapid consumption of the electrolyte.

The silicon spectra show that there is a stronger signal of silicon on the surface of the silicon anode, indicating that the SEI layer cracks repetitively during the charge and discharge process, leading to the continuous exposure of fresh silicon surfaces, which would cause the build-up of additional SEI.

After cycling, the characteristic value of the binding energy for the C 1s peak is maximized at 284.8 eV, mainly corresponding to the C–C and C-H bonds, as shown in Figure 8a. The second local maximum is at 289.7 eV, which is due to lithium carbonate and different semi-organic lithium carbonate. These results demonstrate the presence of typical SEI components such as $(CH_2OCO_2Li)_2$, CH_3OCO_2Li, CH_3OLi, and Li_2CO_3 in the alkyl carbonate. The signal intensity is proportional to the degree of electrolyte decomposition. For the Si/D/Si/D anode, the main signal shifts to 289.2 eV, which proves that the SEI layer on this anode is thinner [29].

The Si 2p spectra in Figure 8b show signals with maxima at around 101 eV, which corresponds to Li_xSiO_y. It is the initial reaction product of lithium with SiO_2. SiO_2 is present in a higher concentration on the original anode. HF is a by-product of the decomposition reaction of PF_6- ions with trace amounts of water and FEC additives. It can be clearly seen from reactions (1)–(3) that a large amount of Li_2O and Li_4SiO_4 are formed [31]. For the silicon anode, we observed a shoulder peak at 105 eV, corresponding to fluorinated silicon suboxide (SiO_xF_y) [31,32].

$$SiO_2 + 4Li \rightarrow Si + 2Li_2O \quad (1)$$

$$2SiO_2 + 4Li \rightarrow Si + Li_4SiO_4 \quad (2)$$

$$SiO_2 + HF \rightarrow SiO_xF_y + H_2O \quad (3)$$

The P2p spectrum in Figure 8c shows a binding energy of 134.0 eV, proving that mainly phosphates and fluorine-substituted phosphates are present. The presence of FEC leads to the hydrolyzing of a large amount of $LiPF_6$ into (fluorinated) phosphates, which are integrated into the inorganic part of the SEI. The general reaction path for phosphate formation has been described in the literature [32]. Therefore, more water is generated in the Si/D/Si/D anode. This is probably due to the reaction of silicon and Li_2O with HF. Similar results were reported by Elazari et al. [33]. Hydrolysis is considered the main factor for the irreversible capacity loss in LIBs, but it appears to have a positive effect on the electrochemical performance. This behavior is consistent with the report by Dalavi et al., who observed a positive effect on the electrochemical performance when decomposed $LiPF_6$ was included in the SEI [34].

The F 1s spectra of both anodes shown in Figure 8d are composed of low-intensity signals at 687.7 eV and a significant signal at 685.5 eV. These are from trace amounts of $LiPF_6$, $Li_xPO_yF_z$, and LiF. The phosphorus species can be confirmed by the P 2p spectrum (see Figure 8c). The signal at 137.5 eV indicates trace amounts of conductive salts, while the signal at 134.5 eV points to partially fluorinated phosphates. The presence of fluorinated phosphates after cycling in conventional carbonate electrolytes as parts of the SEI was reported [32,35].

Figure S1 shows that the crystalline diamond nanoparticles of 3.6 nm adhering on the surface of a silicon thin film improves its hydrophilicity for the uniform wetting of the anode by the electrolyte. Uniform SEI growth on the electrode and the current spreading effects of diamond nanoparticles, due to the slow diffusion of lithium through diamond, are attributed to the prevention of excessive internal stress due to uneven volume expansion across the anode and a smaller increase in anode thickness after 200 cycles of charge–discharge operation. The damages to the anode with diamond nanoparticles are thus reduced and the much better cycling life time and capacity retention are achieved by the Si/D/Si/D anode in comparison with the pure silicon anode. This is further confirmed by the significantly smaller electrochemical impedance enabled by diamond nanoparticles, as shown in Figure S2 and Table S1. Cycling performance of additional silicon–diamond structures are presented in Figure S3.

4. Conclusions

We have discovered that the surface and interfacial layers of single-crystal diamond nanoparticles significantly improve the performance of thin-film silicon-based anodes for lithium-ion batteries. Diamond nanoparticles of 3.6 nm-sized crystals adhere to silicon thin films in the ambient environments without needing vacuum-based diamond CVD processes. The double-layer Si/D/Si/D anode exhibits the uniform alloying of the silicon with lithium and the formation of SEI on the surface, reducing the internal stress and preventing the anode from fracture and pulverization. It results in the desired physical integrity of the anode, including a more stable SEI layer. Significant improvements in the electrochemical stability and reversible capacity retention under high-rate and long-term cycling conditions were demonstrated in comparison with the thin-film silicon anode without diamond additives. The X-ray photoelectron spectroscopy (XPS) analysis reveals that the addition of crystalline nano-diamonds (NDs) leads to the composition of the SEI to contain more carbon and fluorine. This indicates that crystalline nano-diamonds (NDs) play important roles in the decomposition and regeneration processes of the SEI. A facile process of adhering crystalline diamond nanoparticles on thin-film silicon surfaces as an anode material with significantly improved cycling performance for lithium-ion batteries has been reported.

Supplementary Materials: The following supporting information can be downloaded at: https://www.mdpi.com/article/10.3390/batteries10090321/s1, Figure S1: Water on anode surface for comparison in hydrophilicity of Si thin-film anode and Si/D/Si/D thin-film anode [36]; Figure S2: Electrochemical impedance spectra of Si thin-film anode and Si/D/Si/D thin-film anode after the 200th electrochemical cycling: (a) typical Nyquist plot; (b) the equivalent circuit used to fit impedance data. R1 is the cell internal resistance. R2 is the charge transfer resistance. CPE is the double-layer pseudo-capacity. Warburg Element is the solid-state diffusion element [37]; Figure S3: Cycling performance of Si, Si/D/Si, Si/D/Si/D, and Si/D/Si/D/Si/D thin-film anode under test current 0.05 mA/cm^2. Table S1: Parameter values of equivalent circuits based on EIS spectra of anodes of Si thin film and Si/D/Si/D thin film after 200 cycles of charge–discharge.

Author Contributions: Y.T.: Conception, funding, data analysis, interpretation, final manuscript preparation; C.-Y.J.: Material characterization, data analysis, draft manuscript preparation; S.-H.S.: Device fabrication, data collection, initial data analysis, report preparation; Y.-Y.C.: Experimental assistance, data collection and verification, assistance to report preparation. All authors have read and agreed to the published version of the manuscript.

Funding: This work is partially supported by the National Science and Technology Council in Taiwan under the grant number NSTC-112-2221-E-006-075-.

Data Availability Statement: No new data were created.

Acknowledgments: The authors gratefully acknowledge the use of the scanning electron micro-scope (SEM, Hitachi-SU8000, Taipei, Taiwan) and X-ray Photoelectron Spectrometer (PHI Versa Probe 4, Taipei, Taiwan) belonging to the Core Facility Center of the National Cheng Kung University.

Conflicts of Interest: The authors declare that they have no competing financial interests or personal relationships that appear to have influenced the work reported in this paper.

References

1. Li, L.; Wu, Z.; Yuan, S.; Zhang, X.B. Advances and challenges for flexible energy storage and conversion devices and systems. *Energy Environ. Sci.* **2014**, *7*, 2101–2122. [CrossRef]
2. Thackeray, M.M.; Wolverton, C.; Isaacs, E.D. Electrical energy storage for transportation—Approaching the limits of, and going beyond, lithium-ion batteries. *Energy Environ. Sci.* **2012**, *5*, 7854–7863. [CrossRef]
3. Bai, S.; Chen, S.; Shen, X.; Zhua, G.; Wang, G. Nanocomposites of hematite (α-Fe_2O_3) nanospindles with crumpled reduced graphene oxide nanosheets as high-performance anode material for lithium-ion batteries. *RSC Adv.* **2012**, *2*, 10977–10984. [CrossRef]
4. Ge, M.; Cao, C.; Biesold, G.M.; Sewell, C.D.; Hao, S.M.; Huang, J.; Lin, Z. Recent advances in silicon-based electrodes: From fundamental research toward practical applications. *Adv. Mater.* **2021**, *33*, 2004577. [CrossRef] [PubMed]

5. Li, L.; Deng, Y.; Hu, K.; Xu, B.; Wang, N.; Bai, Z.; Yang, J. Nanostructure designing and hybridizing of high-capacity silicon-based anode for lithium-ion batteries. *Prog. Nat. Sci. Mater. Int.* **2023**, *33*, 16–36. [CrossRef]
6. Chen, X.; Li, H.; Yan, Z.; Cheng, F.; Chen, J. Structure design and mechanism analysis of silicon anode for lithium-ion batteries. *Sci. China Mater.* **2019**, *62*, 1515–1536. [CrossRef]
7. Kim, H.; Seo, M.; Park, M.H.; Cho, J. A critical size of silicon nano-anodes for lithium rechargeable batteries. *Angew. Chem. Int. Ed.* **2010**, *49*, 2146–2149. [CrossRef]
8. Wu, H.; Cui, Y. Designing nanostructured Si anodes for high energy lithium ion batteries. *Nano Today* **2012**, *7*, 414–429. [CrossRef]
9. Liu, L.; Lyu, J.; Li, T.; Zhao, T. Well-constructed silicon-based materials as high-performance lithium-ion battery anodes. *Nanoscale* **2016**, *8*, 701–722. [CrossRef]
10. Jiang, S.; Hu, B.; Sahore, R.; Zhang, L.; Liu, H.; Zhang, L.; Zhang, Z. Surface-functionalized silicon nanoparticles as anode material for lithium-ion battery. *ACS Appl. Mater. Interfaces* **2018**, *10*, 44924–44931. [CrossRef]
11. Zamfir, M.R.; Nguyen, H.T.; Moyen, E.; Lee, Y.H.; Pribat, D. Silicon nanowires for Li-based battery anodes: A review. *J. Mater. Chem. A* **2013**, *1*, 9566–9586. [CrossRef]
12. Wen, Z.; Lu, G.; Mao, S.; Kim, H.; Cui, S.; Yu, K.; Chen, J. Silicon nanotube anode for lithium-ion batteries. *Electrochem. Commun.* **2013**, *29*, 67–70. [CrossRef]
13. Salah, M.; Murphy, P.; Hall, C.; Francis, C.; Kerr, R.; Fabretto, M. Pure silicon thin-film anodes for lithium-ion batteries: A review. *J. Power Sources* **2019**, *414*, 48–67. [CrossRef]
14. Tong, L.; Wang, P.; Chen, A.; Qiu, F.; Fang, W.; Yang, J.; Wang, C.; Yang, Y. Improved Electrochemical Performance of Binder-Free Multi-Layered Silicon/Carbon Thin Film Electrode for Lithium-Ion Batteries. *Carbon* **2019**, *153*, 592–601. [CrossRef]
15. Chen, Y.X.; Liao, H.C.; Cheng, Y.W.; Huang, J.H.; Liu, C.P. Scalable Interlayer Nanostructure Design for High-Rate (10C) Submicron Silicon-Film Electrode by Incorporating Silver Nanoparticle. *ACS Appl. Mater. Interfaces* **2023**, *15*, 18845–18856. [CrossRef] [PubMed]
16. Jhan, C.-Y.; Sung, S.-H.; Tzeng, Y. Silicon–Nanodiamond-Based Anode for a Lithium-Ion Battery. *Nanomaterials* **2024**, *14*, 43. [CrossRef]
17. Jhan, C.-Y.; Wang, P.-S.; Sung, S.-H.; Tzeng, Y. Effects of volume-confinement on lithium-ion battery with silicon-based anode. *Mater. Today Commun.* **2024**, *39*, 108578. [CrossRef]
18. Tzeng, Y.; Jhan, C.Y.; Wu, Y.H. Effects of pyrolysis on high-capacity Si-based anode of lithium ion battery with high coulombic efficiency and long cycling life. *Nanomaterials* **2022**, *12*, 469. [CrossRef]
19. Tzeng, Y.; Jhan, C.Y.; Chiu, K.M.; Wu, Y.C.; Chen, G.Y.; Wang, P.S. Si–Ni-alloy-assisted very high-areal-capacity silicon-based anode on Ni foam for lithium ion battery. *Mater. Today Chem.* **2023**, *30*, 101570. [CrossRef]
20. Tzeng, Y.; Jhan, C.Y.; Wu, Y.C.; Chen, G.Y.; Chiu, K.M.; Guu SY, E. High-ICE and high-capacity retention silicon-based anode for lithium-ion battery. *Nanomaterials* **2022**, *12*, 1387. [CrossRef]
21. Marcins, G.; Butikova, J.; Tale, I.; Polyakov, B.; Kalendarjov, R.; Muhin, A. Crystallization processes of amorphous Si by thermal annealing and pulsed laser processing. *IOP Conf. Ser. Mater. Sci. Eng.* **2011**, *23*, 012035. [CrossRef]
22. Becker, C.; Dogan, P.; Ruske, F.; Gorka, B.; Sánchez-Vicens, A.; Conrad, E.; Hüpkes, J. Solid phase crystallized silicon thin-film solar cells on temperature-stable ZnO: Al contact layers. In Proceedings of the 23rd European Photovoltaic Solar Energy Conference, Valencia, Spain, 1–5 September 2008; pp. 2045–2048.
23. Lee, S.H.; Choi, M.; Jung, Y.I.; Sim, S.J.; Moon, J.K.; Choi, J.; Kim, S. Deposition and characterization of silicon thin film on stainless steel by electron beam evaporation. *Thin Solid Films* **2022**, *756*, 139380. [CrossRef]
24. Wang, J.W.; He, Y.; Fan, F.; Liu, X.H.; Xia, S.; Liu, Y.; Zhu, T. Two-phase electrochemical lithiation in amorphous silicon. *Nano Lett.* **2013**, *13*, 709–715. [CrossRef]
25. Adhitama, E.; Van Wickeren, S.; Neuhaus, K.; Frankenstein, L.; Demelash, F.; Javed, A.; Placke, T. Revealing the role, mechanism, and impact of AlF3 coatings on the interphase of silicon thin film anodes. *Adv. Energy Mater.* **2022**, *12*, 2201859. [CrossRef]
26. Tong, L.; Wang, P.; Fang, W.; Guo, X.; Bao, W.; Yang, Y.; Qiu, F. Interface engineering of silicon/carbon thin-film anodes for high-rate lithium-ion batteries. *ACS Appl. Mater. Interfaces* **2020**, *12*, 29242–29252. [CrossRef] [PubMed]
27. Tamirat, A.G.; Lui, Y.; Dong, X.; Wang, C.; Wang, Y.; Xia, Y. Ultrathin Silicon Nanolayer Implanted NixSi/Ni Nanoparticles as Superlong-Cycle Lithium-Ion Anode Material. *Small Struct.* **2021**, *2*, 2000126. [CrossRef]
28. Mukanova, A.; Jetybayeva, A.; Myung, S.T.; Kim, S.S.; Bakenov, Z. A mini-review on the development of Si-based thin film anodes for Li-ion batteries. *Mater. Today Energy* **2018**, *9*, 49–66. [CrossRef]
29. Jaumann, T.; Balach, J.; Klose, M.; Oswald, S.; Langklotz, U.; Michaelis, A.; Giebeler, L. SEI-component formation on sub 5 nm sized silicon nanoparticles in Li-ion batteries: The role of electrode preparation, FEC addition and binders. *Phys. Chem. Chem. Phys.* **2015**, *17*, 24956–24967. [CrossRef]
30. Aryanfar, A.; Brooks, D.J.; Colussi, A.J.; Hoffmann, M.R. Quantifying the dependence of dead lithium losses on the cycling period in lithium metal batteries. *Phys. Chem. Chem. Phys.* **2014**, *16*, 24965–24970. [CrossRef]
31. Philippe, B.; Dedryvère, R.; Allouche, J.; Lindgren, F.; Gorgoi, M.; Rensmo, H.; Edström, K. Nanosilicon electrodes for lithium-ion batteries: Interfacial mechanisms studied by hard and soft X-ray photoelectron spectroscopy. *Chem. Mater.* **2012**, *24*, 1107–1115. [CrossRef]
32. Philippe, B.; Dedryvère, R.; Gorgoi, M.; Rensmo, H.; Gonbeau, D.; Edström, K. Role of the LiPF6 salt for the long-term stability of silicon electrodes in Li-ion batteries–A photoelectron spectroscopy study. *Chem. Mater.* **2013**, *25*, 394–404. [CrossRef]

33. Elazari, R.; Salitra, G.; Gershinsky, G.; Garsuch, A.; Panchenko, A.; Aurbach, D. Li ion cells comprising lithiated columnar silicon film anodes, TiS_2 cathodes and fluoroethyene carbonate (FEC) as a critically important component. *J. Electrochem. Soc.* **2012**, *159*, A1440. [CrossRef]
34. Dalavi, S.; Guduru, P.; Lucht, B.L. Performance enhancing electrolyte additives for lithium ion batteries with silicon anodes. *J. Electrochem. Soc.* **2012**, *159*, A642. [CrossRef]
35. Etacheri, V.; Haik, O.; Goffer, Y.; Roberts, G.A.; Stefan, I.C.; Fasching, R.; Aurbach, D. Effect of fluoroethylene carbonate (FEC) on the performance and surface chemistry of Si-nanowire Li-ion battery anodes. *Langmuir* **2012**, *28*, 965–976. [CrossRef]
36. Sun, X.; Wang, C.; Zhang, X.; Zhai, X.; Feng, J.; Li, H. Nanodiamond-modified separators and optimized graphene anodes for highly enhanced performance of lithium-ion batteries. *J. Alloys Compd.* **2022**, *929*, 167204. [CrossRef]
37. Wu, J.J.; Bennett, W.R. Fundamental investigation of Si anode in Li-ion cells. In Proceedings of the 2012 IEEE Energytech, Cleveland, OH, USA, 29–31 May 2012; pp. 1–5.

batteries

MDPI

Article

First Principles Study of the Phase Stability, the Li Ionic Diffusion, and the Conductivity of the $Li_{10}Ge_xMo_{1-x}P_2S_{12}$ of Superionic Conductors

Yifang Wu [1], Yuanzhen Chen [2] and Shaokun Chong [3,*]

1 Northwest Institute for Nonferrous Metal Research, Xi'an 710016, China; wyf7777v@126.com
2 State Key Laboratory for Mechanical Behavior of Materials, School of Materials Science and Engineering, Xi'an Jiaotong University, Xi'an 710049, China; cyz1984@xjtu.edu.cn
3 Frontiers Science Center for Flexible Electronics, Institute of Flexible Electronics, Northwestern Polytechnical University, Xi'an 710072, China
* Correspondence: iamskchong@nwpu.edu.cn

Abstract: Using first-principles density functional theory (DFT) calculations and ab initio molecular dynamics (AIMD) simulations, we performed this study on the phase stability, the intrinsic redox stability, and the Li^+ conductivity of $Li_{10}Ge_xMo_{1-x}P_2S_{12}$ (x = 0~1) superionic conductors. Molybdenum (Mo) is expected to replace expensive germanium (Ge) to lower tmaterial costs, reduce sensitivity to ambient water and oxygen, and achieve acceptable Li^+ conductivity. The ab initio first principle molecular dynamics simulations show that room-temperature Li^+ conductivity is 1.12 $mS{\cdot}cm^{-1}$ for the $Li_{10}Ge_{0.75}Mo_{0.25}P_2S_{12}$ compound, which is comparable to the theoretical value of 6.81 $mS{\cdot}cm^{-1}$ and the experimental measured one of 12 $mS{\cdot}cm^{-1}$ of the $Li_{10}GeP_2S_{12}$ (LGPS) structure. For $Li_{10}Ge_xMo_{1-x}P_2S_{12}$ (x = 0, 0.25, 0.5 and 1) compounds, the density of states and the projection fractional wave state density were calculated. It was found that when Ge atoms were partially replaced by Mo atoms, the band gap remained unchanged at 2.5 eV, but deep level defects appeared in Mo-substituted compounds. Fortunately, this deep level defect is difficult to ionize at room temperature, so it has no effect on the electronic conductivity of Mo substitute compounds, making Mo substitution a suitable solution for electrolyte materials. The projection fractional wave state density calculation shows that the covalent bond between Mo and S is stronger than that between Ge and S, which reduces the sensitivity of Mo-substituted compounds to water and oxygen contents in the air. In addition, the partial state density coincidence curve between Li and S elements disappears in the 25% Mo-substituted compound with energies of 4–5 eV, indicating that the Li_2S by-product is decreased.

Keywords: first principles; ab initio molecular dynamics; superionic conductors; solid electrolytes; lithium-ion battery

1. Introduction

Lithium-ion batteries (LIBs) have experienced rapid development in the past decade and are widely used in various fields such as medicine, aerospace, and power storage [1]. More stringent criteria for electrolyte materials have been put in place due to the ongoing push for high performance lithium-ion batteries [2]. The highest Li^+ conductivity ever recorded has been found in the $Li_{10}GeP_2S_{12}$ (LGPS) superionic conductor among solid lithium electrolytes, which has the conductivity of 12 $mS{\cdot}cm^{-1}$ at room temperature [3–6].

Despite its cutting-edge Li^+ conductivity and high electrochemical performance, the use of LGPS as a solid electrolyte material is still hindered by three significant issues. First, there is the practical problem of using highly expensive and scarce germanium in LGPS, which would prevent the widespread usage of the substance. Second, the air and moisture sensitivity of sulfide-based compounds can make their manufacture and application more expensive. Third, lithium metal is too prone to chemically react with sulfur elements in LGPS through reduction reactions and generate Li_2S by-products [7].

Researchers have developed a number of strategies to improve the air/water stability of LGPS, including isovalent cation substitutions of Ge^{4+} by Si and Sn and aliovalent cation substitutions by Al [8]. These substituting strategies mainly involve substituting with elements of the same group as Ge or elements adjacent to Ge in the periodic table. However, these elements are hard acids, which is hard to bond with the soft base of S. According to Hard and Soft Acid and Bases (HSAB) theory [9,10], hard acids and bases generally have smaller radii and polarizabilities than soft acids and bases. Hard acids and hard bases are preferentially bound mainly by coulomb forces, while soft acids and soft bases are preferentially bound mainly by covalent bonds. To date, the use of soft acids as substituting elements has not been reported in the literature. Soft acids are substances that can easily accept electron pairs, usually with high polarizability and easily deformable electron orbitals, but have a weak polarization to distort the electron cloud of other atoms or ions. Soft acids include a series of metal ions such as Cu^{+}, Cu^{2+}, Ag^{+}, Cd^{2+}, Hg^{2+}, Au^{+}, Au^{3+}, Pt^{2+}, Pd^{2+}, Mo^{2+}, and Mo^{4+}, etc. The softer acid Mo^{4+} was specially chosen in this work because isovalent cation substitution is expected to have a small effect on the phase stability, electrochemical stability, and lithium-ion diffusivity as for LGPS. Based on HSAB theory, S is a softer base than O and prefers to bind with soft acids such as Mo^{2+} and Mo^{4+}. Therefore, Mo-element substitution in LGPS could not only reduce the material cost but also effectively mitigate the erosion of O in the air, thus helping to maintain the good electrochemical properties of superionic conductors.

In this work, we investigate the phase stability, intrinsic redox stability, and Li^{+} conductivity of $Li_{10}Ge_xMo_{1-x}P_2S_{12}$ (x = 0~1) superionic conductors using first principle calculations. Our aim was to search for an effective element substitution with a similar structural makeup to LGPS, which can strike a better balance between electrochemical performance, cost, and other features.

2. Computational Details

The Vienna Ab initio Simulation Package (VASP) and the projector augmented-wave (PAW) method were used for all calculations in this work. Considering the accuracy and computational cost of different requirements in this paper, we have carefully chosen the suitable functionals and methods for each technique as listed in the following sections.

2.1. Modeling and Structure Optimization

Due to the co-occupation of the Li#1-4, Ge#1, and P#1 atoms in the $Li_{10}GeP_2S_{12}$ (LGPS) structure, it is difficult to conduct simulation calculations. Hence, it was a heavy task to convert a co-occupancy model to full occupancy ones. Eventually, sixty-one full occupancy $Li_{20}Ge_2P_4S_{24}$ models were constructed to identify which atoms fit in which spots. Similarly, sixty-one full occupancy $Li_{20}Mo_2P_4S_{24}$ models were constructed. The PBE generalized-gradient approximation (GGA) functional was employed by DFT calculations to find out the lowest energy of various $Li_{20}Ge_2P_4S_{24}$ and $Li_{20}Mo_2P_4S_{24}$ models to obtain optimized relaxation structures. Structure substitution models for the partial substitution of Mo, namely the $Li_{20}GeMoP_4S_{24}$ model and the $Li_{40}Ge_3MoP_8S_{48}$ supercell model, were implemented on the $Li_{20}Ge_2P_4S_{24}$ model with the lowest energy structure. After Mo substitution, optimal relaxation structures were still obtained using the DFT calculation.

2.2. Binding Energy

When we design a new material through element substitution, the first thing we need to do is to determine whether the material is stable, which is a very key link in the field of material design.

Binding energy refers to the energy released by atoms from their free state to form compounds. For a simple binary compound A_mB_n (where A and B are the two elements contained in the compound, and m and n are the number of corresponding atoms in the chemical formula), its binding energy can be calculated according to the following formula:

$$E_b = \frac{\mathrm{E}(\mathrm{A}_m\mathrm{B}_n) - \mathrm{m} \times \mathrm{E}(\mathrm{A}) - \mathrm{n} \times \mathrm{E}(\mathrm{B})}{\mathrm{m} + \mathrm{n}} \quad (1)$$

where $\mathrm{E}(\mathrm{A}_m\mathrm{B}_n)$ is the ground-state energy of $\mathrm{A}_m\mathrm{B}_n$, and E(A) and E(B) are the energies of free atoms A and B, respectively. All of these energies are calculated in units of eV (electron volts). The more negative the value of E_b, the more stable it is.

In this paper, the PBE generalized-gradient approximation (GGA) functional was employed by DFT calculations to evaluate ground-state energies of various $Li_{10}Ge_xMo_{1-x}P_2S_{12}$ (x = 0~1) models and free energies of Li, Ge, Mo, P, and S single atoms.

2.3. Intrinsic Redox Stability

By figuring out the material's band gap, we evaluated the $Li_{10}Ge_xMo_{1-x}P_2S_{12}$ (x = 0~1) solid electrolyte's inherent stability with regard to inert electrodes. Due to the substantial underestimation of band gaps by standard semi-local DFT, the density of states (DOS) of $Li_{10}Ge_xMo_{1-x}P_2S_{12}$ (x = 0~1) compounds were calculated using the Heyd–Scuseria–Ernzerhof (HSE06) screened hybrid functional [11], which has been tested to give relatively accurate band gaps for a wide range of materials [12,13].

Non-spin-polarized calculations were carried out since $Li_{10}Ge_xMo_{1-x}P_2S_{12}$ compounds do not contain 3D transition metal ions (V to Ni) or other atoms with f-shell outer electrons. The bandgap is not a precise indicator of the electrochemical stability on inert electrodes because it is unknown how it aligns to an external reference potential. However, it might be viewed as the upper limit of the electrochemical window [14].

2.4. Lithium-Ion Transport Capabilities

Using ab initio molecular dynamics (AIMD) simulations, we evaluated the Li^+ diffusivity and conductivity in $Li_{10}Ge_xMo_{1-x}P_2S_{12}$ (x = 0~1) compounds. The Brillouin zone was precisely integrated in the reciprocal space using the γ-point scheme with $3 \times 3 \times 2$ k-meshes. The generalized gradient approximation (GGA) parametrized by Perdew, Burke, and Ernzerhof (PBE) was employed for the exchange−correlation functional, and the PBE configured projected-augmented wave (PAW) method was applied to describe interactions among core electrons. Plane wave basis functions for the Kohn−Sham equation were reduced to an energy cutoff of 259 eV to keep the computing cost at a manageable level. All calculations were performed using a non-spin-polarized. The Verlet algorithm was employed in VASP to integrate Newton's equation. The molecular dynamics time step was set as 2 fs. The AIMD simulations were performed as follows:

1. The $Li_{10}Ge_xMo_{1-x}P_2S_{12}$ (x = 0~1) samples are assigned an initial temperature of 100 K based on a Boltzmann distribution at the start of the MD simulations.
2. The samples are then heated to the required temperature (600 to 1200 K) by velocity scaling over 1000 time steps (2 ps) in the NVT ensemble with a constant volume and a Nosé–Hoover thermostat, and then equilibrated at the equilibrium temperature for 5000 time steps (10 ps) in the NVT ensemble with a constant volume and a Nosé–Hoover thermostat.
3. The MD simulations for diffusion are then run in the NVT ensemble for 40 ps to 100 ps until the diffusion coefficient converges.

To get the diffusion coefficient D, the mean square displacements are utilized over time:

$$D = \frac{1}{2dt}\left\langle [r(t)]^2 \right\rangle \quad (2)$$

where d equals 3, the dimension of the lattice in which diffusion occurs. The t refers to time, counted in unit of fs. The average mean square displacement $\left\langle [r(t)]^2 \right\rangle$ was determined as follows:

$$\left\langle [r(t)]^2 \right\rangle = \frac{1}{N}\sum_i \left\langle \left[r_i(t+t_o)^2 \right] - [r_i(t_o)]^2 \right\rangle \tag{3}$$

where $r_i(t)$ represents the displacement of the i-th Li ion at time t. The calculated displacement $r_i(t)$ represents the displacement of a single Li atom, calculated in the unit of Å. The diffusion coefficient D is determined in $Å^2 \cdot fs^{-1}$ and can be converted to $cm^2 \cdot s^{-1}$ by a multiple of 10^{-1}.

The Nernst–Einstein relation was used to calculate the Li ionic conductivity σ in units of $mS \cdot cm^{-1}$.

$$\sigma = D\frac{\rho Z^2 F^2}{RT} \tag{4}$$

where D is the diffusion coefficient of Li ions in units of $cm^2 \cdot s^{-1}$ in a unit cell. The ρ, Z, F, R, and T are the molar density in units of mol/L, the mean valence electronic charge of a Li ion (dimensionless), the Faraday constant in units of KJ/(V·mol), the gas constant in units of J/(mol·K), and the absolute temperature, respectively.

The Arrhenius equation was used to calculate the average diffusion activation energy of lithium ion E_a in units of eV.

$$\sigma = A \cdot exp\left(-\frac{E_a}{kT}\right) \tag{5}$$

where σ is the Li ionic conductivity. A represents predigital factor. E_a, k, and T are the average diffusion activation energy of lithium ion, the Boltzmann constant in units of eV/K, and the absolute temperature, respectively.

3. Results and Discussion

3.1. Relaxed Structural and Parameters

The optimal relaxation structures of the $Li_{10}Ge_xMo_{1-x}P_2S_{12}$ (x = 0~1) models are shown in Figure 1, in which Figure 1a–d correspond to the $Li_{20}Ge_2P_4S_{24}$, $Li_{20}GeMoP_4S_{24}$, $Li_{40}Ge_3MoP_8S_{48}$, and $Li_{20}Mo_2P_4S_{24}$ structures, respectively. The changes in lattice parameters after Mo substitution are shown in Table 1. For comparison, for the $Li_{40}Ge_3MoP_8S_{48}$ model, both the c-axis lattice parameter and the cell volume are divided by 2.

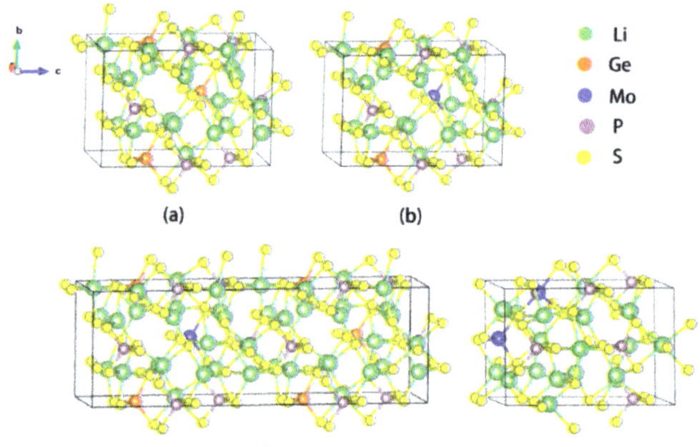

Figure 1. The optimal relaxation structures of $Li_{10}Ge_xMo_{1-x}P_2S_{12}$ (x = 0~1) models. (**a**) $Li_{20}Ge_2P_4S_{24}$, (**b**) $Li_{20}GeMoP_4S_{24}$, (**c**) $Li_{40}Ge_3MoP_8S_{48}$, and (**d**) $Li_{20}Mo_2P_4S_{24}$.

Table 1. The lattice parameters of relaxed $Li_{10}Ge_xMo_{1-x}P_2S_{12}$ (x = 0~1) models.

	$Li_{20}Ge_2P_4S_{24}$	$Li_{40}Ge_3MoP_8S_{48}$	$Li_{20}GeMoP_4S_{24}$	$Li_{20}Mo_2P_4S_{24}$
a (Å)	8.6025	8.5905	8.5899	8.7922
b (Å)	8.6489	8.6606	8.6735	8.6187
c (Å)	12.8824	12.9232	12.9114	12.5916
α	89.1560	89.0088	88.9790	90.9771
β	90.2020	90.2727	90.3518	90.5599
γ	89.2361	89.1494	89.1528	89.6798
V (Å3)	958.2773	961.2006	961.6789	953.9658

3.2. Phase Stability

The binding energies of $Li_{10}Ge_xMo_{1-x}P_2S_{12}$ (x = 0~1) compounds are shown in Table 2. It shows that when the amount of Mo substitution grows, the binding energy becomes more and more negative, indicating improved phase stability.

Table 2. The binding energies of $Li_{10}Ge_xMo_{1-x}P_2S_{12}$ (x = 0~1) compounds.

	$Li_{20}Ge_2P_4S_{24}$	$Li_{40}Ge_3MoP_8S_{48}$	$Li_{20}GeMoP_4S_{24}$	$Li_{20}Mo_2P_4S_{24}$
binding energy (eV)	−4.29	−4.34	−4.39	−4.50

3.3. Bandgaps and Intrinsic Redox Stability

We estimated the density of states (DOS) of all $Li_{10}Ge_xMo_{1-x}P_2S_{12}$ (x = 0~1) compounds using the HSE screened hybrid functional (HSE06) to analyze the intrinsic redox stability of the various $Li_{10}Ge_xMo_{1-x}P_2S_{12}$ (x = 0~1) compounds. The calculated DOS for $Li_{10}Ge_xMo_{1-x}P_2S_{12}$ (x = 0~1) compounds are shown in Figure 2. Figure 2 indicates that bandgaps remain unchanged as Ge atoms are partially substituted by Mo, but the defect state energy level appears in Mo-substituted compounds. These defect state energy levels are located in the center of the forbidden band, far from the bottom of the conduction band or the top of the valence band, which belong to deep level defects. They are difficult to ionize at room temperature and do not affect the electronic conductivity of Mo-substituted compounds.

In $Li_{10}Ge_xMo_{1-x}P_2S_{12}$ (x = 0.5 and 0.75) crystal structures, the Mo atoms are adjacent to S atoms. The degree of coincidence of *DOS* between Mo and S elements near bandgaps at an energy of 1–2 eV are higher than that between Ge and S elements, indicating stronger binding forces between Mo and S elements, which is in accordance with the HSAB theory [15,16]. It can be seen from the projected DOS in the inset of Figure 2b at an energy of 1–2 eV that the binding energy between Mo and S is mainly contributed by the d orbital of Mo and the p orbital of S. Furthermore, it is noteworthy that for the $Li_{10}Ge_xMo_{1-x}P_2S_{12}$ (x = 0.75) compound, the coincidence curve of the partial wave state density between Li and S elements disappears when reaching an energy of 4~5 eV, indicating a decreased probability of forming Li_2S decomposition products of this compound.

The DOS results predict that the bandgap of $Li_{10}GeP_2S_{12}$ is about 2.5 eV, which would be sufficient for preventing electrical conduction. It suggests that the observed electrochemical window of >5 V for this material is likely the result of a passivation phenomenon, where either Li_2S or P_2S_5 is formed as a decomposition product [8]. As can be seen from Figure 2, the positions of the conduction band and the valence band of the Mo-substituted compounds shift down. This leads to an increase in the reduction potential of the compound and a decrease in the oxidation potential, which usually means that these electrolytes remain stable over a wider potential range, thereby improving the electrochemical stability of these electrolytes.

For 0.25 and 0.5 Mo-substituted compounds, bandgaps maintain unchanged except that medium-depth deep-level defects are induced. Generally, the ionization energy of medium-depth deep-level defects is usually between 0.1 eV and 0.5 eV. This energy range is sufficient to make the defect resistant to ionization at room temperature but can be significantly ionized at higher temperatures, such as 100 °C to 300 °C. Obviously, even for batteries at elevated temperatures, which is far less than 100 °C, medium-depth deep-level defects in Mo-substituted compounds would have no impact on electronic conductivities of batteries. By contrast, in the case of the compound with complete Mo-substitution, there is an obvious contraction of the bandgap to 2.0 eV around. Generally speaking, this bandgap is big enough not to become a potential leakage channel.

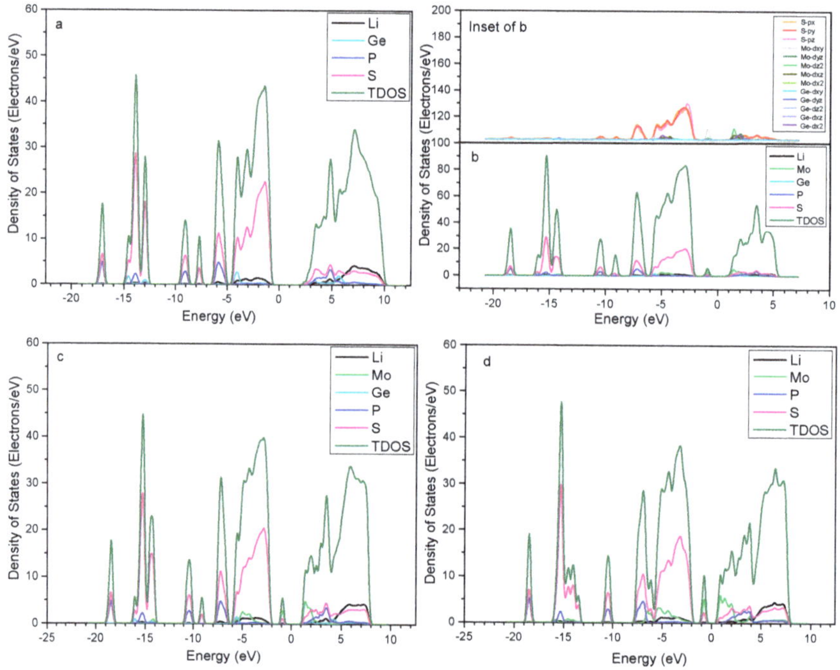

Figure 2. The calculated DOS for $Li_{10}Ge_xMo_{1-x}P_2S_{12}$ (x = 0~1) compounds. (**a**) $Li_{20}Ge_2P_4S_{24}$, (**b**) $Li_{40}Ge_3MoP_8S_{48}$, (**c**) $Li_{20}GeMoP_4S_{24}$, and (**d**) $Li_{20}Mo_2P_4S_{24}$. In all plots, the Fermi level is set to 0 eV and is marked by the dashed line. Inset of b is the projected DOS of the $Li_{40}Ge_3MoP_8S_{48}$ Compound.

3.4. Li^+ Diffusivity and Conductivity

Li ionic diffusion coefficients in $Li_{10}Ge_xMo_{1-x}P_2S_{12}$ (x = 0~1) crystal structures were determined by random walk theory and AIMD simulations as a function of T (=600, 800, 1000, and 1200 K) by vaspkit software 1.3.5 [17], as shown in Figure 3. The lithium-ion diffusion coefficients at room temperature (300 K) were calculated by the linear fitting extrapolation method in Figure 3.

The conductivities and diffusion activation energies of lithium ions were determined by the Nernst equation (See Equation (4)) and Arrhenius equation (See Equation (5)), respectively. The calculated Li ionic diffusion coefficients, conductivities, and diffusion activation energies of lithium ion in $Li_{10}Ge_xMo_{1-x}P_2S_{12}$ (x = 0~1) crystal structures are listed in Table 3. It demonstrates that when Ge is replaced by 25% Mo, lithium-ion conductivity still maintains 1.12 $mS{\cdot}cm^{-1}$, which is in the same order of magnitude as the theoretical predicted value of 6.81 $mS{\cdot}cm^{-1}$ in the LGPS structure (the experimental measured value is 12 $mS{\cdot}cm^{-1}$). The value of 6.81 $mS{\cdot}cm^{-1}$ is very similar to the calculated 6.43 $mS{\cdot}cm^{-1}$ in the recent reference [18]. As a result, appropriate Mo substitution can significantly cut raw material costs while just marginally lowering material conductivity.

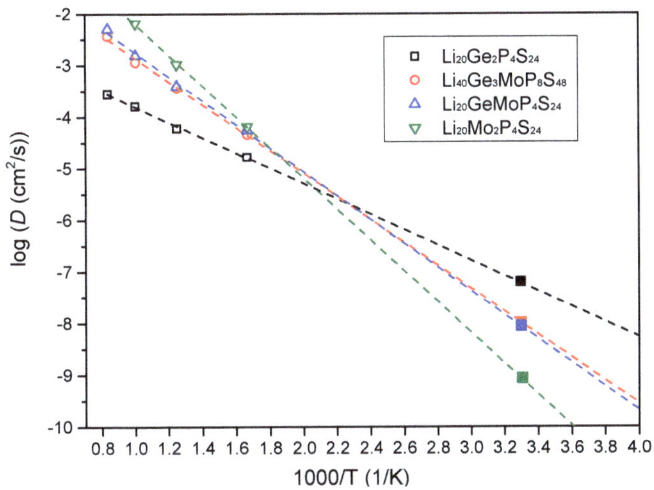

Figure 3. The Li ionic diffusion coefficients in $Li_{10}Ge_xMo_{1-x}P_2S_{12}$ (x = 0~1) crystal structures. The filled black square corresponds to the linearly extrapolated (dashed line) conductivity at 300 K from high-temperature conductivity.

Table 3. The calculated Li ionic diffusion coefficient, conductivity, and diffusion activation energy of lithium ion in $Li_{10}Ge_xMo_{1-x}P_2S_{12}$ (x = 0~1) crystal structures.

	Li Ionic Diffusion Coefficients ($cm^2 \cdot s^{-1}$)	Conductivities of Lithium Ions ($mS \cdot cm^{-1}$)	Diffusion Activation Energies of Lithium Ions (eV)
$Li_{20}Ge_2P_4S_{24}$	5.26×10^{-8}	6.81	0.23
$Li_{40}Ge_3MoP_8S_{48}$	8.66×10^{-9}	1.12	0.37
$Li_{20}GeMoP_4S_{24}$	7.13×10^{-9}	0.92	0.39
$Li_{20}Mo_2P_4S_{24}$	6.65×10^{-10}	0.09	0.53

The calculated activation energy of Li^+ diffusion for totally Mo-substituted $Li_{20}Mo_2P_4S_{24}$ (LMPS) of 0.53 eV is much greater than the 0.23 eV of LGPS, which fits in well with the literature [3,19–21], indicating a much higher diffusion barrier. As a result, the Li^+ conductivity at 300 K is only 0.09 $mS \cdot cm^{-1}$ for LMPS, which is two orders of magnitude lower than LGPS. For compounds with 0.25 and 0.5 Mo substitutes for Ge, calculated activation energies of Li^+ diffusion are between the values of LGPS and LMPS, indicating moderate diffusion barriers.

In fact, for compounds with 0.25 and 0.5 Mo substitutes for Ge, the diffusion of lithium ions remains in the 1D channel along the c-axis, supplemented by a-b plane diffusion pathways [20] because the crystal structure of these compounds does not change compared with LGPS. For these two compounds, their cell volumes are slightly increased, as shown in Table 1, which is favorable for the diffusion of lithium ions as indicated by the literature [8]. However, the diffusion channel of Li ions in these crystals may gradually shrink by an increased amount of Mo due to a higher atomic radius of Mo than Ge. There is an obvious contraction in the cell volume for the compound that Ge is completely replaced by Mo as shown in Table 1, which would lead to a severely constricted lithium-ion diffusion channel and an enormous decrease in the conductivity of lithium ions as indicated in Table 3.

3.5. Discussion

By executing Mo substitutions to create $Li_{10}Ge_xMo_{1-x}P_2S_{12}$ (x = 0~1) compounds, we investigated the characteristics that influence structural, phase, and redox stability and Li ionic diffusivity.

For 0.25 and 0.5 Mo-substituted compounds, their cell volumes are slightly increased while crystal structures remain unchanged. In contrast, there is a distinct contraction in the cell volume for the compound with 100% Mo-substitution.

Our calculation demonstrates that as the content of Mo substitution grows in $Li_{10}Ge_x Mo_{1-x}P_2S_{12}$ (x = 0~1) compounds, the binding energy becomes increasingly negative, indicating increased phase stability.

In terms of redox stability, the calculated DOS suggests that the redox stability remains unchanged for $Li_{10}Ge_xMo_{1-x}P_2S_{12}$ (x = 0.5 and 0.75) compounds. The shift down of calculated conduction bands and valence bands indicates an increase in the reduction potential of the compound and a decrease in the oxidation potential, which means a larger stable electrochemical window. However, when reaching complete Mo-substitution, the bandgap starts to contract, indicating a decline in the inherent redox stability of the compound.

Additionally, the coincidence curve of the partial wave state density between the Li and S elements vanishes at an energy of 4~5 eV for the 0.25 Mo-substituted compound, indicating the decreased probability of forming the Li_2S decomposition product of this compound.

The LMPS compound has a two-orders-of-magnitude lower diffusivity than the LGPS compound, which makes it significantly less interesting than LGPS. This decrease in diffusivity in the LMPS compound is most likely due to the significantly reduced cell volume. The calculated activation energy of Li^+ diffusion for LMPS of 0.53 eV is also significantly higher than 0.23 eV of LGPS.

By contrast, $Li_{10}Ge_{0.5}Mo_{0.5}P_2S_{12}$ and $Li_{10}Ge_{0.75}Mo_{0.25}P_2S_{12}$ compounds have slightly increased cell volume. Hence, higher conductivities of lithium ions are anticipated in these two compounds, which are also projected to lower the cost of raw materials. Our calculation shows that the preferred substitution rate for Ge by Mo is less than 25%. At this rate, lithium-ion conductivity remains at 1.12 $mS{\cdot}cm^{-1}$, which is worthy of comparison with the calculated value of 6.81 $mS{\cdot}cm^{-1}$ of the LGPS structure. The activation energy of the compound is also only 0.14 eV higher than that of LGPS.

In general, for the solid-state synthesis of LGPS, the synthetic raw materials are Li_2S, P_2S_5, and GeS_2. When partial Mo substitution for Ge is adopted, it is relatively easy in synthesis as long as the stoichiometric ratio is well controlled and part of the GeS_2 in the raw materials is replaced by MoS_2. Only by reducing the cost of materials and improving the chemical stability and electrochemical stability of materials, can the large-scale application of new materials be achieved.

4. Conclusions

The necessity to address three key constraints of the LGPS superionic conductor, namely the expensive cost of germanium, the air and moisture sensitivity of a sulfide-based chemistry, and the reduction reaction tendency of lithium metal with sulfur elements, have spurred our exploration of Mo-substituted $Li_{10}Ge_xMo_{1-x}P_2S_{12}$ (x = 0~1) compounds.

Our calculations show that the substitution of 0.25 Mo for Ge is the optimal choice for compounds with this structure, which not only lowers the material cost by adding an appropriate amount of Mo but also reduces the sensitivity of the compound to water and oxygen in the air by improving the binding force between Mo and S, with improved electrochemical stability and suppressed Li_2S by-products. But all of these are at the expense of a slightly decreased conductivity of lithium ions at 1.12 $mS{\cdot}cm^{-1}$ for this compound, which is comparable to the predicted value of 6.81 $mS{\cdot}cm^{-1}$ for the LGPS structure.

The calculated DOS show that bandgaps remain unchanged as Ge atom are partially substituted by Mo, however, the deep defect state energy level appears in these compounds. Fortunately, it is difficult to ionize at ambient temperature and make no impact on their electronic conductivities. The shift down of the conduction band and valence band of the partially Mo-substituted compounds leads to an increase in the reduction potential of the compound and a decrease in the oxidation potential, suggesting a wider potential range and an improved electrochemical stability of compounds.

In conclusion, we believe that replacing Ge with an appropriate Mo atom is an effective strategy to lower material cost and improve chemical stability and electrochemical stability. It is worth noting that the first-principles simulation can only give a theoretical guide, and exactly suitable Mo doping or substitution amounts need to be verified by subsequent experiments. However, theoretical simulations have demonstrated the feasibility of Mo substituting Ge in synthesis and large-scale applications.

Author Contributions: Methodology, Y.W.; Software, Y.C.; Formal analysis, Y.W.; Writing—original draft, Y.W.; Writing—review & editing, S.C.; Funding acquisition, Y.C. and S.C. All authors have read and agreed to the published version of the manuscript.

Funding: This work was supported by National Natural Science Foundation of China (52207248), Guangdong Basic and Applied Basic Research Foundation (2022A1515010208 and 2021A1515110164), and the Open Testing Foundation of Analytical & Testing Center of Northwestern Polytechnical University (2022T024), the Joint Funds of the Natural Science Basic Research Project of Shaanxi Province (No. 2021JLM-23), and the University Joint Project of Shaanxi Province (2021GXLH-Z-067).

Data Availability Statement: The raw data supporting the conclusions of this article will be made available by the authors on request.

Acknowledgments: The authors gratefully acknowledge the High Performance Computing of Network Information Center, Xi'an Jiaotong University for first-principles simulations.

Conflicts of Interest: The authors declare no conflict of interest.

References

1. Zhang, S.Y.; Gu, K.H.; Lu, B.A.; Han, J.W.; Zhou, J. Hydrometallurgical Processes on Recycling of Spent Lithium-Ion Battery Cathode: Advances and Applications in Sustainable Technologies. *Acta Phys.-Chim. Sin.* **2024**, *40*, 2309028. [CrossRef]
2. Zhang, J.K.; Lv, J.C.; Lu, W.; Li, X.; Liu, Y.; Lang, J.H.; Liu, J.; Wang, Z.; Lu, M.; Sun, H. $Li_{10}GeP_2S_{12}$-based solid electrolyte induced by high pressure for all-solid-state batteries: A facile strategy of low-grain-boundary-resistance. *Chem. Eng. J.* **2024**, *487*, 150452. [CrossRef]
3. Kamaya, N.; Homma, K.; Yamakawa, Y.; Hirayama, M.; Kanno, R.; Yonemura, M.; Kamiyama, T.; Kato, Y.; Hama, S.; Kawamoto, K.; et al. A lithium superionic conductor. *Nat. Mater.* **2011**, *10*, 682–686. [CrossRef] [PubMed]
4. Hwang, T.; Conlin, P.; Cho, M.; Cho, K. Electrochemical Stability and Li Ion Diffusion Kinetics of Grain Boundaries in $Li_{10}GeP_2S_{12}$ Solid Electrolyte. *J. Phys. Chem. C* **2023**, *127*, 7528–7535. [CrossRef]
5. Zhang, J.; Huang, L.; Gu, X. Failure mechanism of solid-state electrolyte $Li_{10}GeP_2S_{12}$ in a moist atmosphere: A first-principles study. *Mater. Adv.* **2022**, *3*, 3143–3150. [CrossRef]
6. Wang, Q.; Liu, D.; Ma, X.; Liu, Q.; Zhou, X.; Lei, Z. Sb-doped $Li_{10}GeP_2S_{12}$-type electrolyte $Li_{10}SnP_{2-x}Sb_xS_{12}$ with enhanced ionic conductivity and lower lithium-ion migration barrier. *J. Colloid Interface Sci.* **2022**, *627*, 1039–1046. [CrossRef]
7. Zhao, J.; Zhao, C.; Zhu, J.P.; Liu, X.S.; Yao, J.M.; Wang, B.; Dai, Q.S.; Wang, Z.F.; Chen, J.Z.; Jia, P.; et al. Size-Dependent Chemomechanical Failure of Sulfide Solid Electrolyte Particles during Electrochemical Reaction with Lithium. *Nano Lett.* **2022**, *22*, 411–418. [CrossRef]
8. Ong, S.P.; Mo, Y.F. William Davidson Richards, Lincoln Miara, Hyo Sug Leeb and Gerbrand Ceder. Phase stability, electrochemical stability and ionic conductivity of the (M=Ge, Si, Sn, Al or P, and X=O, S or Se) family of superionic conductors. *Energy Environ. Sci.* **2013**, *6*, 148–156. [CrossRef]
9. Sahu, G.; Lin, Z.; Liu, J.; Dudney, N.; Liang, C. Air-stable, high-conduction solid electrolytes of arsenic-substituted Li_4SnS_4. *Energy Environ. Sci.* **2014**, *7*, 1053–1058. [CrossRef]
10. Yang, M.; Chen, L.Q.; Li, H.; Wu, F. Air/Water Stability Problems and Solutions for Lithium Batteries. *Energy Mater. Adv.* **2022**, *2022*, 9842651.
11. Heyd, J.; Scuseria, G.E.; Ernzerhof, M. Hybrid functionals based on a screened Coulomb potential. *J. Chem. Phys.* **2006**, *124*, 219906, Erratum in *J. Chem. Phys.* **2003**, *118*, 8207–8215. [CrossRef]
12. Heyd, J.; Scuseria, G.E. Efficient hybrid density functional calculations in solids: Assessment of the Heyd-Scuseria-Ernzerhof screened Coulomb hybrid functional. *J. Chem. Phys.* **2004**, *121*, 1187–1192. [CrossRef] [PubMed]
13. Henderson, T.M.; Paier, J.; Scuseria, G.E. Accurate treatment of solids with the HSE screened hybrid. *Phys. Status Solidi B* **2011**, *248*, 767–774. [CrossRef]
14. Ong, S.P.; Andreussi, O.; Wu, Y.; Marzari, N.; Ceder, G. Electrochemical windows of room-temperature ionic liquids from molecular dynamics and density functional theory calculations. *Chem. Mater.* **2011**, *23*, 2979–2986. [CrossRef]
15. Jian, Y.; Huang, Z.; Xing, J.; Sun, L.; Liu, Y.; Gao, P. Phase stability, mechanical properties and electronic structures of TiAl binary compounds by first principles calculations. *Mater. Chem. Phys.* **2019**, *221*, 311–321. [CrossRef]

16. Yang, J.; Huang, J.; Ye, Z.; Fan, D.; Chen, S.; Zhao, Y. First-principles calculations on structural energetics of Cu-Ti binary system intermetallic compounds in Ag-Cu-Ti and Cu-Ni-Ti active filler metals. *Ceram. Int.* **2017**, *43*, 7751–7761. [CrossRef]
17. Wang, V.; Xu, N.; Liu, J.-C.; Tang, G.; Geng, W.-T. VASPKIT: A User-Friendly Interface Facilitating High-Throughput Computing and Analysis Using VASP Code. *Comput. Phys. Commun.* **2021**, *267*, 108033. [CrossRef]
18. Nachimuthu, S.; Cheng, H.-J.; Lai, H.-J.; Cheng, Y.-H.; Kuo, R.T.; Zeier, W.G.; Hwang, B.J.; Jiang, J.-C. First-principles study on selenium-doped $Li_{10}GeP_2S_{12}$ solid electrolyte: Effects of doping on moisture stability and Li-ion transport properties. *Mater. Today Chem.* **2022**, *26*, 101223. [CrossRef]
19. Kuhn, A.; Duppel, V.; Lotsch, B.V. Tetragonal $Li_{10}GeP_2S_{12}$ and Li_7GePS_8-exploring the Li ion dynamics in LGPS Li electrolytes. *Energy Environ. Sci.* **2013**, *6*, 3548–3552. [CrossRef]
20. Mo, Y.F.; Ong, S.P.; Ceder, G. First Principles Study of the $Li_{10}GeP_2S_{12}$ Lithium Super Ionic Conductor Material. *Chem. Mater.* **2012**, *24*, 15–17. [CrossRef]
21. Wang, Z.; Shao, G. Theoretical design of solid electrolytes with superb ionic conductivity: Alloying effect on Li^+ transportation in cubic Li6PA5X chalcogenides. *J. Mater. Chem. A* **2017**, *5*, 21846–21857. [CrossRef]

batteries

MDPI

Article

Improving the Performance of $LiFePO_4$ Cathodes with a Sulfur-Modified Carbon Layer

Su-hyun Kwak and Yong Joon Park *

Department of Advanced Materials Engineering, Graduate School Kyonggi University, 154-42, Gwanggyosan-ro, Yeongtong-gu, Suwon-si 16227, Republic of Korea; sh3517@kyonggi.ac.kr
* Correspondence: yjpark2006@kyonggi.ac.kr

Abstract: $LiFePO_4$ (LFP) cathodes are popular due to their safety and cyclic performance, despite limitations in lithium-ion diffusion and conductivity. These can be improved with carbon coating, but further advancements are possible despite commercial success. In this study, we modified the carbon coating layer using sulfur to enhance the electronic conductivity and stabilize the carbon surface layer via two methods: 1-step and 2-step processes. In the 1-step process, sulfur powder was mixed with cellulose followed by heat treatment to form a coating layer; in the 2-step process, an additional coating layer was applied on top of the carbon coating layer. Electrochemical measurements demonstrated that the 1-step sulfur-modified LFP significantly improved the discharge capacity (~152 $mAh \cdot g^{-1}$ at 0.5 C rate) and rate capability compared to pristine LFP. Raman analyses indicated that sulfur mixed with a carbon source increases the graphitization of the carbon layer. Although the 2-step sulfur modification did not exceed the 1-step process in enhancing rate capability, it improved the storage characteristics of LFP at high temperatures. The residual sulfur elements apparently protected the surface. These findings confirm that sulfur modification of the carbon layer is effective for improving LFP cathode properties, offering a promising approach to enhance the performance and stability of LFP-based lithium-ion batteries.

Keywords: surface modification; cathode; coating; lithium battery; $LiFePO_4$

1. Introduction

The increasing popularity of electric vehicles (EVs) has significantly driven the demand for large-scale lithium-ion batteries (LIBs), fueling the LIB industry. Among the cathodes employed in EV applications, the $LiNi_xCo_yMn_zO_2$ (NCM) series are popular owing to their high energy density and excellent rate capability, which support superior acceleration and extended driving ranges [1–8]. However, the recent interest in lithium iron phosphate ($LiFePO_4$, LFP) cathodes has surged owing to heightened price competition in the EV market and the unstable supply chains for transition metals such as cobalt (Co) and nickel (Ni) [9–13]. LFP cathodes, leveraging the low-cost and abundant iron (Fe), offer several compelling advantages. The stable phase of the material, derived from the robust P-O covalent bond in $(PO_4)^{3-}$ polyanion, ensures superior cyclic performance and enhanced safety compared to NCM-based cathodes, making LFP a promising alternative, particularly for cost-effective and reliable solutions amidst fluctuating raw material markets. However, the low lithium-ion diffusion coefficient, ranging only from 10^{-14} to 10^{-16} $cm^2 \cdot s^{-1}$, coupled with a poor electronic conductivity of 10^{-9} $S \cdot cm^{-1}$ or lower, significantly hinders the rate capability of LFP [14–17], which hinders its applicability.

Forming a carbon coating layer on the surface of LFP is effective for improving its deficient rate capability because it can reduce the Li-ion diffusion length to within a particle and facilitate electron transfer [18–22]. Various organic materials such as citric acid, glucose, and polyethylene glycol have been utilized as carbon sources, and sometimes materials such as carbon black or graphene have been directly complexed with the cathode [16–22]. The commercial success of LFP is largely attributed to the enhanced rate capability due to the carbon coating. However, continuing efforts have been made to modify the carbon to form a more effective carbon layer. Nitrogen doping of carbon to form a more effective electron-transfer carbon coating layer has been used to improve the electrochemical performance of LFPs [23–25]. Additionally, F-doped carbon has been utilized to create better carbon coating layers [26]. Metal oxide coatings on LFP have also been explored, occasionally combined with carbon coatings [27–32].

In this study, we extended these prior efforts by modifying the carbon coating layer with sulfur. Sulfur has been previously employed to enhance the properties of LFP cathodes focusing on doping these cathodes using lithium sulfide and benzyl disulfide via sol–gel methods [33,34]. However, our study focuses on the modification of the carbon coating layer using sulfur, employing two novel methods. In the first method, sulfur powder is mixed with cellulose, a carbon source, to form a coating layer, referred to as the 1-step process. In this approach, sulfur is used as an additive to improve the properties of the carbon during its formation, while most of the sulfur is expected to evaporate during the heat treatment process; however, a small amount of the remaining sulfur acts as a dopant of LFP. In the second method, referred to as the 2-step process, a carbon coating layer is first formed using cellulose. Subsequently, an additional coating layer is applied in the gas phase using sulfur vapor sublimated from a sulfur source. This approach is anticipated to yield a significantly thin and uniform sulfur layer on top of the carbon coating layer. The additional sulfur-related coating is expected to mitigate the side reactions between the carbon layer and the electrolyte, particularly under high-temperature conditions. To the best of our knowledge, these methods have not been explored thus far. These innovative sulfur-related modifications aim to enhance the electrochemical performance of LFP by improving its rate capability and controlling the side reactions, thereby contributing to the development of more efficient and durable LIBs. To determine the effectiveness of our two approaches for carbon modification using sulfur, the electrochemical properties of pristine LFP (with a carbon coating layer) and sulfur-modified carbon-coated LFP were compared and analyzed using various tools such as Raman spectroscopy, X-ray photoelectron spectroscopy (XPS), transmission electron microscopy (TEM), and time-of-flight secondary ion mass spectrometry (TOF-SIMS).

2. Materials and Methods

In this study, LFP was synthesized using a $FePO_4$ precursor provided by Ecopro BM, and mixed with Li_2CO_3 (99.9%, Aldrich, St. Louis, MI, USA) and 5 wt.% cellulose (20 μm, Aldrich) to form a carbon coating layer during cathode synthesis. The precursor and Li_2CO_3 were combined in a molar ratio of 1:1.05. This mixture was then subjected to a milling process using a planetary mono mill (Pulverisette 6, Fritsch, Idar-Oberstein, Germany) at 300 rpm for 3 h with 15 min intervals and a 10 min rest to prevent overheating. The milled mixture was sieved through a 300-mesh sieve. For the 1-step sulfur modification, sulfur (99.998%, Aldrich) was added to the mixed powder in varying amounts (1500, 2000, and 2500 ppm by weight of mixed powder). The mixture was then heat-treated in two stages: first, at 350 °C for 2 h, followed by 700 °C for 6 h under an Ar flow rate of 2 °C/min. This resulted in a 1-step sulfur-modified LFP, while the mixture without the sulfur was the pristine LFP (carbon-coated). A control LFP sample without any carbon coating was also prepared using the same process but without cellulose and sulfur. For the 2-step sulfur modification, pristine LFP was mixed with sulfur (250, 500, 750 ppm by weight of pristine LFP) using a mixer grinder at 30 Hz for 3 min. This mixture was then heat-treated at 300 °C for 4 h in a closed tube under an Ar atmosphere.

The surface morphology after heat treatment was observed using TEM (JEM-2100F, Cs corrector, Japan). The carbon structure was analyzed with a high-resolution Raman spectrometer (Jobin Yvon, LabRam HR Evolution, Japan) to observe the intensity ratio of the disorder peak (D-band at 1342.8 cm^{-1}) and the graphitic peak (G band at 1590.7 cm^{-1}). XPS (K-Alpha+) was utilized to investigate the carbon/sulfur bonds, and the spectra were analyzed using the Avantage Data System and calibrated with the C-S peak in the S 2p intervals. A TOF-SIMS (TOF-SIMS-5, Bi^{1+}) analysis was also conducted to identify the ionic species. The crystal structures of the differently prepared LFP samples were analyzed via X-ray diffraction (XRD, D8 Discover, Bruker, Billerica, MA, USA).

For the electrochemical testing, a homogeneous slurry was prepared by combining the active cathode material, carbon nanotube (CNTs), Super-P, and polyvinylidene fluoride (PVDF) at a precise weight ratio of 80:10:5:5, utilizing anhydrous N-methyl-2-pyrrolidone (NMP, Aldrich) as the solvent. The resulting slurry was uniformly cast onto aluminum foil, serving as the current collector, and subsequently dried under vacuum conditions at 80 °C for 24 h. The electrochemical performance was evaluated using 2032 coin-type cells, which were assembled with a Li metal anode, a polypropylene (Celgard 2400, USA) separator, and the prepared cathode. For cell testing at the standard conditions of 30 °C, 1M $LiPF_6$ dissolved in EC/DMC (1:1 vol%, Enchem Co., Ltd., Chungbuk, Republic of Korea) was used, while for elevated-temperature testing at 45 °C, 1M $LiPF_6$ in EC/EMC (3:7 vol%, Enchem Co., Ltd.) was utilized. The cells underwent cycling over a voltage range of 2.5–4.2 V at various C rates (0.05, 0.1, 0.3, 0.5, and 1.0 C), with 1 C defined as 150 $mA \cdot g^{-1}$, using a Won A Tech voltammetry system. Impedance measurements were conducted in the charged state after the 1st and 100th cycles at a current density of 150 $mA \cdot g^{-1}$, employing an Ametek VersaSTAT 3 electrochemical workstation. An alternating current voltage of 10 mV was applied over a frequency range of 0.1–100 kHz, and Nyquist plot fitting was performed using the ZSimpWin 3.60 software. The galvanostatic intermittent titration technique (GITT) was utilized to evaluate the Li-ion diffusion characteristics following the charge–discharge cycling, intermittently applying a pulse current of 0.1 C for 10 min. The Li^+ ion-diffusion coefficient (D_{Li+}) was calculated using the following equation:

$$D_{Li^+} = \frac{4}{\pi\tau}\left(\frac{n_m V_m}{S}\right)^2\left(\frac{\Delta E_s}{\Delta E_t}\right)^2$$

where τ represents the pulse duration (600 s), n_m denotes the molecular weight of the active material, V_m signifies the molar volume of the active material, S denotes the contact area between the electrolyte and the active material, ΔE_t represents the transient voltage change, and ΔE_s signifies the steady-state voltage change.

3. Results and Discussion

To assess the changes in the electrochemical properties resulting from the 1-step and 2-step sulfur modification processes, the charge–discharge profiles and discharge capacities of the sulfur-modified LFP were compared to those of the pristine LFP. The results, measured at 30 °C, are shown in Figure 1 and Table 1.

Table 1. Discharge capacities of pristine LFP and sulfur-modified LFP (1-step and 2-step processes) at 0.05, 0.1, 0.3, 0.5, and 1 C rates, and their capacity retentions (measured at 30 °C).

	Discharge Capacity ($mAh \cdot g^{-1}$)						Capacity Retention (%)
	0.05 C (1st Cycle)	0.1 C (2nd Cycle)	0.3 C (5th Cycle)	0.5 C (10th Cycle)	1 C (15th Cycle)	0.05 C (20th Cycle)	
Pristine LFP	142.2	130.1	107.8	94.2	76.2	137.9	53.6
1-step S1500ppm	146.1	133.7	109.9	96.4	78.6	140.8	53.8
1-step S2000ppm	152.2	141.8	120.8	106.3	88.5	145.2	58.1
1-step S2500ppm	151.3	140.2	119.5	106.0	88.6	146.7	58.5
2-step S250ppm	143.9	130.2	105.3	91.7	75.1	138.1	52.2
2-step S500ppm	152.8	140.9	118.2	101.6	83.0	147.6	54.3
2-step S750ppm	146.9	134.8	112.5	98.2	80.5	142.2	54.8

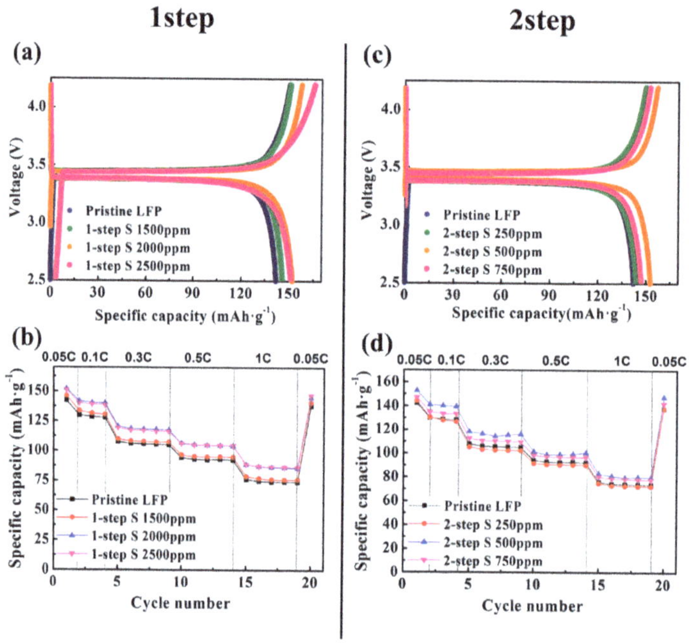

Figure 1. (**a**) Initial charge–discharge profiles and (**b**) discharge capacities at 0.05, 0.1, 0.3, 0.5, and 1 C rates for the pristine LFP and 1-step sulfur-modified LFPs. (**c**) Initial charge–discharge profiles and (**d**) discharge capacities at 0.05, 0.1, 0.3, 0.5, and 1 C rates for the pristine LFP and 2-step sulfur-modified LFPs.

The pristine LFP, containing a carbon coating layer derived from a cellulose source (5wt.%), served as the control. As illustrated in Figure 1a,b, despite certain variations depending on the amount of sulfur source used in the sulfur modification process (adjusted to 1500, 2000, and 2500 ppm), the 1-step sulfur-modified LFP exhibited a notable increase in the discharge capacity. In particular, the best performance was achieved with a sulfur source with a concentration of 2000 ppm. The discharge capacities of the pristine LFP were approximately 142 (0.05 C), 130 (0.1 C), 108 (0.3 C), 94 (0.5 C), and 76 (1 C) mAh·g^{-1}, as summarized in Table 1. Conversely, the 2000 ppm 1-step sulfur-modified LFP demonstrated improved capacities of approximately 152 (0.05 C), 142 (0.1 C), 121 (0.3 C), 106 (0.5 C), and 89 (1 C) mAh·g^{-1}. Furthermore, the rate characteristics also demonstrated a significant improvement with the 1-step sulfur modification. The capacity retention, defined as the percentage of capacity at a 1 C rate compared to that at a 0.05 C rate, improved from 53.6% for pristine LFP to over 58% with the 2000 and 2500 ppm 1-step sulfur modification. Figure S1 and Table S1 (Supplemental Information) present the electrochemical performance of the 1-step sulfur-modified LFP measured at 45 °C, demonstrating a significant increase in the discharge capacity and an improvement in the rate capability, which is consistent with the results at 30 °C. Given that the carbon coating sources for both the pristine and 1-step sulfur-modified LFPs were identical (5 wt.% cellulose), the 1-step sulfur modification significantly enhanced the electrochemical properties of LFP.

Figure 1c,d illustrate the electrochemical performance of the 2-step sulfur-modified LFP measured at 30 °C. When 250 ppm of the sulfur source was used, no improvement in the properties was observed compared to the pristine LFP; however, using 500 and 750 ppm of sulfur clearly improved the discharge capacity. Specifically, the apparently optimal 500 ppm 2-step sulfur-modified LFP exhibited capacities of approximately 153 (0.05 C), 141 (0.1 C), 118 (0.3 C), 102 (0.5 C), and 83 (1 C) mAh·g^{-1}, which were higher than those of the pristine LFP (Table 1). As shown in Figure S2 and Table S1 (Supplemental Information), the 2-step sulfur modification also effectively increased the discharge capacity of the LFP measured at 45 °C. However, the rate capability did not improve as significantly as that with the 1-step sulfur modification. As indicated in Table 1 and Table S1, the capacity retention of the 500 ppm 2-step sulfur-modified LFP was 54.3% at 30 °C and 68.2% at 45 °C, indicating an improvement compared to the pristine LFP. However, these values were lower than the capacity retention observed for the 2000 ppm 1-step sulfur-modified LFP, which were 58.1% at 30 °C and 76.0% at 45 °C. This suggests that the 1-step process was more effective than the 2-step process in enhancing the rate capability. Based on these measurements, the optimal amounts of the sulfur source for the 1-step and 2-step sulfur modifications under our experimental conditions were 2000 ppm and 500 ppm, respectively. Hereafter, the 2000 ppm 1-step sulfur-modified LFP will be referred to as the "1-step SLFP" and the 500 ppm 2-step sulfur-modified LFP will be referred to as the "2-step SLFP," both of which will be compared to the pristine LFP.

Figure S3 compares the XRD patterns of the 1-step and 2-step sulfur-modified SLFPs with that of the pristine LFP. The similarity in the peaks among these samples indicates that the sulfur modification process did not significantly alter the LFP phase. Table 2 summarizes the lattice parameters of the pristine LFP, 1-step SLFP, and 2-step SLFP, as determined by the Rietveld refinement of their XRD patterns. The lattice parameters demonstrated slight variations among the samples. Although the difference between the pristine LFP and 2-step SLFP was minimal, the lattice volume of the 1-step SLFP (291.35 Å^3) was greater than that of the pristine LFP (290.85 Å), and more pronounced changes in the lattice parameters were observed. This suggests that the high-temperature heat treatment process for the 1-step SLFP may have resulted in the doping of certain sulfur elements into the LFP structure, causing the changes observed in the lattice parameters. Conversely, the sulfur elements for the 2-step SLFP were supplied at a lower temperature of 300 °C after the carbon coating layer was already formed. Consequently, the sulfur was likely positioned on the surface rather than being doped into the LFP structure, making any doping effect unlikely. The surface coating layer was examined via a TEM analysis, as illustrated in Figure 2.

Table 2. Lattice parameters obtained from XRD patterns of the pristine LFP, 1-step sulfur-modified SLFP, and 2-step sulfur-modified SLFP via Rietveld refinement.

	a (Å)	b (Å)	c (Å)	V (Å^3)	R_{wp}	GoF
Pristine LFP	10.3253	6.0190	4.6903	290.8579	2.0026	1.7312
1-step SLFP	10.3320	6.0100	4.6920	291.3512	2.6242	2.4149
2-step SLFP	10.3255	6.0046	4.6912	290.873	2.1400	1.9005

For comparison, TEM images of the carbon-uncoated LFP were also captured (Figure 2a), which did not present a distinct surface layer. Conversely, the pristine LFP (carbon-coated) exhibited a clearly defined 2–3 nm thick surface layer, which was identified as the carbon coating layer (Figure 2b). Both the 1-step and 2-step sulfur-modified SLFPs demonstrated a similar thickness of the carbon coating layer and no significant changes were observed owing to the sulfur modification (Figure 2c,d).

Figure 3 compares the Raman spectra of the 1-step and 2-step sulfur-modified SLFP with that of the pristine LFP. Raman spectroscopy is an effective technique for characterizing the carbon structure (including disorders and crystallite formation) of the coating layer owing to the strong scattering properties of carbon, with two E2g modes, predicted to be Raman active. The carbon structure of the coating layer significantly influences the LFP properties, as a higher proportion of graphitic carbon than the disordered carbon enhances the electronic conductivity. Figure 3 demonstrates broad peaks located at approximately 1342.8 and 1590.7 cm^{-1}, which indicate the D and G bands of the carbon coating, respectively. The extent of graphitization and its ratio to disordered carbon is typically characterized by the I_D/I_G (intensity ratio of D and G bands) ratio in the Raman spectrum [35–37]. As shown in Figure 3a,b, the intensity of the G band in the Raman spectrum of the 1-step SLFP is slightly larger than that of the pristine sample.

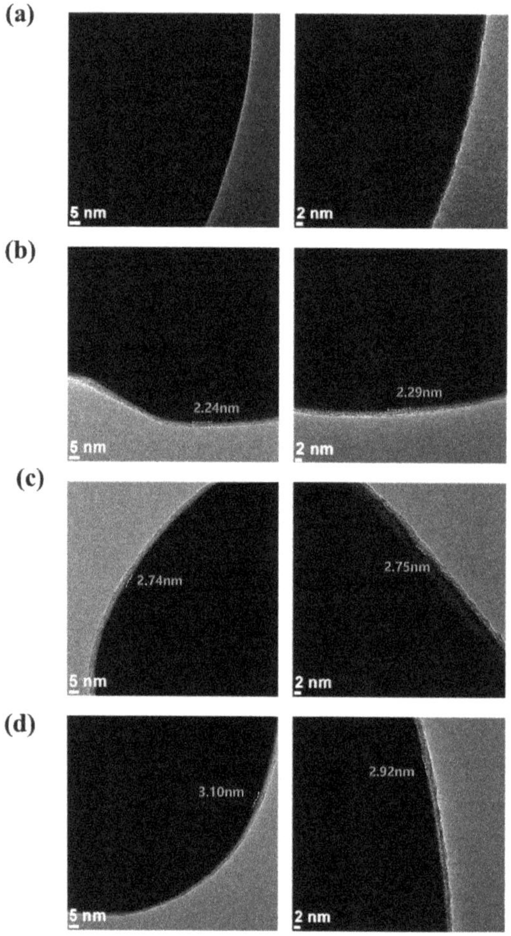

Figure 2. TEM images of the (**a**) carbon-uncoated LFP, (**b**) pristine LFP (carbon-coated), (**c**) 1-step sulfur-modified SLFP, and (**d**) 2-step sulfur-modified SLFP.

The calculated I_D/I_G ratio for the 1-step SLFP (0.90) was relatively lower than that of the pristine sample (0.95), suggesting that the 1-step sulfur modification increased the proportion of graphitic carbon compared to the disordered carbon, resulting in an improved electronic conductivity. The improved rate capability observed in Figure 1 with the 1-step sulfur modification can be attributed to this increased graphitization. The calculated I_D/I_G ratio for the 2-step SLFP was 0.93, which was also lower than that of the pristine LFP (Figure 3c), indicating that the 2-step sulfur modification process also increased the proportion of graphitic carbon, although to a lesser extent than that of the 1-step sulfur modification. The relatively small proportion of graphite carbon likely caused the 2-step sulfur modification to be relatively less successful compared to the 1-step sulfur modification in enhancing the rate capability.

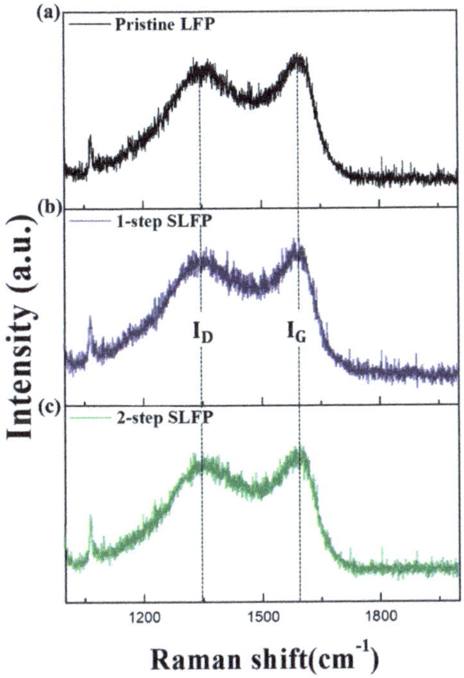

Figure 3. Raman spectra of the (**a**) pristine LFP, (**b**) 1-step sulfur-modified SLFP, and (**c**) 2-step sulfur-modified SLFP.

The Raman analysis confirmed that the carbon structure was altered to a certain extent via sulfur modification. Additionally, the sulfur element remaining in the carbon coating layer may have played a crucial role in protecting the cathode interface from the electrolyte. Given that both the 1-step and 2-step processes involved heat treatment, the sulfur element may have been reduced compared to the initial amount supplied. XPS and TOF-SIMS analyses were conducted to verify the amount of sulfur retained in the surface coating layer. As shown in Figure S4 (Supplemental Information), distinguishing the sulfur-related peaks of the pristine LFP and 1-step SLFP in the XPS spectra was challenging. However, in the 2-step SLFP spectrum, a faint sulfur-related peak was observed at approximately 162–165 eV, indicating that the 2-step SLFP contained more residual sulfur than the 1-step SLFP; however, the peak intensity was too low for a definitive comparison between the samples. This low intensity was likely owing to the minimal amount of the sulfur-containing carbon layer on the cathode surface. Therefore, sulfur-modified carbon samples (excluding the cathode) were prepared utilizing the same method used for the 2-step SLFP and were analyzed via XPS. As shown in Figure 4d, distinct peaks related to the C-S-C bond near 163 and 164.2 eV were identified, indicating that the 2-step SLFP with the same sulfur-modified carbon as the coating layer contains a significant amount of sulfur.

TOF-SIMS was also used for a comparison of the sulfur content on the surface. As shown in Figure 4, the amount of S^- detected in the pristine LFP was negligible (Figure 4a). The 1-step SLFP exhibited a slightly higher amount of S^- than the pristine LFP, indicating that a certain amount of the sulfur element was retained despite the high-temperature (700 °C) heat treatment (Figure 4b). However, the S^- intensity in the 2-step SLFP was relatively higher than in the 1-step SLFP (Figure 4c). The comparison in Figure 4d demonstrates that despite using 2000 ppm of the sulfur source in the 1-step process and only 500 ppm in the 2-step process, the residual sulfur element was greater in the 2-step SLFP, which underwent heat treatment at a lower temperature (300 °C). This suggests that the high-temperature treatment in the 1-step process led to considerable sulfur losses, but it also likely resulted in a certain amount of sulfur doping into the bulk. Additionally, the S element in the 1-step process may have contributed to the increased proportion of graphitic carbon by influencing the carbon formation process.

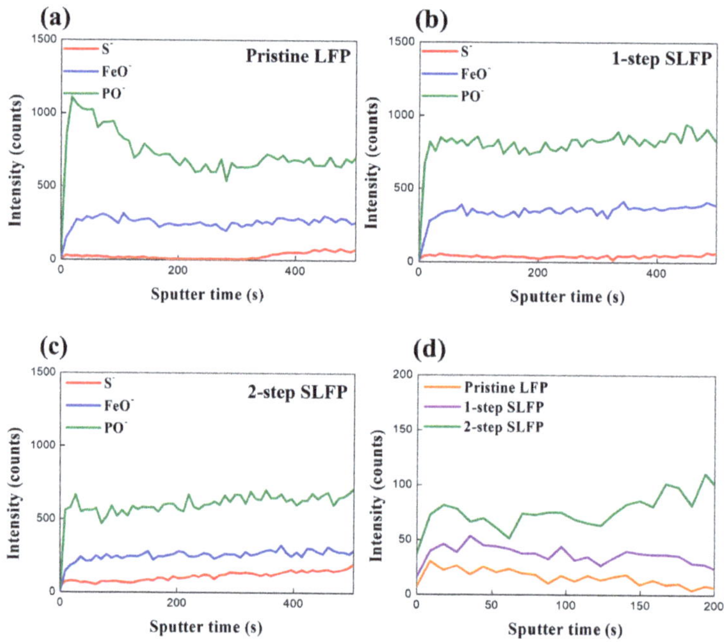

Figure 4. TOF-SIMS spectra of the surface of (**a**) pristine LFP, (**b**) 1-step sulfur-modified SLFP, (**c**) 2-step sulfur-modified SLFP, and (**d**) comparison of their S^- intensity.

To verify the change in the interfacial resistance owing to sulfur modification, the impedance of cells containing the pristine LFP, 1-step, and 2-step SLFP were measured. Figure 5 presents the Nyquist plots of the cells after the 1st and 100th cycles. As illustrated, the size of the semicircle in the Nyquist plots significantly decreased with sulfur modification, indicating a decreased impedance. For a more precise analysis, the Nyquist plots were fitted using the equivalent circuit shown in Figure S5 (Supplemental Information), and the obtained impedance values are summarized in Table 3. The Nyquist plots of the cells comprise the bulk resistance (R_b), resistance of the solid-state interface layer formed on the electrode surface (R_{SEI}), charge transfer resistance between the active material particles and the electrolyte (R_{CT}), and the Warburg impedance (W) [38,39]. As shown in Table 3, R_{CT} was reduced by sulfur modification after both the 1st and 100th cycles. Notably, after the 1st cycle, the 1-step SLFP exhibited lower impedance values (~13 Ω) than those of the 2-step SLFP (~30 Ω). The larger amount of carbon, improved carbon structure, and slight sulfur doping effect may have contributed to these low impedance values in the 1-step SLFP. However, after 100 cycles, the impedance values for both the 1-step (~35 Ω) and 2-step SLFP (~35.8 Ω) were nearly similar. Despite being lower than that of the pristine LFP (~54 Ω), the impedance values increased with cycling for both modified samples, with the 2-step SLFP demonstrating a relatively smaller increase in impedance compared to that of the 1-step SLFP, which may be owing to the sulfur element in the carbon coating layer. The 2-step SLFP carbon coating contained relatively more sulfur, which helped reduce the side reactions with the electrolyte during cycling, thus resulting in a relatively smaller increase in the impedance values during cycling.

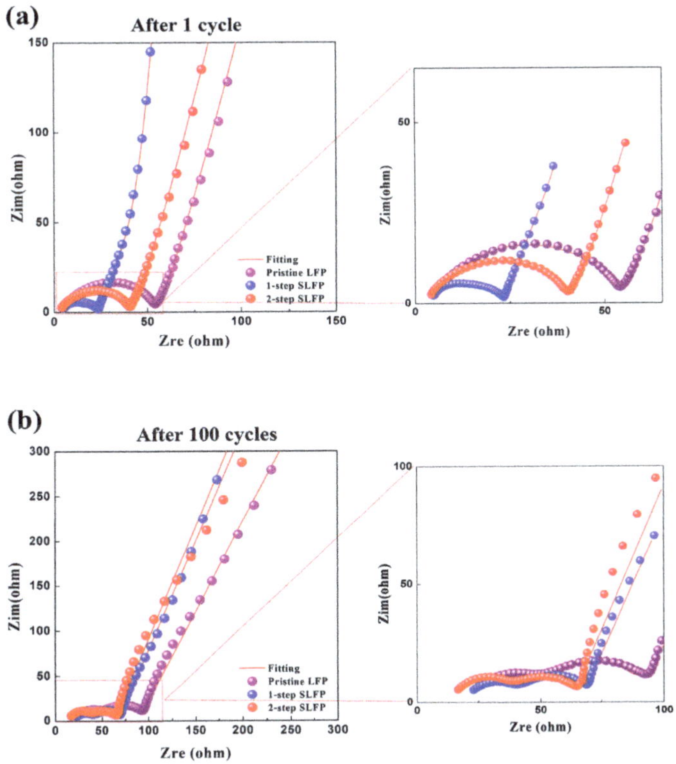

Figure 5. Nyquist plots of the cells containing the pristine LFP, 1-step sulfur-modified SLFP, and 2-step sulfur-modified SLFP after (**a**) 1 cycle and (**b**) 100 cycles.

Table 3. Impedance values of the pristine LFP, 1-step sulfur-modified SLFP, and 2-step sulfur-modified SLFP obtained from the Nyquist plots via fitting.

	After 1 Cycle				After 100 Cycles			
	R_b (Ω)	R_{SEI} (Ω)	R_{CT} (Ω)	R_{total} (Ω)	R_b (Ω)	R_{SEI} (Ω)	R_{CT} (Ω)	R_{total} (Ω)
Pristine LFP	4.96	7.63	43.94	56.54	25.53	16.09	54.28	95.90
1-step SLFP	4.70	6.96	13.05	24.70	22.95	14.66	35.02	70.50
2-step SLFP	4.11	7.90	30.43	42.44	16.75	12.52	35.84	67.25

Figure 6 presents the Li-diffusion coefficient (D_{Li+}) values measured using the GITT method (after 100 cycles), as shown in Figure S6a (Supplemental Information). The results indicate an overall improvement in the D_{Li+} values owing to the sulfur modification. Figure S6b compares the D_{Li}+ values in the plateau range during the discharge process. For the pristine LFP, the D_{Li+} values ranged from 3.96×10^{-14} to 1.16×10^{-13} $cm^2{\cdot}s^{-1}$. Conversely, the 1-step SLFP exhibited D_{Li+} values ranging from 6.87×10^{-14} to 1.41×10^{-13} $cm^2{\cdot}s^{-1}$. Notably, the 2-step SLFP demonstrated slightly higher D_{Li+} values than those of the 1-step SLFP, ranging from 1.62×10^{-13} to 1.83×10^{-13} $cm^2{\cdot}s^{-1}$. Considering this analysis was performed after 100 cycles, the marginally superior D_{Li+} values observed for the 2-step SLFP can be attributed to the interfacial protection effect, which enhances the movement of Li ions. This interfacial protection, provided by the sulfur element in the carbon coating layer, likely reduces the side reactions with the electrolyte and facilitates better ionic conductivity.

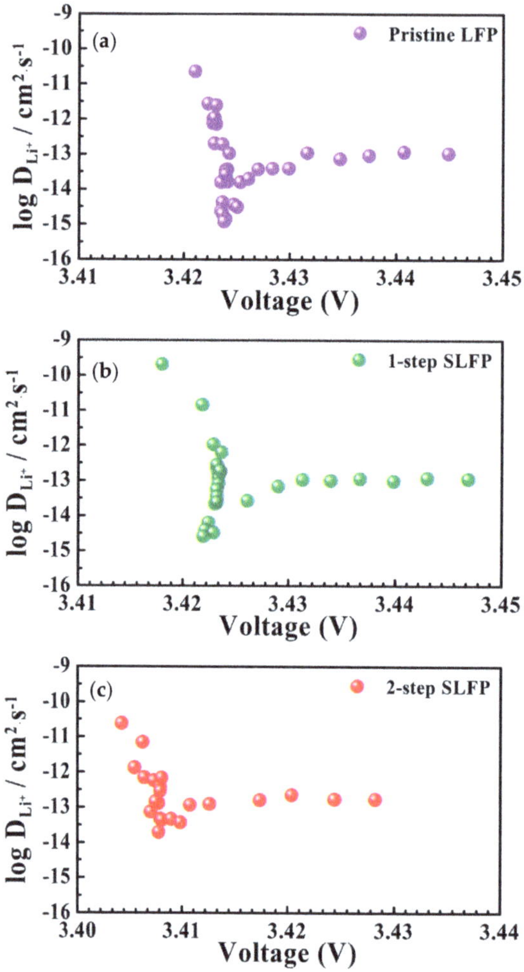

Figure 6. Li-diffusion coefficient (D_{Li+}) values measured using the GITT method for (**a**) pristine LFP, (**b**) 1-step sulfur-modified SLFP, and (**c**) 2-step sulfur-modified SLFP.

To further elucidate the interface-protection effects of the sulfur modification, we conducted a comparative analysis of the electrochemical properties of the cells containing the LFP samples after storage at 60 °C. The elevated temperature accelerated the side reactions at the interface, providing a clear platform to evaluate the interface stabilization imparted by the sulfur modification. As shown in Figure 7a–c, the initial charge–discharge profiles were recorded for the cells stored at 60 °C for 7, 10, and 14 days, respectively.

After 7 days of storage, the differences among the samples were minimal. However, after 10 days, the pristine LFP exhibited a noticeable decrease in capacity, whereas both the 1-step and 2-step SLFP samples demonstrated a minimal deviation from their 7-day performance, indicating enhanced stability. The distinction became more pronounced after 14 days, when the pristine LFP cells ceased to function properly, whereas the 2-step SLFP cells continued to charge and discharge normally with only a slight capacity loss. Conversely, the 1-step SLFP demonstrated a certain amount of degradation; although operational, it exhibited irregularities in its charge/discharge curves (marked with a red circle). Figure 7d–f demonstrate the discharge capacities at 0.05, 0.1, 0.3, 0.5, and 1 C rates post-storage at 60 °C, with the corresponding data summarized in Table 4.

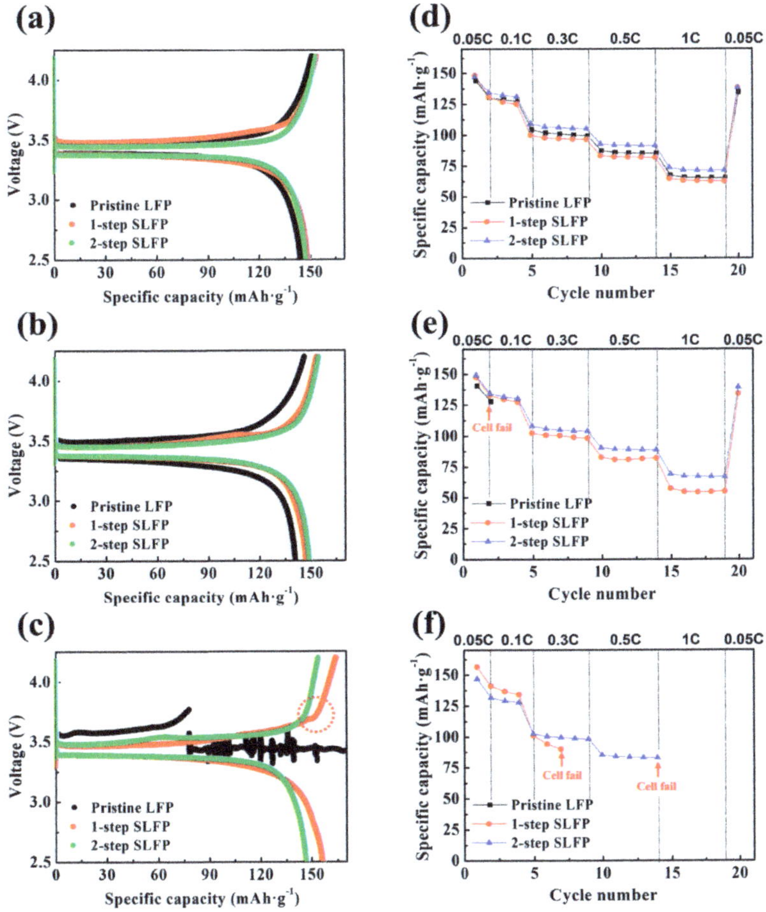

Figure 7. Charge–discharge profiles of the pristine LFP, 1-step sulfur-modified SLFP, and 2-step sulfur-modified SLFP after storage for (**a**) 7 days, (**b**) 10 days, and (**c**) 14 days. Discharge capacities at 0.05, 0.1, 0.3, 0.5, and 1 C rates for the pristine LFP, 1-step sulfur-modified SLFP, and 2-step sulfur-modified SLFP after storage for (**d**) 7 days, (**e**) 10 days, and (**f**) 14 days.

After 7 days, all the cells cycled normally. However, following 10 days of storage, the cell with pristine LFP failed to cycle at a 0.1 C rate. The 1-step SLFP cells remained functional across all C rates, but the capacity retention representing the rate capability significantly diminished to ~39% compared to the 7-day mark (~44%), indicating substantial deterioration. Conversely, the 2-step SLFP cells exhibited a smaller reduction in capacity retention, from ~50% (7 days of storage) to 46% (10 days of storage), and maintained a relatively stable operation. After 14 days, the pristine LFP sample entirely ceased to function, whereas the 1-step SLFP failed to operate at a rate of 0.5 C. Conversely, the 2-step SLFP managed to maintain the performance until it eventually stopped running at a rate of 1 C. These findings indicate that the 2-step sulfur modification process significantly enhances the cathode protection against side reactions with the electrolyte. This improved performance demonstrates the superior stability of sulfur-modified carbon via vaporized sulfur. Although the 1-step sulfur modification also demonstrated improved stability during the storage test compared to the pristine LFP, it was less effective than the 2-step approach. Therefore, the 2-step sulfur modification process offers superior protection against interface side reactions, rendering it more effective in stabilizing the cathode under elevated temperature conditions. Figure 8 schematically summarizes the effects of the 1-step and 2-step sulfur modifications.

Table 4. Discharge capacities of the pristine LFP and sulfur-modified LFP (1-step and 2-step processes) at 0.05, 0.1, 0.3, 0.5, and 1 C rates measured after storage for 7, 10, and 14 days at 60 °C.

	Discharge Capacity (mAh·g^{-1})						Capacity Retention (%)
	0.05 C (1st Cycle)	0.1 C (2nd Cycle)	0.3 C (5th Cycle)	0.5 C (10th Cycle)	1 C (15th Cycle)	0.05 C (20th Cycle)	
Pristine LFP_7days	143.97	130.54	104.41	87.40	67.45	134.31	46.85
1-step SLFP_7days	148.60	130.78	100.27	83.68	65.07	138.50	43.79
2-step SLFP_7days	147.46	134.74	109.58	93.24	73.89	138.23	50.11
Pristine LFP_10days	140.59	127.97	Cell fail				
1-step SLFP_10days	147.54	133.13	102.43	83.08	57.48	113.11	38.96
2-step SLFP_10days_	149.43	134.45	108.45	90.75	69.43	139.60	46.46
Pristine LFP_14days	Cell fail						
1-step SLFP_14days	156.78	141.31	101.47	Cell fail			
2-step SLFP_14days	147.29	132.03	102.90	85.68	Cell fail		

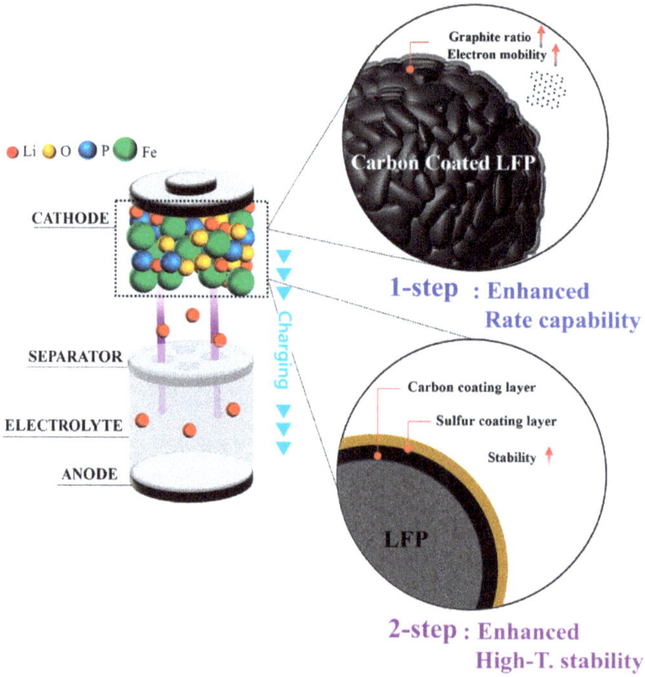

Figure 8. Schematic summarization of the 1-step and 2-step sulfur modification effects.

4. Conclusions

In this study, sulfur modification of the carbon layer was introduced to enhance the performance of the carbon coating layer for LFP cathodes. Two methods were explored: a 1-step process and a 2-step process. In the 1-step process, sulfur powder was mixed with cellulose followed by heat treatment to form a coating layer, with the sulfur element expected to improve the properties of the carbon coating layer during the heating process.

In the 2-step process, an additional coating layer was applied on top of the carbon coating layer using vaporized sulfur during a low-temperature (300 °C) heat treatment, aiming to improve the high-temperature stability of the surface layer. The 1-step sulfur-modified LFP exhibited an enhanced discharge capacity and rate capability compared to the pristine LFP. Under optimal conditions, the sample exhibited a discharge capacity of approximately 152 mAh/g at a 0.05 C rate, with a capacity retention of around 58%, where capacity retention is defined as the percentage of capacity at a 1 C rate compared to that at 0.05 C. Raman spectroscopy indicated increased graphitization in the 1-step SLFP, which enhanced the electronic conductivity. Although the 2-step sulfur modification did not improve the rate capability as effectively as the 1-step process, it better maintained the electrochemical properties during storage at high temperatures (60 °C). TOF-SIMS analyses confirmed the sulfur retention in the surface coating layer, indicating more sulfur retention with the 2-step SLFP owing to the treatment at a lower temperature. The sulfur elements on the surface are expected to enhance the stability of the LFP at high temperatures. Impedance measurements indicated a reduced impedance for the sulfur-modified LFP, and the GITT method confirmed improved Li-diffusion coefficient values owing to the sulfur modification. Based on these results, sulfur modification significantly enhances the electrochemical performance and stability of LFP cathodes, offering a promising solution for improving the performance of LIBs based on LFP. Table 5 provides a comparative summary of the results from this study alongside previous studies focusing on carbon coatings. The data highlight that the current work demonstrates highly competitive properties, with the results suggesting that sulfur modification offers a significant advantage over conventional carbon coatings. It is hoped that further research in this area will continue to explore the potential of sulfur modification, potentially leading to even greater performance improvements for LFP-based LIBs in the future.

Table 5. Discharge capacities of carbon-coated LFP obtained from the previous study.

Cathode	Coating Source	Conditions	Discharge Capacity	Reference
$LiFePO_4$	[Velm]NTf_2	Room temperature 2.5 V–4.2 V	0.1 C 136.4 mAh·g^{-1}	[40]
$LiFePO_4$	Coke	25 °C 2.5 V–4.2 V	0.1 C 145.99 mAh·g^{-1}	[41]
$LiFePO_4$	Graphene nanosheet	Room temperature 2.0 V–4.3 V	0.1 C 145 mAh·g^{-1}	[42]
$LiFePO_4$	Sucrose	2.0 V–4.2 V	0.1 C 128 mAh·g^{-1}	[20]
$LiFePO_4$	Sucrose	2.5 V–4.5 V	0.1 C 125 mAh·g^{-1}	[43]
$LiFePO_4$	Sucrose	25 °C 2.5 V–3.8 V	0.1 C 140 mAh·g^{-1}	[44]
$LiFePO_4$	Sucrose	Room temperature 2.4 V–4.2 V	0.1 C 132 mAh·g^{-1}	[45]
$LiFePO_4$	Cellulose	30 °C 2.5 V–4.2 V	0.1 C 141.8 mAh·g^{-1}	Our study
$LiFePO_4$	Cellulose	30 °C 2.5 V–4.2 V	0.1 C 140.9 mAh·g^{-1}	Our study

Supplementary Materials: The following supporting information can be downloaded at: https://www.mdpi.com/article/10.3390/batteries10100348/s1. Figure S1: (a) Initial charge–discharge profiles and (b) discharge capacities at 0.05, 0.1, 0.3, 0.5, and 1 C rates for the pristine LFP and 1-step sulfur-modified LFPs measured at 45 °C. Figure S2: (a) Initial charge–discharge profiles and (b) discharge capacities at 0.05, 0.1, 0.3, 0.5, and 1 C rates for the pristine LFP and 2-step sulfur-modified LFPs measured at 45 °C. Figure S3: XRD patterns of the pristine LFP, 1-step sulfur-modified SLFP, and 2-step sulfur-modified SLFP. Figure S4: XPS spectra of the (a) pristine LFP, (b) 1-step sulfur-modified SLFP, (c) 2-step sulfur-modified SLFP, and (d) sulfurized carbon using the 2-step process (the same carbon:sulfur ratio as the 2-step sulfur-modified SLFP was used). Figure S5: (a) Equivalent circuit for fitting the Nyquist plots and (b) fitting example using R_b, R_{SEI}, R_{ct}, and W. Figure S6: (a) Measurement profile for the GITT method and (b) comparison of the Li-diffusion coefficient (D_{Li+}) values for the pristine LFP, 1-step sulfur-modified SLFP, and 2-step sulfur-modified SLF in the plateau region. Table S1: Discharge capacities of the pristine LFP and sulfur-modified LFP (1-step and 2-step processes) at 0.05, 0.1, 0.3, 0.5, and 1 C rates, and their capacity retentions (measured at 45 °C). Table S2: Lattice parameters obtained from the XRD patterns of the pristine LFP, 1-step sulfur-modified SLFP, and 2-step sulfur-modified SLFP via Rietveld refinement.

Author Contributions: Conceptualization, investigation, data curation, writing—original draft preparation, S.-h.K.; writing—review and editing, supervision, funding acquisition, Y.J.P. All authors have read and agreed to the published version of the manuscript.

Funding: This work was supported by the Materials and Components Technology Development Program (grant No. 20024249) funded By the Ministry of Trade, Industry and Energy (MOTIE, Republic of Korea), and by the National Research Foundation of Korea (NRF), grant funded by the Korean government (MSIT, No. 2023R1A2C1003330).

Data Availability Statement: Data are contained within the article.

Acknowledgments: This work was supported by the Materials and Components Technology Development Program (grant No. 20024249) funded By the Ministry of Trade, Industry and Energy (MOTIE, Republic of Korea), and by the National Research Foundation of Korea (NRF), grant funded by the Korean government (MSIT, No. 2023R1A2C1003330).

Conflicts of Interest: The authors declare no conflicts of interest.

References

1. Huang, J.; Fan, Z.; Xu, C.; Jiang, F.; Feng, X. Experimental investigation of thermal runaway characteristics of large-format $Li(Ni_{0.8}Co_{0.1}Mn_{0.1})O_2$ Battery under Different Heating Powers and Areas. *Batteries* **2024**, *10*, 241. [CrossRef]
2. Das, D.; Manna, S.; Puravankara, S. Electrolytes, Additives and Binders for NMC cathodes in Li-ion batteries—A review. *Batteries* **2023**, *9*, 193. [CrossRef]
3. Panda, P.K.; Cho, T.S.; Hsieh, C.-T.; Yang, P.C. Cobalt- and Copper-Doped NASICON-Type LATP Polymer Composite Electrolytes Enabling Lithium Titania Electrode for Solid-State Lithium Batteries with High-Rate Capability and Excellent Cyclic Performance. *J. Energy Storage* **2024**, *95*, 112559. [CrossRef]
4. Joo, M.J.; Kim, M.; Chae, S.; Ko, M.; Park, Y.J. Additive-derived surface modification of cathodes in all-solid-state batteries: The effect of lithium difluorophosphate- and lithium Difluoro(Oxalato)Borate-derived coating layers. *ACS Appl. Mater. Interfaces* **2023**, *15*, 59389–59402. [CrossRef]
5. Ji, Y.J.; Noh, S.; Seong, J.Y.; Lee, S.; Park, Y.J. Li_3BO_3-Li_3PO_4 Composites for Efficient Buffer Layer of Sulphide-Based All-Solid-State Batteries. *Batteries* **2024**, *9*, 292. [CrossRef]
6. Madaoui, S.; Vinassa, J.M.; Sabatier, J.; Guillemard, F. An electrothermal model of an NMC lithium-ion prismatic battery cell for temperature distribution assessment. *Batteries* **2023**, *9*, 478. [CrossRef]
7. Joo, M.J.; Park, Y.J. Stabilizing Li_2O-based Cathode /Electrolyte interfaces through succinonitrile addition. *J. Electrochem. Sci. Technol.* **2023**, *14*, 231–242. [CrossRef]
8. Hawley, W.B.; Li, M.; Li, J. Room-temperature eutectic synthesis for upcycling of cathode materials. *Batteries* **2023**, *9*, 498. [CrossRef]
9. Ramasubramanian, B.; Sundarrajan, S.; Chellappan, V.; Reddy, M.V.; Ramakrishna, S.; Zaghib, K. Recent development in carbon-$LiFePO_4$ cathodes for lithium-ion batteries: A mini review. *Batteries* **2022**, *8*, 133. [CrossRef]
10. Mohanty, D.; Chang, M.J.; Hung, I.M. The effect of different amounts of conductive carbon material on the electrochemical performance of the $LiFePO_4$ cathode in Li-ion batteries. *Batteries* **2023**, *9*, 515. [CrossRef]
11. Wu, K.; Hu, N.; Wang, S.; Geng, Z.; Deng, W. Enhancing performance of $LiFePO_4$ battery by using a novel gel composite polymer electrolyte. *Batteries* **2023**, *9*, 51. [CrossRef]
12. Zhang, W.J. Structure and performance of $LiFePO_4$ cathode materials: A review. *J. Power Sources* **2011**, *196*, 2962–2970. [CrossRef]
13. Chen, S.P.; Lv, D.; Chen, J.; Zhang, Y.H.; Shi, F.N. Review on defects and modification methods of $LiFePO_4$ Cathode material for lithium-ion batteries. *Energy Fuels* **2022**, *36*, 1232–1251. [CrossRef]
14. Prosini, P.P.; Lisi, M.; Jane, D.; Pasquali, M. Determination of the Diffusion Coefficient of $LiFePO_4$. *Solid State Ionics* **2002**, *148*, 45–51. [CrossRef]
15. Amin, R.; Balaya, P.; Maier, J. Anisotropy of electronic and ionic transport in $LiFePO_4$ single crystals. *Electrochem. Solid State Lett.* **2007**, *10*, 13–16. [CrossRef]
16. Wang, J.; Sun, X. Understanding and recent development of carbon coating on $LiFePO_4$ cathode materials for lithium-ion batteries. *Energy Environ. Sci.* **2012**, *5*, 5163–5185. [CrossRef]

17. Moon, H.; Kim, D.; Park, G.; Shin, K.; Cho, Y.; Gong, C.; Lee, Y.S.; Nam, H.; Hong, S.; Choi, N.S. Balancing ionic and electronic conduction at the $LiFePO_4$ cathode–electrolyte interface and regulating solid electrolyte interphase in lithium-ion batteries. *Adv. Funct. Mater.* **2024**, *34*, 2403261. [CrossRef]
18. Ni, H.; Liu, J.; Fan, L.Z. Carbon-coated $LiFePO_4$-porous carbon composites as cathode materials for lithium ion batteries. *Nanoscale* **2013**, *5*, 2164–2168. [CrossRef] [PubMed]
19. Wang, C.; Yuan, X.; Tan, H.; Jian, S.; Ma, Z.; Zhao, J.; Wang, X.; Chen, D.; Dong, Y. Three-dimensional carbon-coated $LiFePO_4$ cathode with improved Li-ion battery performance. *Coatings* **2021**, *11*, 1137. [CrossRef]
20. Qi, M.; Liu, Y.; Xu, M.; Feng, M.; Gu, J.; Liu, Y.; Wang, L. Improved electrochemical performances of carbon-coated $LiFePO_4$ microspheres for Li-ion battery cathode. *Mater. Res. Express* **2019**, *6*, 115520. [CrossRef]
21. Nien, Y.H.; Carey, J.R.; Chen, J.S. Physical and electrochemical properties of $LiFePO_4/C$ composite cathode prepared from various polymer-containing precursors. *J. Power Sources* **2009**, *193*, 822–827. [CrossRef]
22. Mathur, P.; Shih, J.Y.; Li, Y.J.J.; Hung, T.F.; Thirumalraj, B.; Ramaraj, S.K.; Jose, R.; Karuppiah, C.; Yang, C.C. In situ metal organic framework (ZIF-8) and mechanofusion-assisted MWCNT coating of $LiFePO_4/C$ composite material for lithium-ion batteries. *Batteries* **2023**, *9*, 182. [CrossRef]
23. Yoon, S.; Liao, C.; Sun, X.G.; Bridges, C.A.; Unocic, R.R.; Nanda, J.; Dai, S.; Paranthaman, M.P. Conductive surface modification of $LiFePO_4$ with nitrogen-doped carbon layers for lithium-ion batteries. *J. Mater. Chem.* **2012**, *22*, 4611–4614. [CrossRef]
24. Yang, J.; Wang, J.; Li, X.; Wang, D.; Liu, J.; Liang, G.; Gauthier, M.; Li, Y.; Geng, D.; Li, R.; et al. Hierarchically porous $LiFePO_4$/nitrogen-doped carbon nanotubes composite as a cathode for lithium ion batteries. *J. Mater. Chem.* **2012**, *22*, 7537–7543. [CrossRef]
25. Zhang, J.; Nie, N.; Liu, Y.; Wang, J.; Yu, F.; Gu, J.; Li, W. Boron and Nitrogen Codoped carbon layers of $LiFePO_4$ improve the high-rate electrochemical performance for lithium ion batteries. *ACS Appl. Mater. Interfaces* **2015**, *7*, 20134–20143. [CrossRef] [PubMed]
26. Wang, X.; Feng, Z.; Hou, X.; Liu, L.; He, M.; He, X.; Huang, J.; Wen, Z. Fluorine doped carbon coating of $LiFePO_4$ as a cathode material for lithium-ion batteries. *Chem. Eng. J.* **2020**, *379*, 122371. [CrossRef]
27. Cho, J.; Kim, Y.J.; Park, B. Novel $LiCoO_2$ cathode material with Al_2O_3 coating for a Li ion cell. *Chem. Mater.* **2000**, *12*, 3788–3791. [CrossRef]
28. Chang, H.H.; Chang, C.C.; Su, C.Y.; Wu, H.C.; Yang, M.H.; Wu, N.L. Effects of TiO_2 coating on high-temperature cycle performance of $LiFePO_4$-based lithium-ion batteries. *J. Power Sources* **2008**, *185*, 466–472. [CrossRef]
29. Cui, Y.; Zhao, X.; Guo, R. Enhanced electrochemical properties of $LiFePO_4$ cathode material by CuO and carbon co-coating. *J. Alloys Compd.* **2010**, *490*, 236–240. [CrossRef]
30. Cui, Y.; Zhao, X.; Guo, R. High rate electrochemical performances of nanosized ZnO and carbon co-coated $LiFePO_4$ cathode. *Mater. Res. Bull.* **2010**, *45*, 844–849. [CrossRef]
31. Cho, J.; Kim, Y.J.; Kim, T.J.; Park, B. Zero-strain intercalation cathode for rechargeable Li-ion cell. *Angew. Chem. Int. Ed. Engl.* **2001**, *40*, 3367–3369. [CrossRef] [PubMed]
32. Zhao, S.X.; Ding, H.; Wang, Y.C.; Li, B.H.; Nan, C.W. Improving rate performance of $LiFePO_4$ cathode materials by hybrid coating of nano-Li_3PO_4 and carbon. *J. Alloys Compd.* **2013**, *566*, 206–211. [CrossRef]
33. Lee, S.B.; Cho, S.H.; Aravindan, V.; Kim, H.S.; Lee, Y.S. Improved cycle performance of sulfur doped $LiFePO_4$ material at high temperatures. *Koreascience* **2009**, *30*, 2223–2226. [CrossRef]
34. Xu, D.; Wang, P.; Shen, B. Synthesis and characterization of sulfur-doped carbon decorated $LiFePO_4$ nanocomposite as high performance cathode material for lithium-ion batteries. *Ceram. Int.* **2016**, *42*, 5331–5338. [CrossRef]
35. Wilcox, J.D.; Doeff, M.M.; Marcinek, M.; Kostecki, R. Factors influencing the quality of carbon coatings on $LiFePO_4$. *J. Electrochem. Soc.* **2007**, *154*, A389–A395. [CrossRef]
36. Ait Salah, A.; Mauger, A.; Zaghib, K.; Goodenough, J.B.; Ravet, N.; Gauthier, M.; Gendron, F.; Julien, C.M. Reduction Fe^{3+} of Impurities in $LiFePO_4$ from Pyrolysis of Organic Precursor Used for Carbon Deposition. *J. Electrochem. Soc.* **2006**, *153*, A1692–A1701. [CrossRef]
37. Doeff, M.M.; Wilcox, J.D.; Kostecki, R.; Lau, G. Optimization of carbon coatings on $LiFePO_4$. *J. Power Sources* **2006**, *163*, 180–184. [CrossRef]
38. Bao, S.J.; Liang, Y.Y.; Li, H.L. Synthesis and electrochemical properties of $LiMn_2O_4$ by microwave-assisted sol-gel method. *Mater. Lett.* **2005**, *59*, 3761–3765. [CrossRef]
39. Nara, H.; Morita, K.; Mukoyama, D.; Yokoshima, T.; Momma, T.; Osaka, T. Impedance analysis of $LiNi_{1/3}Mn_{1/3}Co_{1/3}O_2$ cathodes with different secondary-particle size distribution in lithium-ion battery. *Electrochim. Acta* **2017**, *241*, 323–330. [CrossRef]
40. Xia, J.; Zhu, F.; Wang, G.; Wang, L.; Meng, Y.; Zhang, Y. Synthesis of $LiFePO_4/C$ using ionic liquid as carbon source for lithium ion batteries. *Solid State Ion.* **2017**, *308*, 133–138. [CrossRef]
41. Guo, F.; Huang, X.; Li, Y.; Zhang, S.; He, X.; Liu, J.; Yu, Z.; Li, F. In Situ Low-Temperature Carbonization Capping of $LiFePO_4$ with Coke for Enhanced Lithium Battery Performance. *Molecules* **2023**, *28*, 6083. [CrossRef] [PubMed]
42. Fei, H.; Peng, Z.; Yang, Y.; Li, L.; Raji, A.R.O.; Samuel, E.L.G.; Tour, J.M. $LiFePO_4$ nanoparticles encapsulated in graphene nanoshells for high-performance lithium-ion battery cathodes. *Chem. Commun.* **2014**, *50*, 7117–7119. [CrossRef] [PubMed]
43. Pratheeksha, P.M.; Rajeshwari, J.S.; Daniel, P.J.; Rao, T.N.; Anandan, S. Investigation of In-Situ Carbon Coated $LiFePO_4$ as a Superior Cathode Material for Lithium Ion Batteries. *J. Nanosci. Nanotechnol.* **2018**, *19*, 3002–3011. [CrossRef] [PubMed]

44. Wu, S.; Luo, E.; Ouyang, J.; Lu, Q.; Zhang, X.; Wei, D.; Han, W.K.; Xu, X.; Wei, L. Tuning the graphitization of the carbon coating layer on $LiFePO_4$ Enables Superior Properties. *Int. J. Electrochem. Sci.* **2024**, *19*, 100450. [CrossRef]
45. Chen, C.; Luo, C.; Jin, Y.; Li, J.; Zhao, Q.; Yang, W. Short-Process Spray-Drying Synthesis of Lithium Iron Phosphate@Carbon Composite for Lithium-Ion Batteries. *ACS Sustain. Chem. Eng.* **2024**, *12*, 14077–14086. [CrossRef]

batteries

MDPI

Article

Simulation and Optimization of a Hybrid Photovoltaic/Li-Ion Battery System

Xiaoxiao Yu [1,2,3], Juntao Fan [1], Zihua Wu [1,2,3,*], Haiping Hong [1,*], Huaqing Xie [1,2,3], Lan Dong [1,2,3] and Yihuai Li [1,2,3]

1 School of Energy and Materials, Shanghai Polytechnic University, Shanghai 201209, China; yuxx@sspu.edu.cn (X.Y.); 20221516118@stu.sspu.edu.cn (J.F.); hqxie@sspu.edu.cn (H.X.); donglan@sspu.edu.cn (L.D.); yhli@sspu.edu.cn (Y.L.)
2 Shanghai Engineering Research Center of Advanced Thermal Functional Materials, Shanghai 201209, China
3 Shanghai Thermophysical Properties Big Data Professional Technical Service Platform, Shanghai Polytechnic University, Shanghai 201209, China
* Correspondence: wuzihua@sspu.edu.cn (Z.W.); hphong@sspu.edu.cn (H.H.)

Abstract: The coupling of solar cells and Li-ion batteries is an efficient method of energy storage, but solar power suffers from the disadvantages of randomness, intermittency and fluctuation, which cause the low conversion efficiency from solar energy into electric energy. In this paper, a circuit model for the coupling system with PV cells and a charge controller for a Li-ion battery is presented in the MATLAB/Simulink environment. A new three-stage charging strategy is proposed to explore the changing performance of the Li-ion battery, comprising constant-current charging, maximum power point tracker (MPPT) charging and constant-voltage charging stages, among which the MPPT charging stage can achieve the fastest maximum power point (MPP) capture and, therefore, improve battery charging efficiency. Furthermore, the charge controller can improve the lifetime of the battery through the constant-current and constant-voltage charging scheme. The simulation results indicate that the three-stage charging strategy can achieve an improvement in the maximum power tracking efficiency of 99.9%, and the average charge controller efficiency can reach 96.25%, which is higher than that of commercial chargers. This work efficiently matches PV cells and Li-ion batteries to enhance solar energy storages, and provides a new optimization idea for hybrid PV/Li-ion systems.

Keywords: MPPT; buck circuit; charger controller; Li-ion battery; photovoltaic cells; MATLAB/Simulink

1. Introduction

Currently, the world mainly uses non-renewable energy sources such as coal, oil, natural gas and nuclear energy. The use of non-renewable energy sources releases many gases and pollutants such as carbon dioxide, sulfur dioxide, etc., which cause serious environmental problems such as fog and frost, and land desertification and contribute to the greenhouse effect, while solar power generation is less harmful and more economic [1]. As a result, solar energy has been developed on a large scale as a renewable energy source, particularly photovoltaic (PV) power generation [2–4]. Recently, researchers have conducted extensive research on PV cells, for example, in material improvement, and have achieved the outstanding efficiency of ~23% [5–10]. However, weather conditions and round-the-clock changes affect the stability of PV cell power generation [11]. At present, coupled systems of PV cells and storage devices can reduce the impact from the environment; these include PV–battery systems and PV–phase-change material systems [12,13]. Among many solar energy utilization technologies, the combination of PV cells and batteries can transfer electric energy from PV cells into battery storage, and then the batteries generate electric energy to compensate for the lack of PV power generation caused by poor generation conditions [14,15]. Therefore, it is worth considering how to improve the match between PV cell and batteries.

Li-ion batteries have the advantages of cleanliness, light weight and high power density. Wang et al. [16] found that the charging time of $LiNi_{0.8}Mn_{0.1}Co_{0.1}O_2$ (NMC811) | | graphite lithium-ion batteries could be less than 15 mins, with a voltage higher than 4.5 V. Therefore, Li-ion batteries are considered as great electric energy storage devices and have been utilized for solar PV cell storages [17,18]. Guo et al. [19] proposed a three-stage charging control strategy based on a DC-DC converter to store solar energy in Li-ion batteries. These authors found that the improvement resulting from this strategy significantly enhanced the efficiency of converting solar energy into electrical energy. López et al. [20] utilized a simple MPPT method for a series DC-DC converter PV power system module to measure the PV cell generation voltage in real time and adjust the output power to keep the PV cells operating at maximum power. Bhan et al. [21] investigated the difference between charging systems with and without MPPT charging based on the MATLAB/Simulink environment, and found that a PV system with MPPT charging can generate more power. Szczepaniak et al. [22] proved that MPPT charging can improve generation efficiency. A coupling system combining a MPPT controller a PV generation system can enhance the stability of power generation, especially when the irradiation intensity changes [23,24]. However, the ability to adjust the PV maximum power point (MPP) voltage is still lacking, which reduces the conversion efficiency and shortens its lifetime [25]. Aljarhizi et al. [26] proposed a Li-ion battery charging method based on MPPT charging and a constant-voltage charging control, and found that the lifetime of a Li-ion battery can be extended without overloading. Even though the coupling of PV cells and a Li-ion battery with the MPPT charging method can improve the solar-to-electric efficiency and operating stability, the match of PV cells and Li-ion batteries still needs further optimization and improvement to ensure that the output current of PV cells at MPP meets the charging requirements of lithium-ion batteries; however, this issue is currently rarely studied.

This paper presents a model of a hybrid PV/Li-ion battery system in the MATLAB/Simulink environment. The new Li-ion battery charge controller utilizes an algorithm based on the combination of MPPT charging and Li-ion battery charging technology. The three-stage charging mode comprising constant-current charging, MPPT charging and constant-voltage charging, is utilized for improving the PV/Li-ion battery charging efficiency and effectively extending battery life.

2. System Modeling

Figure 1a shows the model diagram of the coupling system between PV cells and a Li-ion battery charge controller in the MATLAB/Simulink environment. It consists of a PV array, a DC-DC buck circuit, a Li-ion battery and a charge control module. Figure 1b illustrates that the charge control module comprises a Perturb and Observe (P&O) algorithm and a Li-ion battery charge control algorithm. The charge control module generates PWMs for switching the power switching tubes of the DC/DC buck circuit. The PV array module and Li-ion battery module are taken directly from the Simulink Simscape Electrical blockset library. The model uses a 2.7 kW PV array matched with a 12 V/10 Ah Li-ion battery. It was evaluated and simulated using MATLAB/Simulink. The circuitry model and charge controller block are explained in detail in the following sections.

2.1. DC-DC Buck Converter

As shown in Figure 1, a buck circuit is used in this design since it is more reliable and has less complexity [27–30]. The buck circuit comprises a MOSFET, a Schottky diode, an inductor, a resistor and two capacitors (C1, C2). The switching frequency (f_s) is equal to 1000 Hz. Equations (1)–(3) can be used to obtain the values of the buck circuit parameters L and C [31,32]. The capacitor C1 can stabilize the PV array's power output, and the value is set to 1 μF. The value of the capacitor C2 is set to 625 μF. The resistor R is a current-limiting resistor, and its parameter is set to 2 Ω. The forward voltage of the diode is set to 0.7 V.

$$D = \frac{U_o - 0.7}{U_i} \tag{1}$$

$$L = \frac{U_o * (1 - D)}{f_s * \Delta I_b} \tag{2}$$

$$C = \frac{U_i * (1 - D)}{8 * L * f_s * \Delta U_o} \tag{3}$$

where U_o represents the output voltage of the buck circuit, U_i represents the input DC voltage, ΔI_b represents the ripple of the inductor current, and ΔU_o represents the ripple in output voltage.

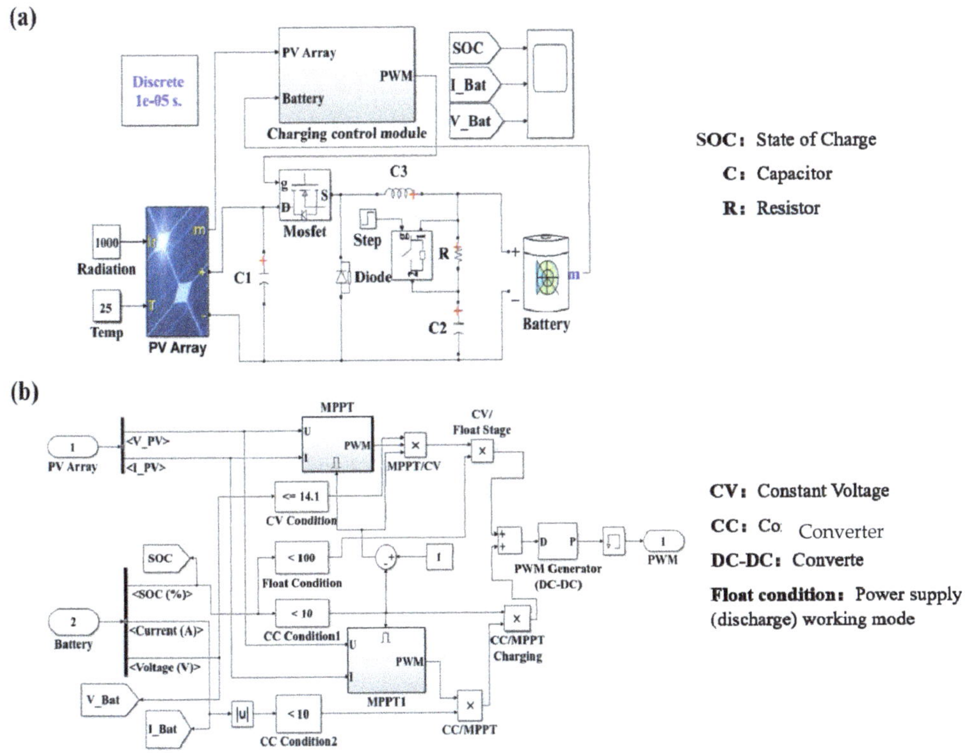

Figure 1. Charging control module: (**a**) solar PV/Li-ion battery charge controller module; (**b**) charge control algorithm module.

2.2. MPPT Algorithm

The power output from the PV panels changes as the irradiance value changes, and there is only one maximum power corresponding to each irradiance value. A model PV panel is simulated in the MATLAB/Simulink environment to analyze MPP, and its voltage–power (U-P) characteristic curves at different solar irradiance intensities are shown in Figure 2. It can be seen that the output power of the PV cell fluctuates with solar irradiance, and the MPP is the inflection point of the U-P curve, which indicates the highest generation power of the PV cells.

At present, the most commonly used MPPT algorithm is the Perturb and Observe (P&O) algorithm, and the flowchart of the P&O algorithm is shown in Figure 3a. The P&O algorithm is utilized to change the output power of PV cells through altering the equivalent load at the output of the cells and constantly adjusting the output power until the maximum output power [33–35]. Figure 3b is the flowchart of the P&O algorithm in the MATLAB/Simulink environment. It can be shown that the P&O algorithm is calculated through voltage U(k), current I(k), which is read from the PV array, and output power

P(k). Firstly, the unit delay block performs the previous sample (k − 1) function. Then, the values of U(k) − U(k − 1) and P(k) − P(k − 1) are calculated by the unit delay block U(k − 1) and P(k − 1). The signal module output 1 or −1 determines the positive or negative state of the duty cycle. Of course, the duty cycle can be changed in steps of 0.01. The duty cycle signal is then connected to the buck section.

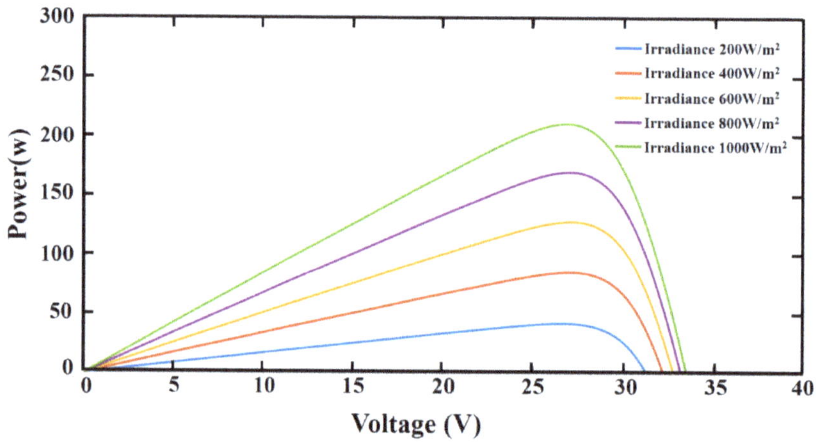

Figure 2. U-P characteristic curves of a solar cell.

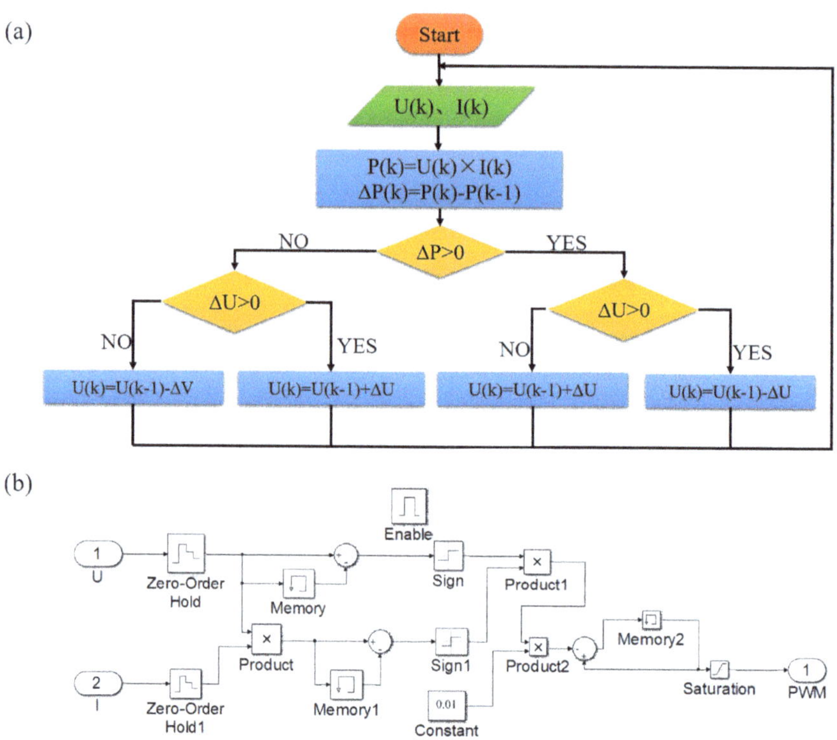

Figure 3. Implementation of the P&O algorithm. (**a**) Flowchart of the P&O algorithm. (**b**) P&O algorithm implementation in Simulink.

2.3. Li-Ion Battery Charging Control Method

Furthermore, the procedure of the Li-ion battery charging algorithm is shown in Figure 4. For SoC (State of Charge) $\leq$ 10%, the first stage maintains constant-current charging and MPPT charging, and the battery is charged at a constant current of 2 A (0.2 C). In particular, the charger switches to MPPT charging when the battery current is below the constant-current set value; otherwise, MPPT charging is prevented and the constant-current charging stage starts. For 10% < SoC $\leq$ 95%, the second stage enters MPPT charging, in which the battery is charged under a fluctuating current. Considering the effect of the weather on the output power of a PV array, it is important to match the models of the PV array and Li-ion battery to ensure that the MPPT charging current does not exceed the maximum charging current of the battery. Finally, for 95% < SoC $\leq$ 100%, the battery is charged at a constant voltage (14.1 V) during the third stage. The charger converts to MPPT charging if the battery voltage falls below the constant-voltage set value; otherwise, the constant-voltage charging continues.

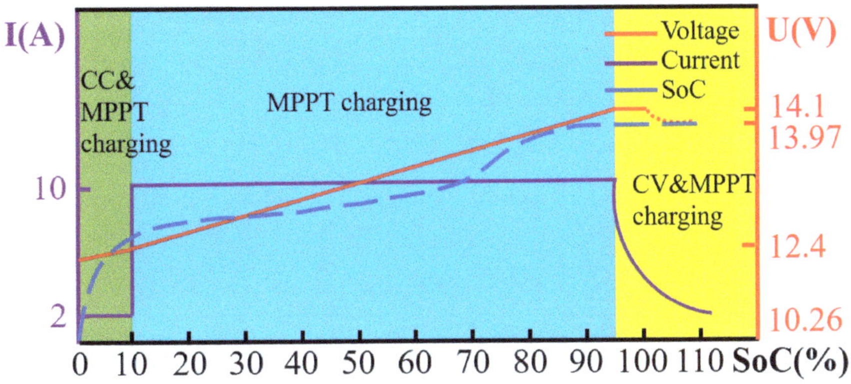

Figure 4. The charging process of the Li-ion battery.

3. Results and Discussion

In the MATLAB/Simulink environment, the PV–battery coupled system performance is simulated and examined. The discrete simulation type is established with a sample duration of 10 µs per sample, and the Simulink model is configured using automated solver selection with a variable step. This paper discusses the charging performance of Li-ion batteries, the tracking performance of MPPT, the overall efficiency performance, and the validation using conventional charge algorithms.

3.1. Li-Ion Battery Charging Performance

The parameters of the PV array and Li-ion battery in the model are shown in Table 1. The PV array consists of four panels in series and produces 2.7 kW of power. The intensity of solar irradiation is set at 1000 W/m^2, the temperature is set to 25 °C and the time is set at 6000 s.

The charging process of the Li-ion battery is shown in Figure 5. It can be seen that the Li-ion battery is initially charged by the battery charge controller during the constant-current charging stage when SoC of the Li-ion battery is less than 10%. Furthermore, The MPPT charging stage begins when the battery SoC reaches 10% at 1800 s, and is maintained when the SoC reaches 95% at 4800 s. However, the MPPT charging is switched into the constant-voltage charging stage once the voltage of the Li-ion battery reaches 14.1 V. Moreover, SoC of the Li-ion battery at 5250 s reaches 100%, and then the charge controller switches to the float stage. Meanwhile, the Li-ion battery maintains a floating voltage of 14.0 V. Considering with the ideal charging process of Li-ion batteries reported in previous research [36], it is demonstrated that the Li-ion battery charging voltage and current in this work meet the charge requirements as the current curve is almost identical to the ideal current curve.

Table 1. Parameters of photovoltaic arrays and Li-ion batteries.

	Parameters	Values
Li-ion battery	Rated capacity	50 Ah
	Nominal voltage	12 V
	Full charge voltage	13.97 V
	Constant-current charging current	10 A
	MPPT charging current	50 A
	Constant-voltage charging voltage	14.1 V
	Float charging voltage	13.97 V
Solar panel	Maximum output power (P_{max})	174 W
	Open circuit voltage (V_{oc})	36.3 V
	Maximum power point voltage (V_{mp})	29 V
	Short-circuit current (I_{sc})	6.5 A
	Maximum power point current (I_{mp})	6 A

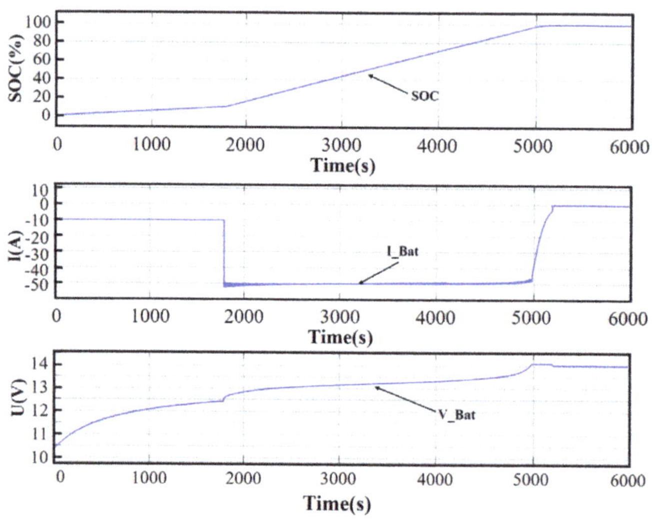

Figure 5. The process of the Li-ion battery charge.

3.2. MPPT Charging Efficiency

In this work, MPPT charging efficiency is defined by the ratio of the PV array output power to MPP power. Here, the effect of the solar irradiation intensity on the efficiency performance is further discussed. Considering the actual outdoor irradiation conditions, clouds affect the solar irradiation intensity, as shown in Figure 6a. In particular, the slower a cloud moves, the smaller the light intensity fluctuation is, and the better the tracking ability of the MPPT control method. Figure 6b clearly shows the efficiency and stability of the MPPT charging method, and the MPPT efficiency can reach 99.9% under the solar irradiation of 200–550 W/m^2.

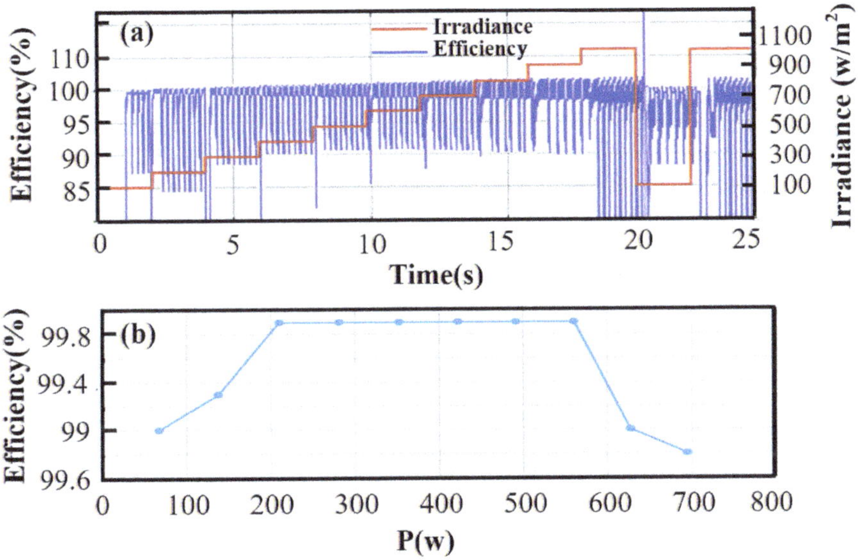

Figure 6. The performance analysis of MPPT charging. (**a**) Solar irradiance and MPPT charging real-time efficiency. (**b**) MPPT charging efficiency.

3.3. Overall Charging Efficiency of Li-Ions

Li-ion charger losses can drastically lower the overall efficiency of the energy storage. Here the overall efficiency of an Li-ion battery is defined by the ratio of the battery input power to the PV array output power. And the charging overall efficiencies of Li-ion batteries under three charging stages are discussed. According to Figure 7a, for 0% < SoC $\leq$ 10%, the constant-current charging efficiency can reach 95.5% at a solar irradiance of 1000 W/m^2. In addition, the MPPT charging efficiency can reach 95.5% at a solar irradiance of 100 W/m^2.

Aiming to highlight the characteristics of this work, the charging mode proposed in this paper and the traditional constant-current charging mode are compared through the same simulation method. Firstly, this study compares constant-current charging combined with MPPT charging to constant-current charging alone. The difference between the two charging methods in the early stage of charging are shown in Figure 7b,c. It can be seen that the two charge controllers have same charging performance under the constant-current charging stage. However, when the irradiation intensity decreases by 100 W/m^2, the three-stage charging mode can switch to the MPPT charging stage, and the charging efficiency can reach 0.25% at 60s, which is 0.7% SoC higher than the traditional charging mode.

For 10% < SoC $\leq$ 95%, the overall charging efficiency during the MPPT charging stage remains stable under uniform growth in the irradiation intensity, which indicates the MPPT method can efficiently achieve the maximum power charging. Moreover, the highest charging efficiency increases with the solar irradiation intensity, and reaches 97% when the irradiation intensity is 1000 W/m^2, as shown in Figure 8a. In particular, the MPPT charging method can rapidly resume a charging ability of 97% when the irradiation intensity undergoes large changes from 100 W/m^2 to 1000 W/m^2.

Furthermore, the charging performance of the MPPT mode is compared with the constant-current charging mode in Figure 8b,c. It is easily seen that the MPPT charging mode results in a more stable charging state than the traditional mode when the irradiation intensity decreases from 1000 W/m^2 to 800 W/m^2. Under the same charging time, the MPPT charging mode can achieve 11.6% SoC, which is 0.3% SoC higher than the constant-current charging mode alone.

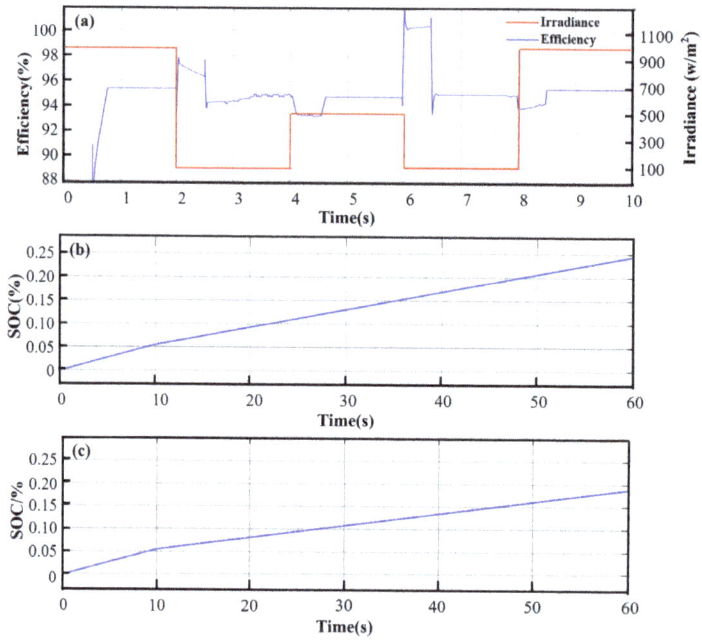

Figure 7. The overall performance analysis of the first-stage charging. (**a**) Solar irradiance and overall efficiency performance. (**b**) SoC variation with new charge controller. (**c**) SoC variation with traditional charge controller.

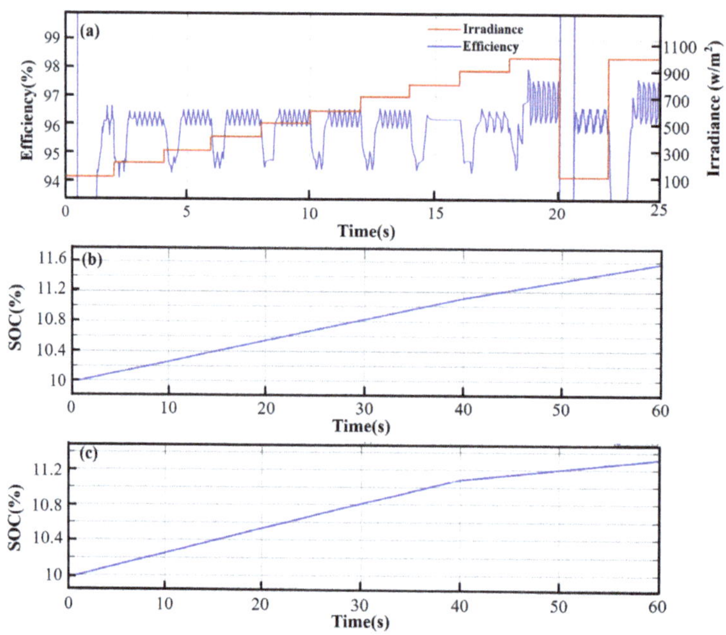

Figure 8. The overall performance analysis of the second-stage charging. (**a**) Light change process and overall conversion efficiency. (**b**) SoC variation with new charge controller. (**c**) SoC variation with traditional charge controller.

For 95% < SoC ≤ 100%, as illustrated in Figure 9a, from 0 to 2 s, the system is in constant-voltage charging mode, but the efficiency curve does not reach stability due to a delay in the simulation. From 2 to 4 s, the system switches to MPPT charging mode with an efficiency of approximately 97%. From 4 to 6 s, the system switches back to constant-voltage charging mode, with a charging efficiency of about 97%.

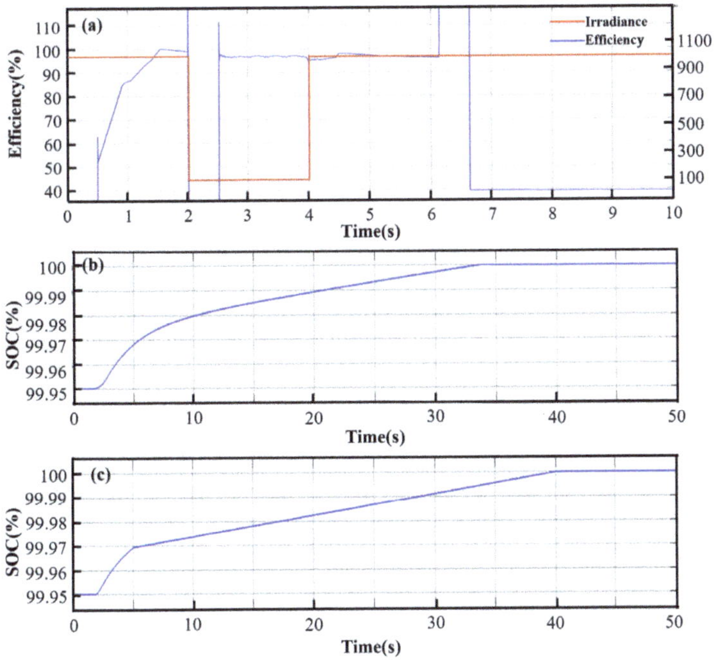

Figure 9. The overall performance analysis of the third-stage of charging. (**a**) Light change process and overall conversion efficiency. (**b**) SoC variation with new charge controller. (**c**) SoC variation with traditional charge controller.

Similarly, this study compares constant-voltage charging combined with MPPT charging to constant-voltage charging alone. Figure 9b,c show the difference between the two charging methods during the end stage of charging. Even though the two charging modes can achieve an SoC of 100%, the charging mode proposed in this work can decrease the charging time of the Li-ion battery to the maximum charging capacity by 5 s.

According to the above simulation results, the average conversion efficiency of the Li-ion battery charge controller is about 96.25%, with the highest efficiency of 97% and lowest efficiency of 95.5%. A comparison of the proposed system's charging efficiency with similar works already published is presented in Table 2.

Table 2. Recent research of nanofluid.

Ref.	System Efficiency
This work	95.5–97%
Palmiro et al. [37]	94–97%
López, J. [23]	85–95%
Zhang, L. [24]	87.48%
Salman et al. [34]	77.85–92.6%

In this section, the efficiency of the improved charging method is verified by comparison with the constant-current and constant-voltage charging methods. For 10% < SOC $\leq$ 90%, MPPT charging is more effective than constant-current charging. Combining constant-current charging and constant-voltage charging with MPPT charging can increase the charging power for Li-ion batteries in environments with varying solar irradiance. This offers faster charging speeds and higher charging efficiencies than charging with constant-current charging only or constant-voltage charging only.

4. Conclusions

An efficient and safe charger system comprising PV cells and Li-ion batteries is crucial for solar energy storage and utilization. This work built a Li-ion battery charge controller model with the MPPT technique in the MATLAB/Simulink environment to explore the charging performance under an unstable surrounding environment. The charging method of the Li-ion battery, the buck circuit, and the maximum power tracking algorithm are all clearly analyzed. The results show that the three-stage charging mode utilized in this work can efficiently enhance battery charging efficiency due to the fastest MPP capture. Meanwhile, the Li-ion battery lifetime is significantly improved. In particular, the three-stage charging method has an average charge controller efficiency of 96.25%. It also shows a higher overall efficiency of 95.5–97% and shorter charging times than the traditional constant-current charging method. The improved charging techniques proposed by this work can enhance solar-driven PV cells–Li-ion battery charging control system performance and offer valuable experiences for further study.

Author Contributions: Conceptualization, Z.W.; methodology, X.Y. and J.F.; software, L.D.; validation, L.D.; formal analysis, L.D.; investigation, X.Y.; resources, Y.L.; data curation, Y.L.; writing—original draft preparation, X.Y. and J.F.; writing—review and editing, H.H.; supervision, Z.W. and H.X.; project administration, Z.W. All authors have read and agreed to the published version of the manuscript.

Funding: This work has been supported by research funding from the National Natural Science Foundation of China (No. 52176081), the Funding Programme Natural Science Foundation of Shanghai (No. 21ZR1424500), the Shanghai Local Capacity Building Programme (No. 22010500500 and No. 21010500700), the Shanghai Sailing Programme (No. 23YF1413500), the Shanghai Rising Star Programme (No. 21QA1403300) and the National Natural Science Foundation of China (No. 52306259).

Data Availability Statement: The original contributions presented in the study are included in the article.

Acknowledgments: We express our sincere gratitude to all those who contributed to this research; in particular, we thank Zihua Wu for his committed instruction and Xiaoxiao Yu for reviewing and editing the manuscript. We also appreciate the advice of Dong Lan, Yi-Huai, and Xie Hui-Qing. We also thank the two reviewers for their insightful and helpful criticism, which helped to strengthen the manuscript.

Conflicts of Interest: The authors declare no conflicts of interest.

References

1. Farghali, M.; Osman, A.I.; Mohamed, I.M.A.; Chen, Z.; Chen, L.; Ihara, I.; Yap, P.-S.; Rooney, D.W. Strategies to save energy in the context of the energy crisis: A review. *Environ. Chem. Lett.* **2023**, *21*, 2003–2039. [CrossRef] [PubMed]
2. Ranalder, L.; Gibb, D. *Renewables in Cities 2019 Global Status Report*; REN21: Paris, France, 2020; Volume 31.
3. Perez, M.; Perez, R. Update 2022—A fundamental look at supply side energy reserves for the planet. *Sol. Energy Adv.* **2022**, *2*, 100014. [CrossRef]
4. Djørup, S.; Thellufsen, J.Z.; Sorknæs, P. The electricity market in a renewable energy system. *Energy* **2018**, *162*, 148–157. [CrossRef]
5. Gao, Y.; Xiao, Z.; Cui, M.; Saidaminov, M.I.; Tan, F.; Shang, L.; Li, W.; Qin, C.; Ding, L. Asymmetric Π-Bridge Engineering Enables High-Permittivity Benzo [1,2-B:4,5-b′] Difuran-Conjugated Polymer for Efficient Organic Solar Cells. *Adv. Mater.* **2024**, *36*, 2306373. [CrossRef]
6. Qi, X.; Song, C.; Zhang, W.; Shi, Y.; Gao, Y.; Liu, H.; Chen, R.; Shang, L.; Tan, H.; Tan, F.; et al. Bidirectional Targeted Therapy Enables Efficient, Stable, and Eco-Friendly Perovskite Solar Cells. *Adv. Funct. Mater.* **2023**, *33*, 2214714. [CrossRef]

7. Gao, Y.; Cui, M.; Qu, S.; Zhao, H.; Shen, Z.; Tan, F.; Dong, Y.; Qin, C.; Wang, Z.; Zhang, W.; et al. Efficient Organic Solar Cells Enabled by Simple Non-Fused Electron Donors with Low Synthetic Complexity. *Small* **2022**, *18*, 2104623. [CrossRef]
8. Huang, X.; Gao, Y.; Li, W.; Wang, J.; Yue, G.; Tan, F.; Wang, H.L. Efficient and Stable Z907-Based Dye-Sensitized Solar Cells Enabled by Suppressed Charge Recombination and Photocatalytic Activity. *ACS Sustain. Chem. Eng.* **2024**, *12*, 13007–13016. [CrossRef]
9. He, S.; Lan, Z.; Zhang, B.; Gao, Y.; Shang, L.; Yue, G.; Chen, S.; Shen, Z.; Tan, F.; Wu, J. Holistically Optimizing Charge Carrier Dynamics Enables High-Performance Dye-Sensitized Solar Cells and Photodetectors. *ACS Appl. Mater. Interfaces* **2022**, *14*, 43576–43585. [CrossRef]
10. Gao, Y.; Shen, Z.; Tan, F.; Yue, G.; Liu, R.; Wang, Z.; Qu, S.; Wang, Z.; Zhang, W. Novel benzo [1,2-b:4,5-b′] difuran-based copolymer enables efficient polymer solar cells with small energy loss and high VOC. *Nano Energy* **2020**, *76*, 104964. [CrossRef]
11. Dubey, S.; Sarvaiya, J.N.; Seshadri, B. Temperature Dependent Photovoltaic (PV) Efficiency and Its Effect on PV Production in the World—A Review. *Energy Procedia* **2013**, *33*, 311–321. [CrossRef]
12. Koutroulis, E.; Kalaitzakis, K. Novel battery charging regulation system for photovoltaic applications. *IEE Proc. Electr. Power Appl.* **2004**, *151*, 191–197. [CrossRef]
13. Stropnik, R. Increasing the efficiency of PV panel with the use of PCM. *Renew. Energy* **2016**, *97*, 671–679. [CrossRef]
14. Dunn, B.; Kamath, H.; Tarascon, J.-M. Electrical Energy Storage for the Grid: A Battery of Choices. *Science* **2011**, *334*, 928–935. [CrossRef] [PubMed]
15. Maheswari, L.; Sornavadivu, R.; Vijayalakshmi, S. Modeling and Simulation of Buck Converter for Charging Battery by Solar Photovoltaic System. *Appl. Mech. Mater.* **2014**, *592*, 2379–2385. [CrossRef]
16. Xu, J.; Zhang, J.; Pollard, T.P.; Li, Q.; Tan, S.; Hou, S.; Wang, C. Electrolyte design for Li-ion batteries under extreme operating conditions. *Nature* **2023**, *614*, 694–700. [CrossRef] [PubMed]
17. Kebede, A.A.; Coosemans, T.; Messagie, M.; Jemal, T.; Behabtu, H.A.; Van Mierlo, J.; Berecibar, M. Techno-economic analysis of lithium-ion and lead-acid batteries in stationary energy storage application. *J. Energy Storage* **2021**, *40*, 102748. [CrossRef]
18. Anuphappharadorn, S.; Sukchai, S.; Sirisamphanwong, C.; Ketjoy, N. Comparison the Economic Analysis of the Battery between Lithium-ion and Lead-acid in PV Stand-alone Application. *Energy Procedia* **2014**, *56*, 352–358. [CrossRef]
19. Guo, L.; Brewer, A.; Speiser, B. Design and Implementation of A Solar Battery Charger. In Proceedings of the 2010 Annual Conference & Exposition, Louisville, KY, USA, 20 June 2010.
20. López, J.; Seleme Jr, S.I.; Donoso, P.F.; Morais LM, F.; Cortizo, P.C.; Severo, M.A. Digital control strategy for a buck converter operating as a battery charger for stand-alone photovoltaic systems. *Sol. Energy* **2016**, *140*, 171–187. [CrossRef]
21. Bhan, V.; Shaikh, S.A.; Khand, Z.H.; Ahmed, T.; Khan, L.A.; Chachar, F.A.; Shaikh, A.M. Performance Evaluation of Perturb and Observe Algorithm for MPPT with Buck–Boost Charge Controller in Photovoltaic Systems. *J. Control Autom. Electr. Syst.* **2021**, *32*, 1652–1662. [CrossRef]
22. Szczepaniak, M.; Otręba, P.; Otręba, P.; Sikora, T. Use of the Maximum Power Point Tracking Method in a Portable Lithium-Ion Solar Battery Charger. *Energies* **2021**, *15*, 26. [CrossRef]
23. Anowar, M.H.; Roy, P. A Modified Incremental Conductance Based Photovoltaic MPPT Charge Controller. In Proceedings of the 2019 International Conference on Electrical, Computer and Communication Engineering (ECCE), Cox's Bazar, Bangladesh, 7–9 February 2019; IEEE: New York, NY, USA, 2019; pp. 1–5. [CrossRef]
24. Zhang, L.; Wang, Z.; Cao, P.; Zhang, S. A Maximum Power Point Tracking Algorithm of Load Current Maximization-Perturbation and Observation Method with Variable Step Size. *Symmetry* **2020**, *12*, 244. [CrossRef]
25. Gibson, T.L.; Kelly, N.A. Solar photovoltaic charging of lithium-ion batteries. *J. Power Sources* **2010**, *195*, 3928–3932. [CrossRef]
26. Aljarhizi, Y.; Hassoune, A.; Al Ibrahmi, E.M. Control Management System of a Lithium-ion Battery Charger Based MPPT algorithm and Voltage Control. In Proceedings of the 2019 5th International Conference on Optimization and Applications (ICOA), Kenitra, Morocco, 25–26 April 2019; IEEE: New York, NY, USA, 2019; pp. 1–6. [CrossRef]
27. Mirzaei, A. Design and construction of a charge controller for stand-alone PV/battery hybrid system by using a new control strategy and power management. *Sol. Energy* **2017**, *149*, 132–144. [CrossRef]
28. Taghvaee, M.H.; Radzi, M.A.M.; Moosavain, S.M.; Hizam, H.; Marhaban, M.H. A current and future study on non-isolated DC–DC converters for photovoltaic applications. *Renew. Sustain. Energy Rev.* **2013**, *17*, 216–227. [CrossRef]
29. Coelho, R.F. Influence of Power Converters on PV Maximum Power Point Tracking Efficiency. In Proceedings of the 2012 10th IEEE/IAS International Conference on Industry Applications, Fortaleza, Brazil, 5–7 November 2012; IEEE: New York, NY, USA, 2012.
30. Venkatramanan, D.; John, V. Dynamic Modeling and Analysis of Buck Converter Based Solar PV Charge Controller for Improved MPPT Performance. *IEEE Trans. Ind. Appl.* **2019**, *55*, 6234–6246. [CrossRef]
31. Pathare, M.; Shetty, V.; Datta, D.; Valunjkar, R.; Sawant, A.; Pai, S. Designing and Implementation of Maximum Power Point Tracking(MPPT) Solar Charge Controller. In Proceedings of the 2017 International Conference on Nascent Technologies in Engineering, Navi Mumbai, India, 27–28 January 2017; p. 5.
32. Tan, R.H.; Er, C.K.; Solanki, S.G. Modeling of Photovoltaic MPPT Lead Acid Battery Charge Controller for Standalone System Applications. *E3S Web Conf.* **2020**, *182*, 03005. [CrossRef]

33. Kaur, T.; Gambhir, J.; Kumar, S. Arduino based solar powered battery charging system for rural SHS. In Proceedings of the 2016 7th India International Conference on Power Electronics (IICPE), Patiala, India, 17–19 November 2016; IEEE: New York, NY, USA, 2016; pp. 1–5. [CrossRef]
34. Salman, S.; Ai, X.; Wu, Z. Design of a P-&-O algorithm based MPPT charge controller for a stand-alone 200W PV system. *Prot. Control Mod. Power Syst.* **2018**, *3*, 25.
35. de Brito, M.A.G.; Galotto, L.; Sampaio, L.P.; e Melo, G.D.A.; Canesin, C.A. Evaluation of the Main MPPT Techniques for Photovoltaic Applications. *IEEE Trans. Ind. Electron.* **2013**, *60*, 1156–1167. [CrossRef]
36. Gao, Y.; Zhang, X.; Cheng, Q.; Guo, B.; Yang, J. Classification and Review of the Charging Strategies for Commercial Lithium-Ion Batteries. *IEEE Access* **2019**, *7*, 43511–43524. [CrossRef]
37. Palmiro, F.; Rayudu, R.; Ford, R. Modelling and simulation of a solar PV lithium ion battery charger for energy kiosks application. In Proceedings of the 2015 IEEE PES Asia-Pacific Power and Energy Engineering Conference (APPEEC), Brisbane, Australia, 15–18 November 2015; IEEE: New York, NY, USA, 2015; pp. 1–5. [CrossRef]

batteries

MDPI

Review

Rechargeable Li-Ion Batteries, Nanocomposite Materials and Applications

Sara El Afia [1], Antonio Cano [1,*], Paul Arévalo [1,2] and Francisco Jurado [1]

[1] Department of Electrical Engineering, University of Jaen, 16990 Jaen, Spain; se000038@red.ujaen.es (S.E.A.); warevalo@ujaen.es (P.A.); fjurado@ujaen.es (F.J.)
[2] Faculty of Engineering, Department of Electrical Engineering, Electronics and Telecommunications (DEET), University of Cuenca, Balzay Campus, Cuenca 010107, Azuay, Ecuador
* Correspondence: acano@ujaen.es

Abstract: Lithium-ion batteries (LIBs) are pivotal in a wide range of applications, including consumer electronics, electric vehicles, and stationary energy storage systems. The broader adoption of LIBs hinges on advancements in their safety, cost-effectiveness, cycle life, energy density, and rate capability. While traditional LIBs already benefit from composite materials in components such as the cathode, anode, and separator, the integration of nanocomposite materials presents significant potential for enhancing these properties. Nanocomposites, including carbon–oxide, polymer–oxide, and silicon-based variants, are engineered to optimize key performance metrics, such as electrical conductivity, structural stability, capacity, and charging/discharging efficiency. Recent research has focused on refining these composites to overcome existing limitations in energy density and cycle life, thus paving the way for the next generation of LIB technologies. Despite these advancements, challenges related to high production costs and scalability remain substantial barriers to the widespread commercial deployment of nanocomposite-enhanced LIBs. Addressing these challenges is essential for realizing the full potential of these advanced materials, thereby driving significant improvements in the performance and practical applications of LIBs across various industries.

Keywords: Li-ion batteries; nanocomposite materials; nanotechnology

1. Introduction

Energy plays a crucial role in the progress of our world, with electricity being the most widely utilized form of energy. It is primarily generated from fossil fuels like coal, oil, and natural gas. In recent years, the substantial use of conventional fossil fuels and their environmental and climate impacts have spurred increased interest in technologies like energy storage to enhance energy efficiency. The battery or chemical energy system, utilizing the conversion from chemical energy to electrochemical energy, has captured considerable interest in the energy storage field [1]. The main technologies utilized in rechargeable battery systems include lithium-ion (Li-ion), lead–acid, nickel–metal hydride (NiMH), and nickel–cadmium (Ni–Cd). Rechargeable batteries constitute a substantial portion of the global battery market.

The Li-ion battery stands out as the most popular and widely used rechargeable battery, attributed to its high gravimetric and volumetric energy density, along with a significant cost reduction over the last decade [2]. The main applications of rechargeable Li-ion batteries include portable electronic devices, electric vehicles, and solar energy storage. Currently, Li-ion batteries already reap benefits from composite materials, with examples including the use of composite materials for the anode, cathode, and separator.

Lithium-ion batteries are an appealing option for power storage systems owing to their high energy density. Despite this advantage, significant polarization during high charging and discharging rates results in low energy efficiency [3]. This polarization occurs due to the slow diffusion of lithium in the active material and an increase in electrolyte resistance with rising charging and discharging rates. Addressing these issues necessitates the creation of nanocomposite electrode materials with extensive surface areas and brief diffusion paths, enabling efficient ionic transport and electronic conduction.

Nanocomposites are composite materials in which one phase has nanoscale morphology such as nanoparticles or nanostructures. They are multiphase materials and, at least, the phases should have dimensions in the range of 10–100 nm [1]. Nanocomposites involve the fusion of a larger matrix and nanoscale phases, distinguished by variations in properties due to differences in both structure and chemistry. There are three categories of nanocomposites classified by their matrix: ceramic matrix, polymer matrix, and metal matrix nanocomposites. Recently, nanocomposite materials have attracted attention due to their remarkable thermal conductivity, mechanical strength, and resistance to solvents.

Consequently, nanocomposite materials can provide ample opportunities for advancing Li-ion batteries. By manipulating these materials at the nanoscale, unprecedented improvements in material properties can be achieved [4].

This review critically examines the advancements in research pertaining to rechargeable lithium-ion batteries (LIBs), emphasizing the significant contributions of nanocomposite materials to their performance enhancement. Given the essential function of LIBs in diverse applications—including consumer electronics, electric vehicles, and stationary energy storage—the incorporation of innovative nanocomposite materials is poised to tackle critical challenges such as energy density, cycle life, and charge/discharge rates.

This review outlines recent progress in various nanocomposite materials, such as carbon–oxide, polymer–oxide, and silicon-based composites, highlighting their specific applications aimed at optimizing LIB performance. Key findings from recent research are presented, focusing on the enhancements in conductivity, stability, and overall efficiency attributed to these nanocomposites. Furthermore, this review addresses the obstacles related to the scalability and cost-effectiveness of these materials, which continue to hinder their wider adoption.

In summary, this review highlights the promising potential of nanocomposite materials in advancing LIB technology. It emphasizes the need for continued research and innovation to overcome existing challenges and unlock new opportunities that could lead to significant improvements in the performance and viability of next-generation rechargeable batteries. By exploring the intersection of nanocomposite materials and LIBs, this work aims to deepen understanding of their capabilities and chart future pathways for development in this critical field.

2. Lithium-Ion Batteries

Lithium-ion batteries were introduced to the industrial marketplace in 1991 [1]. Utilizing carbon and lithium cobalt oxide ($LiCoO^2$) as the electrode's materials. Since their introduction, lithium-ion batteries have made significant progress in various sectors, such as electronic devices, power sources, and energy storage devices. For that, lithium-ion batteries are recognized currently as the prevailing choice in battery chemistry.

Batteries are generally divided into two main types: primary batteries and secondary batteries. Primary batteries, or single-use cells, can be utilized and discharged once before disposal. However, secondary batteries, such as lithium-ion batteries, are designed to be cycled, allowing them to be charged and discharged repeatedly throughout their lifespan.

The lithium-ion battery pack consists of distinct modules, each containing numerous individual cells assembled in either series or parallel configurations within the module. These modules are subsequently assembled in a specific configuration to constitute a complete battery pack. The number of cells depends on the battery application, and the assembly configuration significantly influences the voltage and current output of the battery. Specifically, the series configuration increases the voltage of the battery module, while the parallel configuration enhances the current. Furthermore, a series–parallel combination is employed to achieve a balance between voltage and capacity [2].

Lithium serves as the primary material in lithium-ion batteries owing to its distinctive chemical characteristics, making it a preferred option for battery components. Notably, lithium is the third smallest element after hydrogen and helium, featuring only three protons and three neutrons [5]. Being a highly electropositive material, lithium facilitates the efficient movement of ions between the anode and cathode during the charging and discharging processes. Additionally, this material boasts a higher energy density, signifying its ability to store a substantial amount of energy in a compact size. Furthermore, lithium's lower atomic weight contributes to its lighter nature compared to many other metals. As a result of readily releasing its outer electron, lithium exhibits high reactivity, enabling a smooth flow of power through a cell [6].

Nanostructured materials are used in lithium-ion storage devices because of their high surface area, porosity, etc. These characteristics allow for introducing new active reactions, decreasing the path length for lithium-ion transport, reducing the specific surface current rate, and improving stability and specific capacitance. In addition, designed nanostructured composite materials could decrease the internal resistance of lithium-ion batteries, resulting in higher specific capacities even at high charge/discharge current rates [7].

The potential advantages of nanostructured active electrode materials can be summarized: new reactions can be used that are not possible with bulk materials; a larger electrode/electrolyte contact area, leading to higher charge/discharge rates; short path lengths for both electronic and Li-ion transport (permitting operation even with low electronic or low Li-ion conductivity, or at higher power); etc.

There are some disadvantages, such as a more complex synthesis process for nanomaterials, which will increase the cost of lithium-ion batteries. Therefore, the development of simpler synthesis methods will allow for the large-scale production of nanostructured active materials.

Lithium-ion batteries offer a host of advantages that make them a leading choice in energy storage technology. They exhibit remarkable specific energy, durability, and longevity. Moreover, Li-ion batteries feature a conventional design and operate at elevated voltage levels, further enhancing their overall efficiency and effectiveness. Table 1 presents a comparison of lithium-ion (Li-ion) batteries with other widely used rechargeable battery types, such as lead–acid, Ni-MH, and Ni-Cd. It emphasizes variations in specific power, gravimetric energy density, and lifespan, while also noting the advantages and disadvantages of each. The comparison shows that Li-ion batteries outperform others in terms of energy density, lifespan, and overall performance, although they are more costly and pose greater safety risks when compared to alternatives like lead–acid and Ni-MH batteries. Lithium-ion batteries provide the highest energy density and extended lifespan compared to alternative battery technologies. They demonstrate the highest level (approximately 95%) in terms of energy efficiency, allowing for discharge rates of up to 100%. Additionally, they exhibit a low self-discharge rate, enable rapid charging, and boast various other enhanced performance characteristics, rendering them highly appealing [8]. However, the intricate nature of material synthesis poses numerous challenges in the pursuit of creating new high-energy lithium-ion batteries.

2.1. Basic Concepts of Li-Ion Batteries

The essential components of lithium-ion batteries include the cathode (positively charged electrode), the anode (negatively charged electrode), electrolyte, separator, and current collector. The positive electrode serves to store and release electrons during the battery's operation, while the negative electrode facilitates the movement of electrons [9]. The electrolyte is a conductive substance that sits between the cathode and anode, carrying and transferring the lithium ions between both ends of the battery. The separator acts as a barrier, preventing contact between the two electrodes. The current collector collects the flow of electrons.

Table 1. Li-ion battery attributes.

Battery Attribute	Specific Power (Wkg^{-1})	Gravimetric Energy Density ($Whkg^{-1}$)	Lifespan (Cycles)	Strengths	Weaknesses
Li-ion	500–2000	150–200	1500–4500	High energy density, high voltage operation, no memory effect, low self-discharge; long life cycle	Safety concerns, restricted possibilities for additional size reduction, and limited capacity for further improvement
Lead–acid	30–40	30–40	200–300	Low cost, reliable, widely used	Low energy density, short lifespan, heavy
Ni-MH	250–1000	60–120	500–1000	Moderate energy density, safer than Ni-Cd	Memory effect, higher self-discharge rate
Ni-Cd	150–300	40–60	1000–1500	Long cycle life, operates in low temperatures	Toxic, memory effect, lower energy density

Typically, the anode and cathode are always composed of different materials. The anode is usually constructed from graphite, which is a form of carbon known for its conductivity and stability [6]. On the other hand, the cathode is generally made of lithium metal oxide, such as lithium cobalt oxide or lithium iron phosphate, known for their high energy density and excellent performance. An electrolyte is a solution that transfers ions between the anode and cathode. There are three types of electrolytes in lithium-ion batteries: organic electrolytes, such as dimethyl carbonate, gel polymer electrolytes, such as polyethylene oxide, and solid electrolytes, like lithium ceramic materials. Finally, a separator is generally a porous material made of polyethylene or propylene that prevents direct physical contact between the anode and cathode while facilitating the movement of lithium ions between them. Figure 1 shows the components of rechargeable batteries.

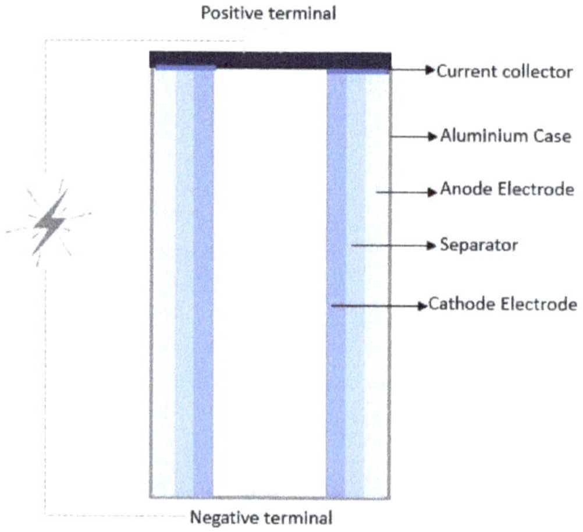

Figure 1. Components of rechargeable batteries.

The core principles and concepts that serve as the foundation for lithium-ion batteries derive from electrochemical mechanisms. This indicates that batteries employ a chemical process to convert stored chemical energy into electric energy. In simpler terms, the stored chemical energy undergoes a conversion into electrical energy. Moreover, during operation, the chemical reaction that produces electricity can be reversed by applying an external current [9].

Redox reactions, which involve reduction and oxidation, are crucial in the charging and discharging mechanisms of lithium-ion batteries. These reactions can be divided into two halves: oxidation occurs at the anode, leading to electron loss, while reduction takes place at the cathode, resulting in the gain in electrons. In the charging phase, lithium ions migrate from the cathode to the anode, accompanied by electron flow in the external circuit, generating electrical energy. Conversely, during discharging, lithium ions move from the anode to the cathode, and electrons flow through the external circuit, releasing electrical energy [6,9]. Figure 2 shows a schematic of the lithium-ion battery.

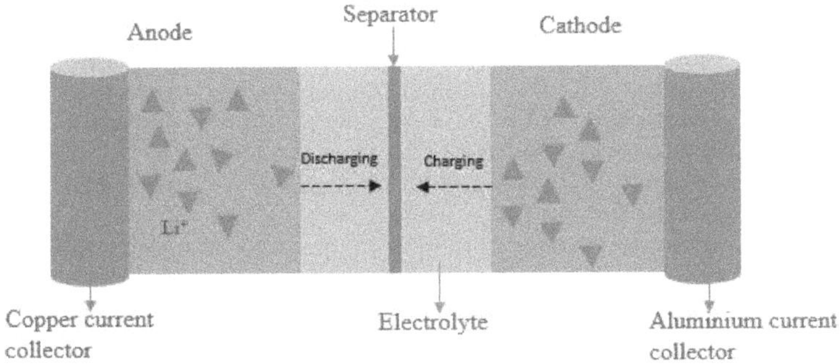

Figure 2. Basic working principle of a lithium-ion battery.

2.2. Electrodes Materials for Lithium Ions Battery

Lithium-ion batteries are widely employed across a diverse range of applications, both small and large, owing to their high energy density and environmentally friendly nature. However, despite these advancements, lithium-ion batteries face certain limitations, particularly in transportation and high-energy storage applications [10]. Moreover, lithium-ion batteries encounter challenges in low-temperature conditions, attributed to the distinctive electronic and ionic conductivities of the anode material, cathode material, and electrolyte solution, resulting in diminished capacity and inefficient charging. This indicates that the effectiveness of lithium-ion batteries is substantially impacted by the selection of materials for their principal components [11].

At low temperatures, the interaction between the anodic and cathodic materials becomes more evident, impacting the overall conductivity and transport of ions within the battery. Consequently, the meticulous selection and optimization of electrode materials can enhance the effectiveness of lithium-ion batteries [10].

Generally, lithium-ion batteries utilize graphite as the anode material due to its low cost, effective conductivity, and outstanding reversibility. Furthermore, the utilization of graphite material comes with certain drawbacks, such as its restricted capacity and potential safety concerns linked to the insertion of Li+ into the anode's structure [12]. On the other hand, the cathode, typically composed of lithium metal oxide, holds significant importance in conventional lithium-ion batteries. It serves as the primary supplier of lithium ions within the battery system, exerting a considerable impact on the capacity of lithium-ion batteries. Consequently, the development of cathode materials with advantageous attributes, including high performance, safety, and large capacity, would significantly enhance the widespread adoption of lithium-ion batteries.

To improve the effectiveness of lithium-ion batteries under low-temperature conditions, multiple approaches have been suggested, such as the development of electrode materials [11]. Scientists are dedicated to designing materials with micro- and nanostructures, alongside composites featuring diverse morphologies, orientations, and particle dimensions. Moreover, the optimization of lithium-ion battery performance heavily relies on the utilization of electrode composite materials and nanocomposites.

A composite material is formed through the combination of two or more substances characterized by distinct physical and chemical properties. This combination results in an innovative composite material that displays enhanced qualities. A composite comprises two primary components: the matrix and the fibre. The matrix, serving as the foundational substance, can be a polymer, ceramic, or metal. It functions as a continuous phase, enveloping and binding the reinforcement materials. This matrix plays a crucial role in providing support, distributing loads, and ensuring cohesion. The reinforcement, typically in fibrous form and fashioned from materials like carbon or glass, strengthens the overall structure. In the realm of lithium-ion batteries, composite materials refer to the amalgamation of a lithium-ion conductive matrix material and a reinforcing substance, such as carbon or metal oxides [13]. The matrix functions as the conduit through which lithium ions can traverse, facilitating the seamless progression of charge and discharge cycles. Concerning the materials used in lithium-ion electrode construction, anode composites consist of a blend of active components like graphite or silicon. This amalgamation seeks to mitigate volume expansion and mechanical stress and enhance cycling stability. Conversely, cathodes incorporate a combination of active materials, such as lithium cobalt oxide, lithium manganese oxide, or lithium iron phosphate, with the objective of boosting the energy density, rate capability, and overall performance of lithium-ion batteries.

Table 2 compares different Li-ion battery technologies using composite materials, focusing on the combination of anode and cathode materials. It highlights that lithium cobalt oxide with graphite anodes delivers high energy density but suffers from overcharging sensitivity and limited thermal stability. Conversely, when paired with lithium iron phosphate or lithium manganese oxide, graphite anodes offer superior thermal stability and increased safety, though at the expense of reduced energy density. This emphasizes the need to balance energy storage, safety, and thermal characteristics when choosing materials for specific uses.

Table 2. Comparative analysis of Li-ion battery technologies.

Anode Material	Cathode Material (Lithium Metal Oxide)	Advantages	Disadvantages
Graphite	Lithium Cobalt Oxide	High energy density	Sensitiveness to overcharge, poor thermal stability
Graphite	Lithium Iron Phosphate	Excellent thermal stability	Low energy density
Graphite	Lithium Manganese Oxide	Excellent thermal stability and elevated safety	Low energy density
Graphite	Nickel Manganese Cobalt (NMC)	High energy density, good thermal stability, longer cycle life	Expensive, environmental concerns related to nickel and cobalt extraction

Composites offer significant advantages in various aspects of material performance, outperforming single-material alternatives, particularly when compared to the isolated use of individual elements [13]. Despite these advantages, there are some drawbacks, such as the interface between electrode materials, which can affect the overall efficiency of the battery. Thus, the use of nanocomposite materials emerges as an interesting solution to address these challenges, leveraging the enhanced properties of nanoscale components to optimize performance and mitigate the drawbacks associated with traditional composites.

In conclusion, this section has provided a foundation for understanding lithium-ion battery technology. The subsequent section will delve into the advancements in nanocomposite materials and their role in improving the performance and efficiency of these batteries.

3. Nanocomposite Materials

Nanotechnology has been a fascinating field for researchers since the last century [14]. It is the science that deals with materials and devices at the nanometer scale [15]. At the nanoscale, materials showcase unique chemical, and physical properties that differ from those at the level of individual atoms, molecules, or bulk matter. These distinct characteristics give rise to novel applications, opening up new possibilities in various scientific and technological domains. A nanometer (nm) is a unit of length, 9-10 m, representing one billionth of a meter. The technology is quantified by the scale, with 1 nm equating to

1/1,000,000 m. Illustratively, human hair has a thickness of 60–80,000 nm, and red blood corpuscles measure 2–500 nm in width.

In recent years, the speedy advancement of nanotechnology has increased the significance of studying nanocomposites for the creation of novel materials tailored for advanced applications. A composite material, composed of multiple components with distinct microscopic characteristics, is classified as a nanocomposite when one of the reinforcing dimensions operates within the nanoscale range [16].

Nanocomposites constitute a unique category of composites characterized by the existence of morphological features at the nanoscale such as nanoparticles, nanotubes, or other nanostructures within at least one of their phases. First introduced in the literature by Blumstein in 1961 [14], these materials display a multiphasic characteristic, where at least one phase has dimensions spanning from 10 to 100 nanometers. In contrast to their micro-composite equivalents, nanocomposites have complex structures that are dependent on a variety of variables, including composition, interfacial interactions, and the unique characteristics of each component. Figure 3 shows a schematic diagram of matrix-reinforced nanocomposite.

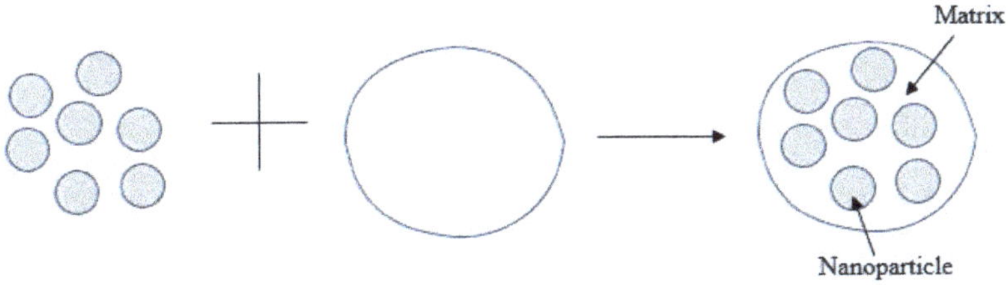

Figure 3. Schematic diagram of matrix-reinforced nanocomposites.

3.1. *Synthesis Method*

A nanocomposite material is a solid material with multiple phases, where at least one phase possesses dimensions in one, two, or three directions that are smaller than 100 nanometers [1]. Alternatively, it may have a structure with a nanoscale repeating distance between the various phases comprising the material. On the other hand, a nanocomposite consists of a matrix, typically a polymer, that encases and integrates nanoscale reinforcements [14].

The creation processes of nanocomposite materials vary, depending on the intended composition and desired properties. Numerous techniques are employed in making nanocomposite materials, such as melt intercalation, solution blending, in situ polymerization and sol–gel. Melt intercalation is a technique utilized in the fabrication of nanocomposite materials [15]. This process entails the fusion of a polymer matrix and the incorporation of nanoscale additives while in a molten state. It is considered the conventional approach for producing nanocomposites with thermoplastic polymers. The fundamental concept of this approach involves heating the matrix until it reaches a molten state, enabling the seamless inclusion of nanoscale reinforcements, such as clay, graphene, silica nanoparticles, or other nanomaterials. This technique improves the distribution of nanomaterials within the polymer matrix, resulting in enhanced mechanical, thermal, and barrier characteristics.

The second method involves solution blending, a process relying on solvents where both the polymer and pre-polymer are soluble. The ability of the polymer and pre-polymer to dissolve causes the clay layers to swell, resulting in the separation of the layered clay into individual layers using a solvent like water, alcohol, and toluene [16]. On the other hand, the intercalation process is aided by annealing, where molecules undergo reorientation. The identification of increased tensile strength and modulus provides evidence for the development of partially intercalated material in specific instances. Essentially, this process results in the creation of a nanocomposite with a mixed immiscible–intercalated structure. The complete procedure generally involves three stages: first, the dispersion of clay in a polymer solution; second, the meticulous removal of the solvent; and lastly, the casting of the composite film.

The third technique, known as in situ polymerization, represents a method applied for the creation of nanocomposite materials. In this process, polymerization unfolds directly within the material's matrix, giving rise to the development of a polymer composite [17]. This method entails the concurrent or sequential polymerization of monomers while incorporating nanoscale fillers or reinforcements. A distinctive feature of in situ polymerization is the simultaneous occurrence of the formation of the polymer matrix and the dispersion of nanoscale fillers during the reaction. Nanoparticles, nanotubes, or nanoclays are frequently utilized as nanofillers in this method. In a standard in situ polymerization procedure, nanoparticles are dispersed within a monomer or monomer solution, and subsequent polymerization of the monomer through conventional polymerization techniques results in the creation of nanocomposite materials.

The fourth technique is sol–gel, which involves transforming a solution (sol) into a gel-like substance and then solidifying it, resulting in the formation of a nanocomposite [18]. The sol–gel methodology proves advantageous in creating nanocomposites characterized by controlled structures, compositions, and properties. At its core, the sol–gel method revolves around generating a homogeneous sol from precursor substances and subsequently converting it into a gel. Sol–gel stands out as an exceptionally adaptable approach to obtaining both the matrix and filler components of a nanocomposite, allowing for chemical adjustments at the interface to optimize overall structure and properties. Furthermore, sol–gel techniques find widespread application in the formulation of nanocomposite materials due to the facile occurrence of these transformations with a diverse range of precursors, and they can be executed at or around room temperature. Sol–gel chemistry enables the formation of a diverse array of host matrices under gentle conditions, employing cost-effective reagents.

Table 3 presents various approaches for synthesizing nanocomposites, such as melt intercalation, solution blending, in situ polymerization, and the sol–gel process. Each method has distinct benefits, including improved dispersion of nanoparticles and better material properties. The choice of technique depends on the specific requirements and intended use of the nanocomposite.

Table 3. Summary of various synthesis methods for producing nanocomposite materials.

Method	Synthesis of Nanocomposite
Melt Intercalation	Polymer matrix is melted, nanoscale additives are introduced into the molten mixture, followed by mixing, cooling, and solidification.
Solution Blending	Nanoparticles are distributed within a solvent, accompanied by the inclusion of polymer or matrix material. The nanocomposite material is achieved through the evaporation of the solvent, ensuring a homogeneous mixture.
In situ Polymerization	Monomers are integrated with nanomaterials, leading to polymerization alongside the simultaneous dispersion of nanoscale additives.
Sol–Gel Process	Conversion of a colloidal solution into a gel (three-dimensional network), followed by additional procedures such as drying and heat treatment to produce the nanocomposite material.

3.2. Classification of Nanocomposite Materials

Nanocomposite materials belong to a category of substances wherein nanoscale fillers or reinforcements are integrated into a matrix material. This integration leads to improved characteristics in comparison to conventional composites [1]. Nanocomposites may be categorized according to the dispersion matrix into two primary groups: polymeric and non-polymeric [19]. The categorization is based on the presence or absence of polymeric material in the composite.

Polymer nanocomposites belong to the nanocomposite family, characterized by the integration of reinforcements. Typically, in the form of nanoparticles or nanofibers, these reinforcements are integrated into a matrix composed of polymers. Moreover, the collaboration between the polymer matrix and the reinforcements at the nanoscale level significantly improves the performance of polymer nanocomposites. Polymer nanocomposites have been widely utilized due to their capacity for amenability to functional modifications and the potential for manufacturing in diverse dimensions [20]. The synthesis of nanocomposites involved the use of several polymers, such as rubber, propylene, styrene, and ethylene vinyl acetate. Thus, incorporating polymeric components into nanocomposites contributes to the enhancement of their mechanical, thermal, and biodegradable properties. This integration plays a crucial role in optimizing the overall performance and sustainability of the composite materials.

Non-polymer nanocomposite materials consist of a matrix made from metals, ceramics, or carbon, which includes nanoparticles or nanofillers. These reinforcements are distributed within the matrix to improve the overall material's properties (mechanical, thermal, electrical. . .). Additionally, they are often referred to as inorganic nanocomposites. These materials can be categorized into metal-based nanocomposites and ceramic-based nanocomposites. Metal-based nanocomposites are innovative materials that involve a metal matrix and ceramic reinforcement. These materials fall into categories of either continuous or non-continuous reinforced materials. They offer a range of improvements in properties compared to monolithic alloys, including super plasticity, heightened strength, and enhanced electrical resistivity [19]. Ceramic-based nanocomposites consist of nanoparticles or nanofibers dispersed within a ceramic matrix. The predominant ceramic component is typically derived from oxide groups like nitrides, borides, or silicide. These materials are characterized by enhanced toughness, increased ductility, and improved strength and hardness. Figure 4 presents a classification of nanocomposite material based on the matrix.

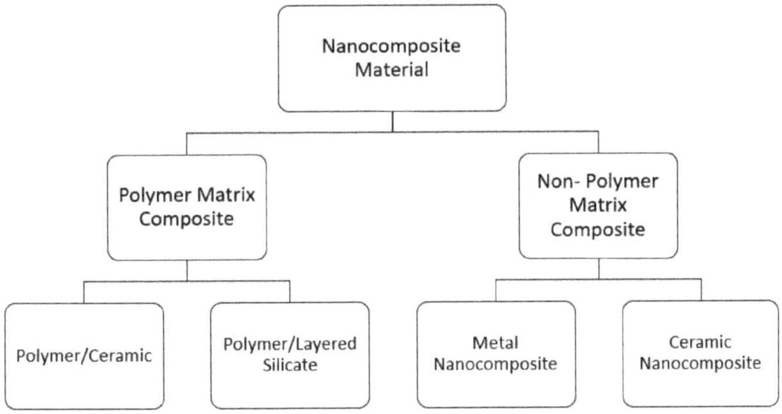

Figure 4. Classification of nanocomposite materials based on the matrix.

3.3. Electrochemical Performance

The electrochemical process involves creating electricity from chemical reactions [21], while electrochemical performance assesses how materials behave in these reactions. Evaluating electrochemical performance includes examining factors like specific capacity, rate capability, energy density, power density, and conductivity. In energy storage technologies, the efficiency of nanocomposite materials is measured by their electrochemical performance, which can be influenced by the unique characteristics they exhibit [19].

Nanocomposite materials are essential for enhancing the electrochemical performance of energy storage technologies, such as batteries. These materials boast promising properties, such as increased energy density or capacity compared to traditional composite materials. Improved rate capability enables faster charge and discharge rates [1]. Additionally, the inclusion of nanoparticles in nanocomposites influences electrical conductivity, leading to improved efficiency in electrochemical processes.

Table 4 contrasts the performance of conventional lithium-ion batteries with those incorporating nanocomposite materials. The table emphasizes the advantages of nanocomposites in mitigating issues such as electrolyte interface barriers, improving energy density, and enhancing charge/discharge rates. By increasing the surface area and conductivity of electrodes, nanocomposite materials contribute to superior overall performance, especially in extreme temperature conditions and high-rate charging/discharging scenarios. This comparison underscores the potential of nanocomposites to address key limitations of traditional Li-ion batteries.

Table 4. Comparison between conventional and nanocomposite Li-ion batteries.

Conventional Lithium-Ion Battery (Weakness)	Nanocomposite Li-Ion Battery (Strength)
At extreme temperatures, the battery electrolyte forms a strong and durable interface barrier.	Utilizing nanocomposite electrodes prevents the formation of an electrolyte interface barrier by promoting a more uniform and stable interaction between the electrode and electrolyte component.
Moderate energy density limits overall performance.	Increased surface area contributes to high energy density by offering more active sites for electrochemical reactions, with additional enhancement from improved conductivity facilitated by nanoscale structures.
The significant distance the Li ion needs to traverse within the battery's electrode material impacts the overall discharge rate.	The utilization of nanocomposite materials reduces the current distance within the electrode material, thereby accelerating both the recharging and discharging rates.

After examining the fundamental properties of nanocomposite materials, we will focus on their specific applications in the development of lithium-ion batteries.

4. Nanocomposite Materials in Li-Ion Battery Development

Lithium-ion batteries have garnered significant attention, especially with the increasing demand for electric vehicles and renewable energy storage applications. In recent years, substantial research has been dedicated to crafting advanced batteries with exceptional conductivity, power density, and both gravimetric and volumetric energy. The electrodes within lithium-ion batteries play a pivotal role in defining the battery's overall performance, lifespan, capacity, and cycle stability [22]. As a result, there is a crucial need to explore novel electrode materials to enhance the electrochemical performance of lithium-ion batteries. Concurrently, the integration of nanocomposite materials is a promising pathway that holds significant potential for the progress and development of lithium-ion batteries.

4.1. Progress in Anode and Cathode Nanocomposite Materials

4.1.1. Nanocomposite Anode Materials for Li-Ion Batteries

The anode electrode is considered as the most significant component of a lithium-ion battery, playing a crucial role in the overall performance of the battery. Generally, the most frequently used material for anode electrodes is graphite. Graphite is a crystalline form of carbon, consisting of stacked layers of graphene where carbon atoms are arranged in a hexagonal lattice structure. It is a pure material composed solely of carbon atoms. Graphite possesses chemical stability within a voltage range of 2.9 to 4.5 V and is cost-effective [23]. Nevertheless, its theoretical capacity is limited to 372 mAh/g [1]. The limited capacity of graphite does not meet the demands of rapidly advancing technology. It is imperative to create novel materials with improved Li storage properties to address this challenge. The main concept involves substituting graphite, either entirely or partially, with mixed metal anode materials.

Currently, Si (silicon) and Sn (tin) are capturing considerable attention due to their outstanding gravimetric and volumetric capacity among various alloy anodes.

Silicon, an economical and abundant material, is widely recognized as a highly promising anode material for lithium-ion batteries (LiBs) due to its high theoretical specific capacity and low discharge potential [20]. Additionally, it boasts an ultra-high theoretical gravimetric capacity of 4200 mAh/g, surpassing that of commercially available graphite anodes by a factor of 10 [24]. Thus, nanocomposites incorporating dispersed silicon nanoparticles within a matrix are designed to enhance capacity and enhance cycling stability, surpassing conventional graphite anodes. Nanocomposite materials with a foundation in silicon exhibit a wide spectrum of compositions, incorporating various nanomaterials with the goal of improving the efficiency of anodes in lithium-ion batteries.

The initial nanocomposite material is silicon–carbon nanotubes (Si-CNTs), which incorporate carbon nanotubes into silicon. This composite comprises carbon microcapsules that enclose silicon nanoparticles and carbon nanotubes. Si-CNT has been fabricated through a surfactant-mediated sol–gel technique followed by carbonization [25]. The integration of Si-CNT microcapsules proved successful as an anode in lithium-ion batteries, exhibiting noteworthy reversible capacity and coulombic efficiency of 80%. The introduction of silica as an intermediary layer demonstrated a substantial enhancement in the capacity retention capability of Si-CNT microcapsules [26]. A different nanocomposite material gaining attention for anodes is silicon–graphene, where graphene, a hexagonally arranged single layer of carbon atoms, is acknowledged for its exceptional electrical conductivity and mechanical robustness. The incorporation of silicon with carbon materials, such as graphene, proves attractive for augmenting the electrochemical efficiency of silicon-based anodes. The enhanced electrochemical performance is ascribed to the supportive role of graphene in dispersing silicon nanostructures and acting as a highly conductive framework, facilitating efficient interaction between them. Moreover, graphene plays a vital role in averting the expansion/contraction of volume and aggregation of silicon nanostructures throughout the lithium charge/discharge process.

Silicon–oxide nanocomposites developed for anode electrodes in lithium-ion batteries consist of silicon nanoparticles combined with a variety of oxide compounds. Examples of these oxides include titanium dioxide, silicon dioxide, aluminium oxide, zinc oxide, and cobalt oxide. These oxide materials play a crucial role in enhancing the electrochemical performance of the resulting nanocomposite anodes [27]. The manufacturing of these materials utilizes various technologies, such as sol–gel synthesis or chemical vapour deposition. Typically, the fabrication process involves creating silicon nanoparticles and incorporating them into an oxide matrix. The benefits provided by these nanocomposites include improved cycling stability and higher capacity. However, the intricacies of fabrication procedures, limited options for oxide selection, and the need to balance capacity with structural stability pose ongoing challenges in the advancement and refinement of these silicon–oxide nanocomposite materials for lithium-ion battery applications. The integration of nanocomposite materials into silicone-based anodes enhances cycling stability, boosts energy density, and accelerates charge/discharge rates in lithium-ion batteries.

On the other hand, tin nanoparticles emerge as a promising alternative for lithium-ion battery anodes, poised to replace carbon materials [28]. This shift is attributed to their remarkable theoretical Li-ion storage capacity, reaching 994 mAh/g, significantly surpassing that of graphite, which stands at 372 mAh/g. Tin, a member of the carbon family and a chemical element, exhibits soft, ductile, and highly crystalline silvery-white metal properties [29]. To address the challenge of Sn material expansion during cycling, nanocomposites based on Sn metal anodes have been developed [30]. These include Sn alloy-based materials, Sn–metal oxide composites, and Sn sulphide-based materials. The incorporation of tin with other metals to form alloys has been explored to enhance overall electrochemical performance. Various alloying elements such as copper (Cu), silver (Ag), antimony (Sb), and molybdenum (Mo) can be combined with tin to create different alloy types. The selection of alloying elements plays a crucial role in determining the final properties of the nanocomposite.

Examples of Sn alloy-based composite anodes encompass Sn-based alloy variations like Sn–Cu alloys, Sn–Co alloys, Sn–Al alloys, and Sn–Mo alloys [31]. The synthesis of Sn alloy-based nanocomposites involves methods such as mechanical alloying, electrodeposition, and sol–gel processes. Additionally, incorporating nanoscale reinforcements such as carbon nanotubes or graphene further enhances the performance of these alloy-based nanocomposites. Metal sulphides based on tin can occur in two stable forms, specifically, SnS_2 (tin selenide) and SnS (tin sulphide) [30,31]. Featuring a distinctive two-dimensional layer structure and significant layer spacing, these materials exhibit a high theoretical specific capacity, making them promising candidates for anode materials in lithium-ion batteries. The layered configurations inherent in tin-based metal sulphides not only offer favourable sites for charge storage but also contribute to improved electron/ion transport, thereby enhancing the efficient insertion of ions [30]. A common strategy involves compounding Sn-based sulphides with carbon materials and Sn-based sulphide anode materials including constructing SnS_2-based heterojunctions and forming hybrids with other Sn-based sulphides, which are promising strategies for enhancing the electrochemical performance.

There exist two primary categories of nanocomposites featuring tin-based oxides, namely, SnO and SnO_2, both considered promising options as active anode materials for advanced lithium-ion batteries (LIBs) [32]. Tin oxides exhibit accelerated lithiation/delithiation kinetics and significantly improved cyclability, making them potential candidates for use in the next generation of LIBs. SnO_2 provides simplicity in designing nanostructures and demonstrates a positive synergistic impact when incorporated with highly conductive materials or transition metal oxides [31]. On the other hand, SnO-based materials exhibit a substantial theoretical capacity [30,32]. Combining SnO with graphene or carbon materials to create hybrids represents an efficient approach for fabricating Sn-based anodes. Furthermore, the appealing aspect of this nanocomposite material is heightened by its low discharge potential, rendering it a more attractive choice as an anode material in lithium-ion batteries [32].

The importance of tin (Sn) in anode nanocomposite materials for lithium-ion batteries cannot be overstated. Its elevated theoretical capacity and adaptability in forming diverse nanocomposites make it an indispensable component [29]. Tin's intrinsic ability to store and release electrical energy, coupled with its compatibility with various materials, facilitates the creation of customized anode structures tailored to specific requirements. This multifaceted nature positions tin nanoparticles as a key player in advancing the efficiency and versatility of lithium-ion batteries.

Table 5 provides a comprehensive overview of nanocomposite anode materials for lithium-ion batteries, emphasizing their characteristics, advantages, challenges, and notable examples [33].

Table 5. Nanocomposite anode materials for Li-ion batteries.

Material	Characteristics	Advantages	Challenges	Examples
Graphite	Crystalline form of carbon with a hexagonal lattice structure.	Chemical stability, cost-effective, widely available.	Limited capacity (372 mAh/g), cannot meet advancing technological demands.	-
Silicon (Si)	Abundant material with ultra-high theoretical capacity (4200 mAh/g).	High capacity, low discharge potential, improved cycling stability in nanocomposites.	Volume expansion/contraction during cycling, aggregation of silicon nanoparticles.	Si-CNT, Si–graphene, Si–oxide composites.

Table 5. *Cont.*

Material	Characteristics	Advantages	Challenges	Examples
Si-CNT Nanocomposite	Carbon nanotubes with silicon nanoparticles encapsulated in carbon microcapsules.	Improved reversible capacity and cycling stability, coulombic efficiency of 80%.	Complex fabrication process, moderate capacity retention without silica intermediary layer.	Si-CNT microcapsules with silica layer.
Si–Graphene Nanocomposite	Combines silicon nanoparticles with a graphene matrix.	Excellent electrical conductivity, mechanical robustness, mitigates volume changes.	Cost of graphene, potential aggregation of nanoparticles.	Silicon–graphene hybrids.
Si–Oxide Nanocomposites	Silicon nanoparticles combined with oxide compounds (e.g., TiO_2, SiO_2, ZnO).	Enhanced cycling stability, higher capacity.	Intricate fabrication processes, limited oxide material options, balancing capacity with stability.	Si-TiO_2, Si-Al_2O_3 composites.
Tin (Sn)	Soft, ductile metal with high theoretical capacity (994 mAh/g).	Elevated storage capacity, adaptable to form various nanocomposites.	Volume expansion during cycling, structural degradation.	Sn alloys (Sn-Cu, Sn-Co), Sn sulphides (SnS, SnS_2), Sn oxides (SnO, SnO_2).
Sn–Alloy Nanocomposites	Sn combined with metals like Cu, Co, Al, Mo, and nanoscale reinforcements like CNT or graphene.	Enhanced electrochemical performance, improved structural stability.	Alloying process complexity, achieving a balance between capacity and cyclability.	Sn-Cu, Sn-Mo alloys with carbon nanotubes.
Sn–Sulphide Nanocomposites	Tin-based sulphides (SnS, SnS_2) with layered structures.	High theoretical capacity, good electron/ion transport, layered structure.	Limited scalability, potential performance degradation under high cycling rates.	SnS_2-based heterojunctions, hybrids with other Sn sulphides.
Sn Oxide Nanocomposites	Nanostructures featuring SnO and SnO_2, often combined with conductive materials.	Accelerated lithiation/delithiation, enhanced cyclability, low discharge potential.	Complex hybrid design, maintaining synergy between oxide and conductive matrix.	SnO–graphene hybrids, SnO_2 with transition metal oxides.

4.1.2. Nanocomposite Cathode Materials for Li-Ion Batteries

The adoption of nanocomposite materials for the cathode of Li-ion batteries stands out as a crucial strategy, presenting unparalleled advantages in terms of energy density, cycling stability, and overall electrochemical performance. Generally, there are three types of cathode materials: layered oxides, spinels, and olivines.

The first category is layered oxide, characterized by its composition of layered structures, usually incorporating transition metal oxides, along with nanoscale reinforcements. $LiCoO_2$ (lithium cobalt oxide) is an example of this type, which is the first Li-ion chemistry that was discovered in 1980 and subsequently introduced to the market by SONY in 1991. On the other hand, the nanocomposite lithium cobalt oxide is essentially a composition of $LiCoO_2$ as the matrix, coupled with nanoscale reinforcements like carbon nanotubes, graphene, or metal oxides. Layered oxides possess Li ion diffusion channels in two dimensions. Various methods, such as sol–gel synthesis, chemical vapour deposition, or physical vapour deposition, can be employed to produce these materials [34]. Each approach provides distinct benefits in terms of regulating composition, structure, and morphology, ultimately impacting the performance of the resultant nanocomposite material. The superiority of the $LiCoO_2$ nanocomposite cathode compared to bulk $LiCoO_2$ is apparent in its enhanced rate capability. Nanostructuring has demonstrated its effectiveness in boosting the performance of positive electrodes in lithium-ion batteries by diminishing the diffusion distances necessary for electrons and lithium ions within nano-sized crystals or particles. Moreover, nanocrystalline samples have exhibited a unique voltage profile marked by a more gradual curve and the lack of a plateau during lithiation. This occurrence was linked to the heightened importance of surface reactions, disordered structure, and the distribution of site energy for reacting with lithium in the nanocomposite cathode. As a result, the $LiCoO_2$ nanocomposite cathode demonstrates exceptional rate capability, particularly under demanding high-rate cycling conditions, solidifying its prominence in the lithium-ion battery market.

The second category is spinels, which integrates nanoscale structures with crystal phases characteristic of spinels. Spinels belong to a group of materials characterized by a distinct arrangement of atoms in their crystal lattice. One example of a nanocomposite spinel cathode for lithium-ion batteries is lithium manganese oxide ($LiMn_2O_4$) with nanoscale modifications. In its conventional form, $LiMn_2O_4$ functions as a spinel cathode material. The nanocomposite material $LiMn_2O_4$ is composed of nanoscale materials or structures, such as carbon nanotubes, metal nanoparticles, or graphene, within $LiMn_2O_4$ as a matrix. The nanochain-structured $LiMn_2O_4$ demonstrated better performance in terms of both rate capability and cycling stability when compared to commercially accessible $LiMn_2O_4$, which consists of aggregated particles at the submicron scale [34]. $LiMn_2O_4$ is traditionally synthesized through the solid-state reaction involving lithium and manganese salts. Nevertheless, these methods often encounter issues like inhomogeneity, irregular morphology, and large particle sizes. To address these challenges, wet chemical techniques, such as the sol–gel method, are recommended for the synthesis of $LiMn_2O_4$ nanocomposite materials. Contrastingly, the robust structural stability and significantly enhanced safety and environmental sustainability features have rendered spinel $LiMn_2O_4$ the most appealing choice as a cathode material for both transportation and large-scale batteries.

The last category relates to olivines, a term encompassing a group of minerals distinguished by their orthosilicate crystal structure containing elements such as iron and magnesium [1]. An example of this crystal structure is found in lithium iron phosphate ($LiFePO_4$). Nanocomposite cathodes utilizing $LiFePO_4$ generally comprise a blend of materials. $LiFePO_4$ (LFP) serves as the primary active material or matrix responsible for supplying lithium ions, alongside nanomaterials such as carbon additives, conductive polymers, or other relevant substances [34].

Nanotechnology has facilitated the utilization of $LiFePO_4$ and other metal phosphates as positive electrodes in lithium-ion batteries. Given their inherently low ionic and electronic conductivity, the integration of nanoparticles or particles coated with nanoscale conductive films becomes imperative to achieve the maximum charge storage capacity. Nevertheless, the compelling attributes of nanostructured LFP, including its cost-effectiveness, outstanding performance, and safety advantages, have positioned it as the preferred phosphate material. Consequently, it has become a focal point for extensive research and development within the industrial sector. Nanostructured materials for the cathode of lithium-ion batteries represent the core of significant advancements in efficient energy storage. Surface processes and transport kinetics play pivotal roles in these fundamental developments. Moreover, nanocomposite materials exhibit additional enhancements in properties when compared to their individual constituent phases.

Table 6 provides a summary of the three primary categories of nanocomposite cathode materials for lithium-ion batteries: layered oxides, spinels, and olivines [35].

4.2. Role of Nanocomposites in Electrolytes and Separators

In lithium-ion batteries, the electrolyte plays a crucial role in enabling the seamless movement of lithium ions between the cathode and anode during electrochemical reactions. Typically, electrolyte materials for lithium-ion batteries can be classified into two categories: solid polymer electrolytes and liquid electrolytes. Solid polymer electrolytes exhibit superior performance compared to liquid electrolytes, yet they encounter processing challenges, primarily linked to potential toxicity issues. Despite their notable advantages, solid polymer electrolytes come with drawbacks such as physic-chemical incompatibility with the anode, resulting in diminished ion conductivity [36]. Liquid electrolytes maintain their dominance in the field of lithium-ion applications, primarily because solid polymer electrolytes face processing challenges related to poor chemical stability and potential toxicity. Despite the advantages offered by solid polymer electrolytes, their drawbacks, such as physic-chemical incompatibility with the anode, contribute to a reduction in ion conductivity, keeping liquid electrolytes as the prevalent choice.

Table 6. Nanocomposite cathode materials for lithium-ion batteries.

Category	Composition	Advantages	Challenges	Examples
Layered Oxides	Transition metal oxides with nanoscale reinforcements (e.g., graphene, carbon nanotubes, metal oxides).	High energy density, excellent rate capability, reduced electron and lithium-ion diffusion distances.	Complex synthesis processes, potential structural disorder, high cost of nanoscale reinforcements.	$LiCoO_2$ nanocomposite with graphene or CNTs.
Spinels	Spinel crystal structures with nanostructures (e.g., carbon nanotubes, graphene, metal nanoparticles).	Robust structural stability, high cycling stability, good safety, and environmental sustainability.	Inhomogeneous morphology, irregular particle sizes, and low conductivity in conventional forms.	Nanochain-structured $LiMn_2O_4$ nanocomposite.
Olivines	Orthosilicate crystal structures (e.g., $LiFePO_4$) blended with carbon additives, conductive polymers, or nanoscale coatings.	Cost-effective, safe, excellent thermal stability, and enhanced ionic and electronic conductivity.	Inherently low conductivity of base materials, requiring nanoscale coatings or additives.	Nanostructured $LiFePO_4$ with carbon or conductive polymer coatings.

Various types of nanocomposite electrolytes exist, one example being the incorporation of ceramic nanopowers (Al_2O_3, SiO_2, and TiO_2) into polyethylene electrolytes [36]. This addition has demonstrated an improved electrical conductivity in lithium-ion batteries. The utilization of smaller particles, when compared to pure polyethylene, could enhance dispersion. Furthermore, the integration of nanostructured additives plays a role in enhancing the physical stability of the solid polymer structure. Nanocomposite liquid electrolytes are a combination of liquid electrolytes with nanomaterials (graphene oxide, carbon nanotubes, Nanostructured Ceramic Particles, and clay nanoparticles) to enhance the overall performance and safety of LIBs. The primary objective of the nanocomposite liquid electrolyte is to tackle challenges commonly linked with traditional liquid electrolytes, notably addressing concerns related to volatility and flammability.

In a notable instance, $Li_{1.3}Al_{0.3}Ti_{1.7}(PO_4)_3$(LATP) nanoparticles are seamlessly integrated with polyethylene through a solvent thermal technique, resulting in the formation of a solid composite electrolyte known as PEOLITFSI-LATP [37]. This amalgamation leads to the enhancement in electrochemical properties within the polyethylene oxide (PEO)-based electrolyte. Additionally, the exploration of methacrylate-functionalized SiO_2 (MA-SiO_2) nanoparticles in cross-linked composite gel polymer electrolytes reveals a significant improvement in cycling performance, particularly under heightened operating temperatures.

Nanomaterials play a crucial role in electrolytes by primarily improving the mass transport essential for the operation of lithium-ion batteries. The separator plays a crucial role in lithium-ion batteries by effectively segregating the anode and cathode electrodes. Simultaneously, it facilitates the movement of ions between these electrodes and promotes essential ionic transport within the battery. Presently, the primary components of lithium-ion separators predominantly comprise polyolefin materials, specifically polyethylene or polypropylene. The utilization of these materials encounters various challenges, primarily due to their low melting points and safety concerns. Polyethylene and polypropylene separators, in particular, exhibit relatively low melting points, with PE at 130 °C and PP at 160 °C [38]. These low melting points pose a significant risk of battery explosions in situations involving overheating or short circuits. Moreover, the hydrophobic nature and limited porosity of the existing separator materials give rise to challenges such as thermal shrinkage and wettability problems. To tackle these challenges, a practical solution involves integrating nanomaterials into LIB separators. This can be accomplished through the development of nanocomposites, which entail combining polymers with ceramics, or by applying nanoceramics coated onto polymer substrates. The use of nanocomposite material for separators improves the durability of LIB systems by reducing physical damage and preventing ion migration via crossover.

There are various nanocomposites used as separators in lithium-ion batteries (LIBs), such as SiO_2 ceramic layers onto polypropylene (PP) separators. This application enhances rate capability, battery safety, coulombic efficiency, and mechanical strength. Additionally, it reduces thermal shrinkage. Another notable example is the Cellulose/PVDF-HFP Composite Non-woven, which serves as an advanced separator for LIBs. It offers high ionic conductivity, cost-effectiveness, and environmental friendliness. Furthermore, the incorporation of a ceramic coating with tailored porosity and engineered surface area is a common practice. Ceramic nanoparticles are coated using methods like dip-coating, automatic machining, and sol–gel techniques.

4.3. Advancements in Nanocomposite Materials for Lithium-Ion Battery Technologies

Currently, investigations into lithium-ion batteries (LIBs) are increasingly directed towards the creation of nanocomposite materials that emphasize multifunctional capabilities, scalability, and sustainability. The advancement of gradient-structured nanocomposites is a promising strategy for enhancing lithium-ion battery (LIB) technologies [39]. These materials exhibit a continuous variation in composition or properties throughout their structure, which optimizes lithium-ion diffusion pathways while simultaneously alleviating mechanical stress at the electrode–electrolyte interface. By customizing the structural and compositional gradients, these innovative nanocomposites can significantly improve electrochemical stability, ionic conductivity, and cycle life, thereby contributing to the development of next-generation high-performance LIB systems [39].

The advancement of lithium-ion batteries (LIBs) is increasingly dependent on the integration of self-healing and hybrid nanocomposites, which are essential for overcoming significant challenges related to durability and multifunctionality. Self-healing nanocomposites utilize materials such as synthetic polymers or metallic structures that can autonomously repair microstructural damage inflicted by mechanical stress or extended charge/discharge cycles. This capability not only restores electrical conductivity but also preserves structural integrity, thereby significantly prolonging the lifespan of LIBs [40]. On the other hand, hybrid nanocomposites combine materials with complementary properties to enhance both electrochemical and thermal performance. This strategic combination not only increases energy storage capacity but also facilitates effective thermal regulation, thereby reducing the risk of thermal runaway incidents [41].

Finally, the development of solid-state lithium-ion batteries (SSBs) represents a promising area for the application of nanocomposites. Solid electrolytes that incorporate both ceramic and polymer phases exhibit exceptional ionic conductivity while retaining mechanical flexibility [42]. Moreover, advanced interfacial nanocomposites designed to mitigate dendrite formation are essential for the successful integration of high-capacity lithium metal anodes within SSBs. These innovations are critical for enhancing the performance and safety of next-generation battery technologies [43].

After investigating the role of nanocomposite materials in improving the performance of lithium-ion batteries, the next section will examine the practical applications of these advanced batteries in various industries.

5. Applications of Li-Ion Batteries Based on Nanocomposite Materials

Nowadays, the integration of nanocomposite materials has attracted considerable interest and stands out as a crucial breakthrough in the field of energy storage, specifically within the domain of lithium-ion batteries [44]. Rechargeable lithium-ion batteries incorporating nanocomposite materials are widely utilized across diverse industries, revolutionizing energy storage solutions. Consequently, the utilization of these materials has transformed the realm of battery technology, heralding a new era of improved performance and efficiency.

The integration of nanocomposite materials into Li-ion batteries has numerous applications, ranging from small energy storage devices to large-scale solutions, emphasizing their extensive applicability across various contemporary industries. Medical Instruments, Mobile Devices, Aerospace Applications, Renewable Energy Storage Systems, and electric vehicles (EVs) exemplify key domains where nanocomposite-enhanced lithium-ion batteries play a vital role [44]. However, in this paragraph, we will specifically delve into the applications of Renewable Energy Storage Systems and electric vehicles (EVs), aiming to provide a detailed examination of their utilization and advancements in these key areas.

Cutting-edge nanocomposite materials have revolutionized the field of renewable energy storage technology, with a particular focus on lithium-ion batteries [45]. These enhanced batteries are recognized as ground-breaking solutions for efficiently storing clean energy, especially in solar energy systems. Lithium-ion batteries play a crucial role in solar energy systems, serving as integral devices in this technology. They perform the essential function of storing excess energy generated during sunny periods. Subsequently, this stored energy is released during cloudy days or night-time, ensuring a continuous and reliable power supply. Furthermore, these batteries contribute to achieving grid independence, offering autonomy in energy supply, especially during grid outages. These batteries provide a prolonged lifespan and significant energy density, offering a dependable and resilient solution for solar energy systems.

On the other hand, lithium-ion batteries used in solar energy systems face specific challenges, notably, cycling instability and restricted rate capability. To tackle these issues, researchers have increasingly explored the potential of nanocomposite materials. These innovative materials aim to improve battery performance by rectifying concerns related to cycling stability and amplifying rate capability by optimizing electrode conductivity. Thus, opting for lithium-ion batteries empowered with nanocomposite materials in solar systems is a prudent choice for several compelling reasons. The incorporation of nanocomposite technology facilitates rapid charging for solar energy systems, reducing the time needed to replenish energy storage. This feature is especially advantageous during periods of intermittent sunlight, ensuring quick adaptation to changing weather conditions in the context of solar energy utilization. Further, the use of nanocomposite materials supports the development of a more streamlined and lighter battery design while maintaining optimal performance. This characteristic simplifies the installation process, rendering the solar system suitable for a range of applications, particularly on residential rooftops. Therefore, the solar system integrates nanocomposite lithium-ion batteries that utilize cutting-edge nanoscale materials. These materials elevate the battery's overall performance, extending its lifespan and boosting energy density.

Nanocomposite technology not only guarantees dependable energy storage but also promotes the sustainability and eco-friendliness of the entire system. In the pursuit of sustainable and eco-friendly transportation solutions, the electric vehicle (EV) sector has garnered significant attention [46]. As the demand for alternative fuel sources intensifies due to global warming concerns and fuel shortages, lithium-ion batteries have become a focal point for enhancing the performance of electric vehicles. The drawbacks of traditional electric vehicles, such as long charging times and large battery sizes, can be mitigated through the incorporation of nanocomposite materials in lithium-ion batteries. Nanomaterials, with their unique physical and chemical properties, hold the key to revolutionizing battery technology. These materials, whether spontaneously formed, synthesized, or engineered for specific tasks, offer increased performance and storage capacity while reducing the overall size of batteries.

Nanotechnology, with its ability to tailor materials to specific needs, has found applications across various sectors. In the realm of electric vehicles, nanomaterials play a crucial role in improving battery efficiency and addressing the challenges associated with EVs, such as long-distance travel and extended recharge periods. Lithium-ion batteries, with their inherent advantages over traditional nickel–metal hydride batteries, benefit from the integration of nanomaterials to enhance their performance. Nanocomposite materials, including carbon nanotubes, titanium dioxide, and vanadium oxide, have demonstrated the potential to optimize lithium-ion battery technology. These materials enable higher concentrations of lithium, resulting in increased power production and improved battery capabilities. The ongoing research in electrode compositions, such as nanowires and nanoparticles, aims to make batteries cost-effective and lightweight. Major automotive players, like Ford Motor Company, are actively exploring nanotechnology to create lighter vehicles, reducing energy consumption. Despite the promising strides in nanotechnology, the research is ongoing, with a focus on reducing costs and ensuring scalability for large-scale commercial applications.

6. Conclusions

In conclusion, the exploration of nanocomposite materials for rechargeable lithium-ion batteries has unveiled a promising avenue for significant advancements in performance parameters. The remarkable characteristics of lithium-ion batteries, with their widespread applications in consumer electronics, electric vehicles, and stationary energy storage, underscore the importance of continuous improvements in safety, cost efficiency, cycle life, energy density, and rate capability. The integration of nanocomposite materials, including carbon–oxide, polymer–oxide, and silicon-based nanocomposites, represents a crucial step towards achieving these goals.

While nanocomposite materials hold great promise, challenges such as high costs and scalability issues in commercial production still impede their widespread adoption. Overcoming these hurdles is imperative to fully harness the potential of nanocomposite-enhanced Li-ion batteries in various applications, including renewable energy systems and electric vehicles.

Nanotechnology not only improves the efficiency of lithium-ion batteries but also contributes to the development of eco-friendly and efficient electric vehicles. The integration of nanomaterials in battery technologies is not only limited to performance enhancement but also addresses environmental concerns, as evidenced by life cycle assessments conducted by the Environmental Protection Agency (EPA).

The future of electric vehicles hinges on the continued advancements in nanocomposite materials, providing a path towards energy-efficient and sustainable transportation solutions. As the automotive industry evolves, the integration of nanotechnology in lithium-ion batteries stands as a pivotal step in the transition towards a greener and more efficient transportation landscape.

Author Contributions: Conceptualization, S.E.A. and P.A.; methodology, S.E.A. and A.C.; software, A.C.; validation, S.E.A. and P.A.; formal analysis, S.E.A.; investigation, P.A.; resources, S.E.A. and A.C.; data curation, P.A.; writing—original draft preparation, S.E.A.; writing—review and editing, A.C.; visualization, S.E.A. and P.A.; supervision, A.C.; project administration, S.E.A. and F.J.; funding acquisition, S.E.A. and F.J. All authors have read and agreed to the published version of this manuscript.

Funding: This research was funded by the Thematic Network 723RT0150 "Red para la integración a gran escala de energías renovables en sistemas eléctricos (RIBIERSE-CYTED)" financed by the call for Thematic Networks of the CYTED (Ibero-American Program of Science and Technology for Development) for 2022.

Data Availability Statement: Data will be made available upon request due to (specify the reason for the restriction: e.g., privacy, legal or ethical reasons).

Acknowledgments: The authors thank Universidad de Cuenca (UCUENCA), Ecuador, for easing access to the facilities of the Micro-Grid Laboratory, Faculty of Engineering, for allowing the use of its equipment, to provide the technical support for the descriptive literature analysis included in this article.

Conflicts of Interest: The authors declare no conflict of interest.

References

1. Demirocak, D.; Srinivasan, S.; Stefanakos, E. A Review on Nanocomposite Materials for Rechargeable Li-Ion Batteries. *Appl. Sci.* **2017**, *7*, 731. [CrossRef]
2. Zhan, H.; Xiao, J.; Nie, Z.; Li, X.; Wang, C.; Zhang, J.-G.; Liu, J. Nanostructured Materials for Rechargeable Batteries: Synthesis, Fundamental Understanding, and Limitations. *Curr. Opin. Chem. Eng.* **2013**, *2*, 151–159. [CrossRef]
3. Cai, C.; Wang, Y. Novel Nanocomposite Materials for Advanced Li-Ion Rechargeable Batteries. *Materials* **2009**, *2*, 1205–1238. [CrossRef]
4. Stephenson, T.; Li, Z.; Olsen, B.; Mitlin, D. Lithium Ion Battery Applications of Molybdenum Disulfide (MoS_2) Nanocomposites. *Energy Environ. Sci.* **2014**, *7*, 209–231. [CrossRef]
5. Palacin, M.R. Battery Materials Design Essentials. *Acc. Mater. Res.* **2021**, *2*, 319–326. [CrossRef]
6. Dash, R.; Kommu, P.; Bhattacharyya, A.S. Electrode Materials in Lithium-Ion Batteries. In *Materials Horizons: From Nature to Nanomaterials*; Swain, B.P., Ed.; Springer Nature Singapore: Singapore, 2023; pp. 77–89. [CrossRef]
7. Jiang, C.; Hosono, E.; Zhou, H. Nanomaterials for lithium-ion batteries. *Nanotoday* **2006**, *1*, 28–33. [CrossRef]
8. Rakhimov, E.; Khoshimov, D.; Sultonov, S.; Jamoldinov, F.; Imyaminov, A.; Omonov, B. Battery technologies: Exploring different types of batteries for energy storage. *BIO Web Conf.* **2024**, *84*, 05034. [CrossRef]
9. Deng, D. Li-Ion Batteries: Basics, Progress, and Challenges. *Energy Sci. Eng.* **2015**, *3*, 385–418. [CrossRef]
10. Arote, S.A. *Lithium-Ion and Lithium–Sulfur Batteries: Fundamentals to Performance*; IOP Publishing: London, UK, 2022. [CrossRef]
11. Huang, Y.; Duan, J.; Zheng, X.; Wen, J.; Dai, Y.; Wang, Z.; Luo, W.; Huang, Y. Lithium Metal-Based Composite: An Emerging Material for Next-Generation Batteries. *Matter* **2020**, *3*, 1009–1030. [CrossRef]
12. Khan, B.M.; Oh, W.C.; Nuengmatch, P.; Ullah, K. Role of Graphene-Based Nanocomposites as Anode Material for Lithium-Ion Batteries. *Mater. Sci. Eng. B* **2023**, *287*, 116141. [CrossRef]
13. Selinis, P.; Farmakis, F. A Review on the Anode and Cathode Materials for Lithium-Ion Batteries with Improved Subzero Temperature Performance. *J. Electrochem. Soc.* **2022**, *169*, 010526. [CrossRef]
14. Le, B.; Khaliq, J.; Huo, D.; Teng, X.; Shyha, I. A Review on Nanocomposites. Part 1: Mechanical Properties. *J. Manuf. Sci. Eng.* **2020**, *142*, 100801. [CrossRef]
15. Mittal, V. *In-Situ Synthesis of Polymer Nanocomposites*; Wiley-VCH Verlag GmbH & Co. KGaA: Weinheim, Germany, 2012.
16. Guo, F.; Aryana, S.; Han, Y.; Jiao, Y. A Review of the Synthesis and Applications of Polymer–Nanoclay Composites. *Appl. Sci.* **2018**, *8*, 1696. [CrossRef]
17. DineshKumar, S.; Purushothaman, S. Synthesis and Characterization of Polymer Nanocomposites for Biomedical Applications—Current Perspectives and Challenges. *Int. J. Res. Eng. Sci.* **2016**, *4*, 1–6.
18. Bokov, D.; Jalil, A.T.; Chupradit, S.; Suksatan, W.; Ansari, M.J.; Shewael, I.H.; Valiev, G.H.; Kianfar, E. Nanomaterial by Sol-Gel Method: Synthesis and Application. *Adv. Mater. Sci. Eng.* **2021**, *2021*, 1–21. [CrossRef]
19. Sen, M. Nanocomposite Materials. In *Nanotechnology and the Environment*; Sen, M., Ed.; IntechOpen: London, UK, 2020. [CrossRef]
20. Khan, I.; Khan, I.; Saeed, K.; Ali, N.; Zada, N.; Khan, A.; Ali, F.; Bilal, M.; Akhter, M.S. Polymer Nanocomposites: An Overview. In *Smart Polymer Nanocomposites*; Elsevier: Amsterdam, The Netherlands, 2023; pp. 167–184. [CrossRef]
21. Shiraishi, S. Electrochemical Performance. In *Materials Science and Engineering of Carbon*; Elsevier: Amsterdam, The Netherlands, 2016; pp. 205–226. [CrossRef]
22. Febrian, R.; Septiani, N.L.W.; Iqbal, M.; Yuliarto, B. Review—Recent Advances of Carbon-Based Nanocomposites as the Anode Materials for Lithium-Ion Batteries: Synthesis and Performance. *J. Electrochem. Soc.* **2021**, *168*, 110520. [CrossRef]
23. Ma, X.; Zhang, P.; Zhao, H.; Wang, Q.; Zhang, G.; Chi, S.-S.; Liu, Z.; Qian, Y.; Wang, J.; Wang, C.; et al. $LiCoO_2$/Graphite Cells with Localized High Concentration Carbonate Electrolytes for Higher Energy Density. *Liquids* **2021**, *1*, 60–74. [CrossRef]
24. Larkin, R.-J.; Willenberg, S.C.; Ross, N. Silicon-Based Anodes Towards Enhanced Cycling Efficiencies for Next-Generation Lithium-Ion Batteries. *Int. J. Electrochem. Sci.* **2023**, *18*, 100158. [CrossRef]
25. Bae, J. Fabrication of Carbon Microcapsules Containing Silicon Nanoparticles–Carbon Nanotubes Nanocomposite by Sol–Gel Method for Anode in Lithium-Ion Battery. *J. Solid State Chem.* **2011**, *184*, 1749–1755. [CrossRef]
26. Sehrawat, P.; Shabir, A.; Abid; Julien, C.M.; Islam, S.S. Recent Trends in Silicon/Graphene Nanocomposite Anodes for Lithium-Ion Batteries. *J. Power Sources* **2021**, *501*, 229709. [CrossRef]
27. Zhou, Z.; Dong, P.; Wang, D.; Liu, M.; Duan, J.; Nayaka, G.P.; Wang, D.; Xu, C.; Hua, Y.; Zhang, Y. Silicon-Titanium Nanocomposite Synthesized via the Direct Electrolysis of SiO_2/TiO_2 Precursor in Molten Salt and Their Performance as the Anode Material for Lithium-Ion Batteries. *J. Alloys Compd.* **2019**, *781*, 362–370. [CrossRef]

28. Wang, J.; Kober, D.; Shao, G.; Epping, J.D.; Görke, O.; Li, S.; Gurlo, A.; Bekheet, M.F. Stable Anodes for Lithium-Ion Batteries Based on Tin-Containing Silicon Oxycarbonitride Ceramic Nanocomposites. *Mater. Today Energy* **2022**, *26*, 100989. [CrossRef]
29. Jamil, M.A.; Ali, G.; Khan, K.I.; Jan Iftikhar, F.; Zaman, S.; Shaikh, S.F.; Pandit, B.; Wali, Q.; Patil, S.A. Highly Efficient Tin Fluoride Nanocomposite with Conductive Carbon as a High Performance Anode for Li-Ion Batteries. *J. Alloys Compd.* **2022**, *900*, 163447. [CrossRef]
30. Liu, H.; Wang, S.; Zhao, J.; Zhang, B.; Liu, L.; Bao, R.; Jing, Z. Sn-Based Anode Materials for Lithium-Ion Batteries: From Mechanism to Modification. *J. Energy Storage* **2024**, *80*, 109862. [CrossRef]
31. Tomboc, G.M.; Wang, Y.; Wang, H.; Li, J.; Lee, K. Sn-Based Metal Oxides and Sulfides Anode Materials for Na-Ion Battery. *Energy Storage Mater.* **2021**, *39*, 21–44. [CrossRef]
32. Zoller, F.; Böhm, D.; Bein, T.; Fattakhova-Rohlfing, D. Tin Oxide Based Nanomaterials and Their Application as Anodes in Lithium-Ion Batteries and Beyond. *ChemSusChem* **2019**, *12*, 4140–4159. [CrossRef]
33. Poorshakoor, E.; Darab, M. Advancements in the development of nanomaterials for lithium-ion batteries: A scientometric review. *J. Energy Storage* **2023**, *75*, 109638. [CrossRef]
34. Sukkurji, P.A. Advanced Anode and Cathode Materials for Li-ion Batteries: Application to Printing Methodology. Ph.D. Thesis, Institut für Nanotechnologie (INT), Eggenstein-Leopoldshafen, Germany, 2021.
35. Julien, C.; Mauger, A.; Zaghib, K.; Groult, H. Comparative Issues of Cathode Materials for Li-Ion Batteries. *Inorganics* **2014**, *2*, 132–154. [CrossRef]
36. Sharma, S.; Pathak, D.; Kumar, R.; Sharma, V.; Arora, N.; Kaur, S.; Sharma, V. Nanocomposite Polymer Electrolytes for Energy Devices. In *Nano Tools and Devices for Enhanced Renewable Energy*; Elsevier: Amsterdam, The Netherlands, 2021; pp. 27–40. [CrossRef]
37. Liu, J.; Liu, T.; Pu, Y.; Guan, M.; Tang, Z.; Ding, F.; Xuc, Z.; Li, Y. Facile Synthesis of NASICON-Type $Li_{1.3}Al_{0.3}Ti_{1.7}(PO_4)_3$ Solid Electrolyte and Its Application for Enhanced Cyclic Performance in Lithium-Ion Batteries Through the Introduction of an Artificial Li_3PO_4 SEI Layer. *RSC Adv.* **2017**, *7*, 46545–46552. [CrossRef]
38. l'Abee, R.; DaRosa, F.; Armstrong, M.J.; Hantel, M.M.; Mourzagh, D. High Temperature Stable Li-Ion Battery Separators Based on Polyetherimides with Improved Electrolyte Compatibility. *J. Power Sources* **2017**, *345*, 202–211. [CrossRef]
39. Paramsothy, M. Nanocomposites: Nanoscience and Nanotechnology (Nanoscale Phenomena) in Advanced Composites. *Nanomaterials* **2021**, *12*, 81. [CrossRef] [PubMed]
40. Wang, C.; Wu, H.; Chen, Z.; McDowell, M.T.; Cui, Y.; Bao, Z. Self-healing chemistry enables the stable operation of silicon microparticle anodes for high-energy lithium-ion batteries. *Nat. Chem.* **2013**, *5*, 1042–1048. [CrossRef] [PubMed]
41. Zhang, L.; Yang, M.; Zhang, S.; Wu, Z.; Amini, A.; Zhang, Y.; Wang, D.; Bao, S.; Lu, Z.; Wang, N.; et al. V_2O_5-C-SnO_2 Hybrid Nanobelts as High Performance Anodes for Lithium-ion Batteries. *Sci. Rep.* **2016**, *6*, 33597. [CrossRef] [PubMed]
42. Wu, Y.; Shuang, W.; Wang, Y.; Chen, F.; Tang, S.; Wu, X.-L.; Bai, Z.; Yang, L.; Zhang, J. Recent Progress in Sodium-Ion Batteries: Advanced Materials, Reaction Mechanisms and Energy Applications. *Electrochem. Energy Rev.* **2024**, *7*, 17. [CrossRef]
43. Chen, Y.; Xu, L.; Yang, X.; Li, Q.; Yao, M.; Wang, G. Nanocomposite Design for Solid-State Lithium Metal Batteries: Progress, Challenge, and Prospects. *Adv. Nanocompos.* **2024**, *1*, 120–143. [CrossRef]
44. Barbosa, J.; Gonçalves, R.; Costa, C.M.; Lanceros-Mendez, S. Nanocomposites for Energy Storage Applications. In *Advances in Nanocomposite Materials for Environmental and Energy Harvesting Applications*; Springer International Publishing: Cham, Switzerland, 2022; pp. 533–565. [CrossRef]
45. Shalan, A.E.; Hamdy Makhlouf, A.S.; Lanceros-Méndez, S. *Advances in Nanocomposite Materials for Environmental and Energy Harvesting Applications*; Springer International Publishing: Cham, Switzerland, 2022.
46. Kurc, B.; Pigłowska, M.; Rymaniak, Ł.; Fuć, P. Modern Nanocomposites and Hybrids as Electrode Materials Used in Energy Carriers. *Nanomaterials* **2021**, *11*, 538. [CrossRef]

batteries

MDPI

Article

Nickel Stabilized Si/Ni/Si/Ni Multi-Layer Thin-Film Anode for Long-Cycling-Life Lithium-Ion Battery

Yonhua Tzeng *, Yu-Yang Chiou and Aurelius Ansel Wilendra

Institute of Microelectronics, National Cheng Kung University, Tainan 70101, Taiwan; q16121135@gs.ncku.edu.tw (Y.-Y.C.); q16125016@gs.ncku.edu.tw (A.A.W.)
* Correspondence: tzengyo@mail.ncku.edu.tw

Abstract: Silicon-based anodes suffer from the loss of physical integrity due to large volume changes during alloying and de-alloying processes with electrolytes. By integrating electrochemically inert, physically strong, ductile nickel layers with a multi-layered thin-film silicon anode, the long-life cycling of the Si/Ni/Si/Ni anode was demonstrated. A capacity retention of 82% after 200 cycles was measured, surpassing the performance of conventional silicon thin-film anodes. This is attributed to the effective suppression of internal local stress induced by nonuniform volume expansion by the nickel layers. These findings offer a promising pathway towards the practical implementation of high-capacity silicon-based anodes in advanced lithium-ion batteries.

Keywords: lithium-ion battery; silicon; multi-layer thin film; anode; nickel; large surface area; protective layer

1. Introduction

Commercial lithium-ion batteries currently rely on graphite-based anodes. However, the surging demand for high-capacity energy storage solutions, particularly in electrical vehicles [1] and other applications, necessitates a departure from conventional graphite anodes, which exhibit a limited specific capacity of 372 mAh/g. Silicon-based anodes, with their exceptional specific capacity of 3579 ($Li_{15}Si_4$)–4200 ($Li_{22}Si_5$) mAh/g and relatively low discharge potential (approximately 0.4 V vs. Li/Li^+), emerge as a compelling alternative to address this capacity limitation.

Despite their promising characteristics, silicon anodes confront several critical challenges that hinder their practical implementation. The inherent low electrical conductivity of silicon and its substantial volume expansion (approximately 400%) during charge–discharge cycles are particularly problematic. These factors lead to increased internal resistance and stress [2], reduced Coulombic efficiency, and ultimately, degraded long-term performance during cycling [3].

To address these challenges, researchers have explored various strategies. Structurally, the use of silicon nanoparticles [4,5], nanotubes [6,7], nanowires [8,9], and thin films [10,11] have been investigated. Additionally, surface modification techniques, such as carbon (C) coating [12] and carbon composites [13], diamond nanoparticles (ND) [14], copper (Cu) [15], and other materials [16,17], have been employed to mitigate the potential damage caused by the volumetric expansion of silicon.

Multi-layer thin-film structures, such as Si-Y multi-layer thin films, as anode materials of high-capacity lithium-ion batteries [18] deliver a high reversible capacity of 2450 mAh/g under a current density of 0.4 C after 50 cycles. The volumetric expansion of silicon can be effectively suppressed by forming a Fe layer between Si layers, which was tested using the constant charge and discharge current of 30 μA/cm^2 between 0 and 1.2 V at 30 °C [19].

Yang et al. [20] reported a multi-layer thin-film structure consisting of carbon/silicon/carbon/silicon (with the outer carbon layer in contact with the electrolyte), fabricated through magnetron sputtering. The protective carbon layers significantly enhance the cycling stability of the silicon anode. The initial discharge specific capacity was approximately 2045.9 mAh/g, which stabilized at around 1500 mAh/g after 200 cycles, resulting in a capacity retention rate of 85.4%. Salah et al. [21] reported silicon/tin thin-film anodes for low- and high-power-density lithium-ion batteries.

Tzeng et al. [14] reported a multi-layer thin-film silicon-based anode with nanoscale diamond particles as an interfacial layer and the surface layer. Diamond particles decorated with silicon thin films result in better uniform lithium-ion flux and an improved wettability of the anode to the electrolyte. Both effects enhanced the physical integrity of the silicon thin-film anode and the cycling performance.

FU et al. [22] reported a multi-layer electrode made of NiO/SnO_2 by alternating magnetron sputtering. Ni nanoparticles (NP) formed from NiO can lower the Li-O bonding energy, thereby enhancing the transition reaction from Sn/Li_2O to SnO_2, further decreasing the Li loss. Additionally, Ni can also enhance the conductivity and alleviate volume expansion. The NiO/SnO_2 multi-layer electrode exhibits a high ICE of 92.3% and retains ~97% capacity at a low test current density. As shown in Table S1, silicon offers a much higher specific capacity and a lower potential versus Li/Li^+ than tin, making it an attractive anode material. However, the substantial volume expansion (300–400%) of silicon during cycling compared to tin (260%) results in internal stress, which may cause structural degradation and capacity loss. Furthermore, in their study, the maximum current density utilized in the C-rate tests was limited to a low current density of 5 A/g. Due to the use of oxide materials, the resistance values obtained from EIS measurements were high and cycling performance at a high test current density left much room for improvement.

In this study, nickel was used as an interfacial layer and a surface layer in a nickel/silicon multi-layer thin-film anode. We employed thermal evaporation to deposit approximately 100 nm of silicon onto a copper foil substrate, followed by the deposition of a 20 nm nickel (Ni) layer directly onto the silicon surface. This process was repeated twice to yield a multi-layer Si/Ni/Si/Ni anode structure. The Ni layer plays a pivotal role in this anode. As the surface layer, it prevents potential side reactions by blocking direct contact between the electrolyte and the silicon. Additionally, it provides mechanical strength to accommodate the volume changes of silicon during cycling. This protective function safeguards the overall structure from damage caused by the silicon's expansion, thereby significantly improving the long-term cycling performance of the battery.

2. Materials and Methods

2.1. Electrode Preparation

All electrodes were fabricated through thermal evaporation to deposit thin films. N-type crystalline silicon (99.999% purity) and a Ni slug (99.995% purity) served as the evaporation sources, while copper foils (14 mm in diameter) were used as current collectors.

The deposition process was conducted under a background pressure of approximately 5×10^{-6} torr, with a substrate temperature set at 250 °C. The deposition rate was controlled between 0.5 and 1 nm/s.

Prior to silicon deposition, the Cu foil was thoroughly cleaned in an ultrasonic bath using acetone, ethanol, and deionized water to remove surface contaminants. Following this cleaning step, thin-film deposition was initiated. For the reference silicon anode, a single 200 nm layer of silicon was directly deposited onto the Cu foil surface. In contrast, the Si/Ni anode was fabricated by sequentially depositing a 100 nm silicon layer and a 20 nm nickel layer. This sequence was repeated twice to form the Si/Ni/Si/Ni multi-layer anode structure (Figure 1).

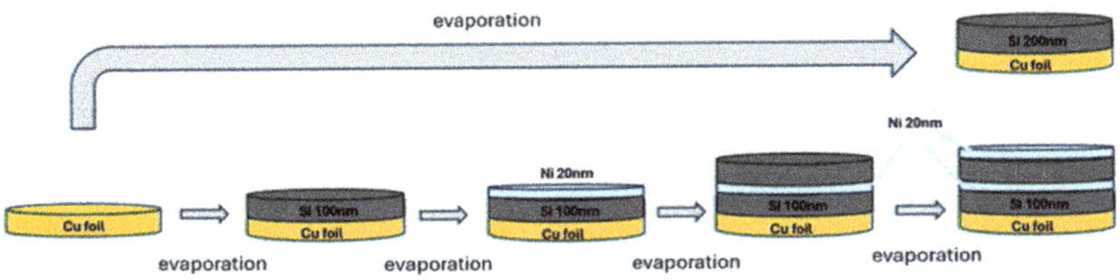

Figure 1. Schematic diagram of the fabrication processes for a Si-based anode and a Si/Ni/Si/Ni thin-film anode.

2.2. Fabrication of Coin Cells

The thin-film electrodes were assembled into CR2032 coin cells paired with lithium metal cathodes within an argon-filled glove box. Rigorous control was maintained to ensure oxygen and moisture levels remained below 0.3 ppm.

The electrolyte comprised a 1 M $LiPF_6$ solution dissolved in a 2:1:2 volume ratio of dimethyl carbonate (DMC), diethyl carbonate (DEC), and ethylene carbonate (EC). Additionally, 10 wt.% of fluoroethylene carbonate (FEC) was incorporated into the electrolyte composition.

2.3. Electrochemical Testing

Galvanostatic discharge/charge tests were performed using a BAT battery testing system within a voltage range of 0.01–1.5 V for the half-cells under the test current density of 0.05 mA/cm^2. For the C-rate stepped tests, the current density was gradually increased from 0.05 mA/cm^2 to 1.5 mA/cm^2 and then reduced back to 0.05 mA/cm^2 to evaluate electrode stability under varied current conditions. Electrochemical impedance spectroscopy (EIS) and cyclic voltammetry (CV) analyses were conducted using an Autolab instrument (Metrohm AUTOLAB BV, Taipei, Taiwan). EIS measurements were performed over a frequency range from 0.1 Hz to 100 kHz, while CV scans were conducted at room temperature with a scan rate of 0.05 mV/s.

2.4. Electrode Characterization

To analyze the electrode surface morphology before and after cycling, the coin cells were disassembled within an argon-filled glove box, and the electrodes were retrieved. The retrieved electrodes were rinsed with diethyl carbonate (DEC) to remove residual electrolyte and air-dried within the glove box. Scanning electron microscopy (SEM, Hitachi-SU8000, Taipei, Taiwan) was employed to examine the surface morphology of the electrode.

3. Results

Figure 2a,b present the XRD analysis showing that the elevated temperature during deposition caused alloying reactions between Cu and Si, as well as between Si and Ni, leading to the formation of alloy layers. Figure 2b shows a magnified XRD spectrum of Figure 2a to display the weaker signals. These alloy layers significantly improved interfacial contact between layers, functioning as an "alloy binder" to enhance the overall bonding strength of the materials. The lack of a distinct Si signal in the XRD pattern is likely due to the amorphous nature of the silicon.

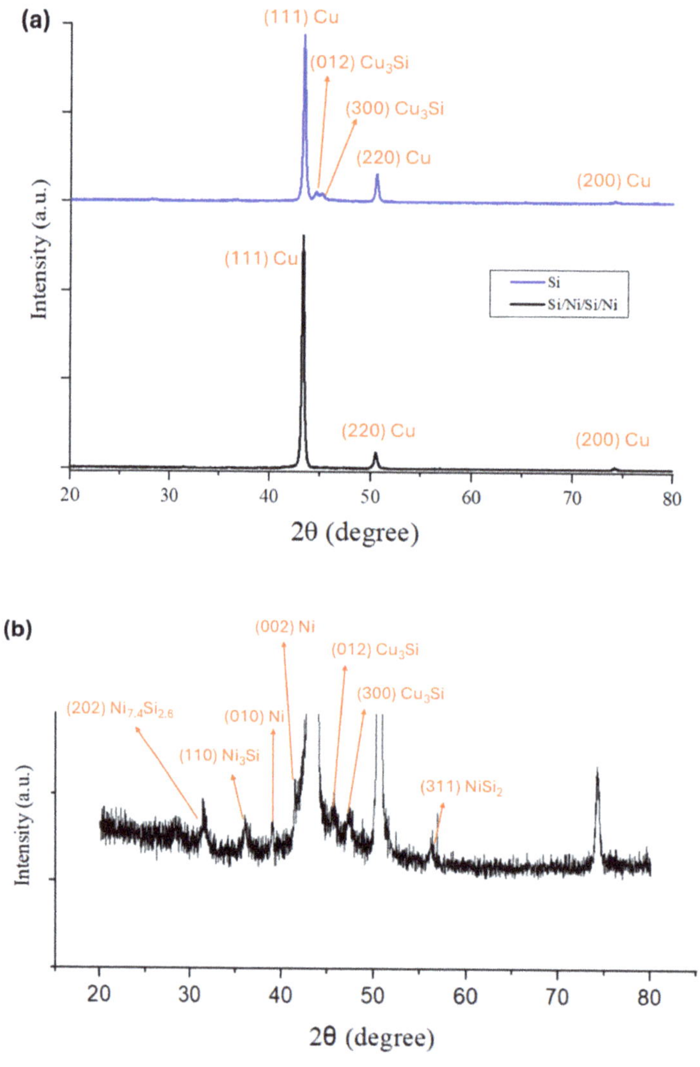

Figure 2. XRD analysis of (**a**) a Si anode and a Si/Ni/Si/Ni thin-film anode; (**b**) magnified intensity of a Si/Ni/Si/Ni thin-film anode.

Figure 3a,b present scanning electron microscopy (SEM) images of a silicon thin-film anode and a Si/Ni/Si/Ni multi-layer thin-film anode, respectively. The Si thin-film anode (Figure 3a) exhibits a nonuniform surface morphology. The Ni top layer (Figure 3b) formed relatively uniform semispherical structures and numerous pores in between and, thus, facilitated more uniform lithium-ion transport. Because nickel and its compound are more ductile than silicon, the surface passivation layer may enhance the physical integrity of the anode. This enhanced surface area with special structures improves the contact area between the electrolyte and the anode [23], facilitating efficient lithium-ion transport under a high current density [24]. Additionally, a uniform and rough surface enhances the adhesion between the solid electrolyte interphase (SEI) and the anode, reducing interfacial resistance to charge transport and contributing to an improved overall performance.

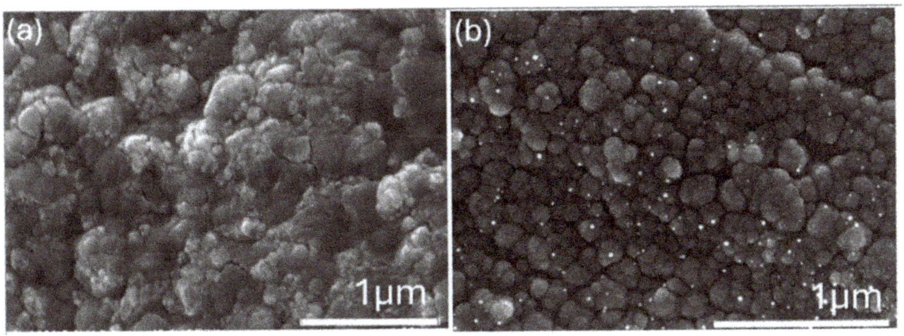

Figure 3. SEM images of the top view of (**a**) a Si thin-film anode and (**b**) a Si/Ni/Si/Ni thin-film anode.

Figure 4a,b show the cyclic voltammetry (CV) profiles of the electrodes, where the cathodic and anodic peaks correspond to lithium-ion insertion and extraction processes, respectively. The high degree of overlap in the Si/Ni/Si/Ni thin-film anode's CV curves after the first cycle indicates the excellent cycling stability and high reversible capacity of this anode. The oxidation peaks observed at approximately 0.3 V and 0.51 V (detailed values are provided in Table S2) during the de-alloying process are attributed to the phase transition from Li_xSi_y to amorphous silicon. The delithiation process of Li_xSi_y is typically a multi-step process. Peak A at 0.3V might correspond to the gradual transition from high-lithium-content alloys (such as $Li_{15}Si_4$ and $Li_{22}Si_5$) to lower-lithium-content phases (such as $Li_{13}Si_4$ and Li_7Si_3). At peak B at 0.51V, the structure gradually transformed into a low-lithium-content silicon phase ($Li_{3.75}Si$) or amorphous silicon.

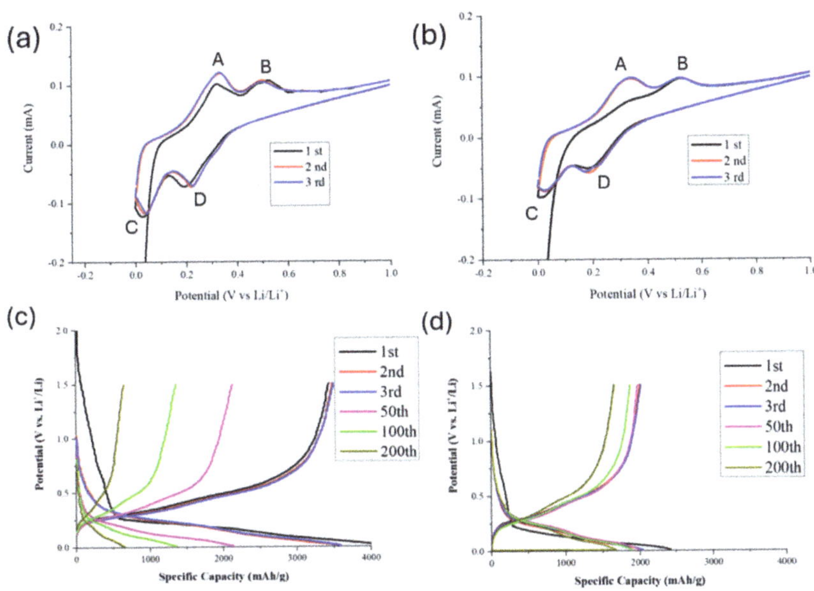

Figure 4. CV curve of (**a**) a Si thin-film anode and (**b**) a Si/Ni/Si/Ni thin-film anode; charge–discharge curves of different cycling numbers for the (**c**) Si thin-film anode and (**d**) the Si/Ni/Si/Ni thin-film anode.

During alloying, lithium ions were inserted into amorphous silicon, resulting in the formation of different amorphous $L_{ix}Si_y$ phases at around 0.03 V and 0.19 V, as indicated by the two reduction peaks. At peak D of 0.19 V, $Li_{12}Si_7$ was formed. Higher Li phases (such as $Li_{15}Si_4$, $Li_{21}Si_5$) were formed at peak C at approximately 0.02V [25].

$$Li_xSi_y \rightarrow xLi^+ + e^- + ySi \quad (1)$$

$$xLi^+ + e^- + ySi \rightarrow Li_xSi_y \quad (2)$$

As shown in Figure 4c,d, the first cycle charge capacity of the Si thin-film anode (3438 mAh/g) was nearly 1.5 times that of the Si/Ni/Si/Ni thin-film anode (1992 mAh/g) (Figure 4d). The reduced capacity in the Si/Ni/Si/Ni anode can be attributed to the presence of Ni layers, which have a higher density (8.9 g/cm³) than Si (2.33g/ cm³). The increased mass loading of Ni reduced the overall specific capacity.

The Ni interfacial layer and the surface layer play significant roles in stabilizing the thin-film Si anode. Zhang et al. reported that a silicon anode having been subjected to prolonged exposure to high potentials led to the formation of a soft and thick SEI layer and made it more prone to cracking after multiple cycles of charging and discharge [26]. Figure 4c,d, show that a slow decline of the first-cycle lithiation curve for a thin-film Si anode is correlated with the substantial growth of a thick SEI layer on the Si thin-film anode. SEI formed by reactions of Si with the electrolyte is mostly irreversible. This gradual voltage drop during the initial lithiation stage is attributed to the significant lithium-ion consumption, which is required for the substantial SEI formation and buildups. This slow decline is in clear contrast with the fast decline of the first-cycle lithiation curve for a Si/Ni/Si/Ni anode, which exhibits a much thinner SEI.

Joshi et al. showed that transition metals accelerate the growth of inorganic components within the SEI layer [27]. Inorganic phases exhibit high elastic moduli [28], and contribute to a more mechanically robust SEI layer, which suppresses crack formation and reduces the exposure of new silicon surfaces. Excessive growth of the SEI on fresh silicon surfaces is, thus, minimized. Moreover, the intermetallic compound formed by Ni with Si acts as a binder to prevent silicon from cracking and an electrochemically inactive passivation layer for protecting silicon from reactions with the electrolyte. This inactive interlayer does not participate in electrochemical reactions with the electrolyte and remains dimensionally stable during repeated electrochemical cycling [29], further improving the physical integrity of the Si/Ni/Si/Ni anode and enhancing the long-term stability.

Figure 5a illustrates the cycling performance of two anodes over 200 cycles. The Si thin-film anode achieved a higher initial Coulombic efficiency (ICE) of 82.6% but suffered substantial capacity fading and retained only 18.7% of its initial capacity after 200 cycles (645 mAh/g). This corresponds to a capacity decay rate of 0.41% per cycle. In contrast, the Si/Ni/Si/Ni thin-film anode (ICE = 82.2%) exhibited a lower initial capacity but exceptional cycling stability and retained 82% of its capacity after 200 cycles (1647 mAh/g) with a significantly lower decay rate of 0.09% per cycle. The superior cycling performance of the Si/Ni/Si/Ni anode can be attributed to the dual function of the Ni layer, which acts as an electrochemical barrier for preventing direct contact between Si and the electrolyte, while also providing mechanical support to mitigate Si volume changes during cycling, thereby preserving the electrode's structural integrity.

Figure 5b presents the C-rate step testing results for the Si and Si/Ni/Si/Ni thin-film anodes. The pristine silicon thin-film anode exhibited a significant capacity reduction of 67% (from 3457 mAh/g to 1120 mAh/g) as the current density increased from 0.25 mA/cm² to 1.5 mA/cm². In comparison, the Si/Ni/Si/Ni multi-layer anode demonstrated a smaller capacity decrease of 49.6% (from 1959 mAh/g to 985 mAh/g). This result highlights the effectiveness of the nickel layers in enhancing the electrode's surface area and uniformity, enabling faster and even lithium-ion insertion and extraction during high-rate cycling and improving the high-rate performance.

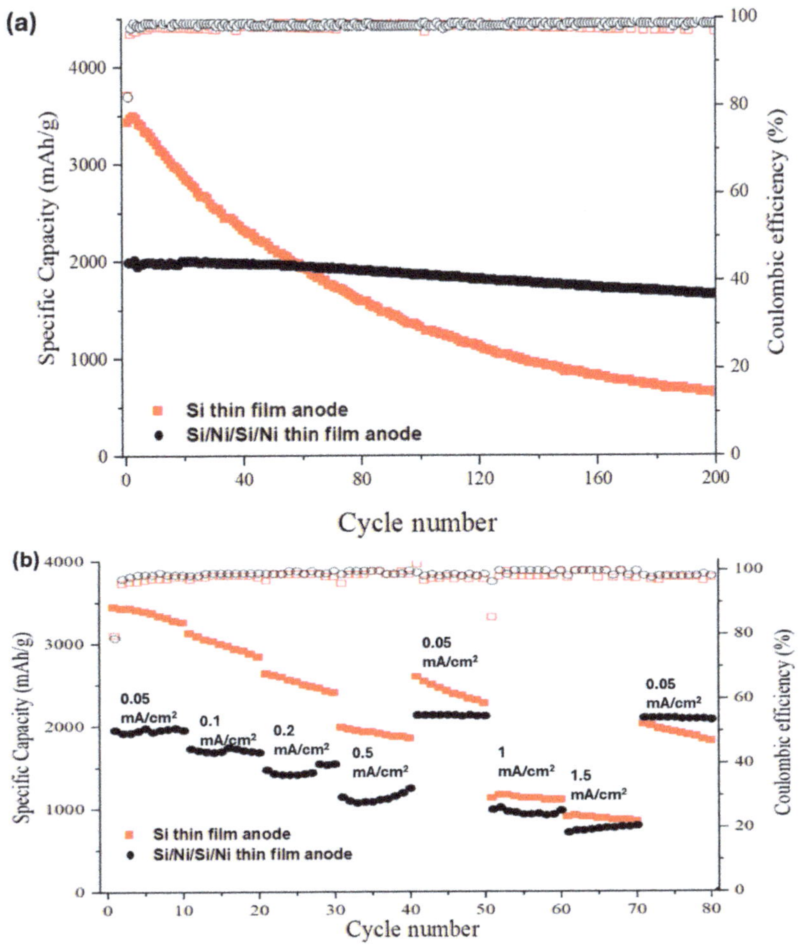

Figure 5. (**a**) Cycle performance of a Si thin-film anode and a Si/Ni/Si/Ni thin-film anode under test current density of 0.05 mA/cm^2; (**b**) C-rate performance of a Si thin-film anode and a Si/Ni/Si/Ni thin-film anode.

When the current density returned to 0.05 mA/cm^2, the Si/Ni/Si/Ni thin-film anode (2076 mAh/g) outperformed the pristine silicon thin-film anode (1823 mAh/g). This confirms that the Si/Ni/Si/Ni multi-layer structure not only suppresses the adverse effects of volume changes but also provides enhanced protection and stability for the electrode.

As depicted in Figure 6a, for the Si thin-film anode, the electrochemical reaction appears uneven, leading to the detachment of the Si film from the current collector. Further evidence is shown in Figures S2 and S3. In contrast, the Si/Ni/Si/Ni thin-film anode (Figure 6b) exhibits uniform electrochemical reactions. The Ni layer effectively improves the uniformity of Li-ion transport and maintains the physical integrity of the anode.

Figure 7a demonstrates the distinct multi-layer structure of the Si/Ni/Si/Ni thin-film anode, with a total thickness of approximately 321 nm. Figure 7b shows the cross-sectional SEM image of the Si/Ni/Si/Ni electrode after 100 cycles of discharge and charge operations under a high current density. Although some localized uneven expansion of the silicon layer was observed, the overall structure remained intact. This indicates that the nickel layer successfully mitigated the stress induced by the volume changes of silicon during cycling.

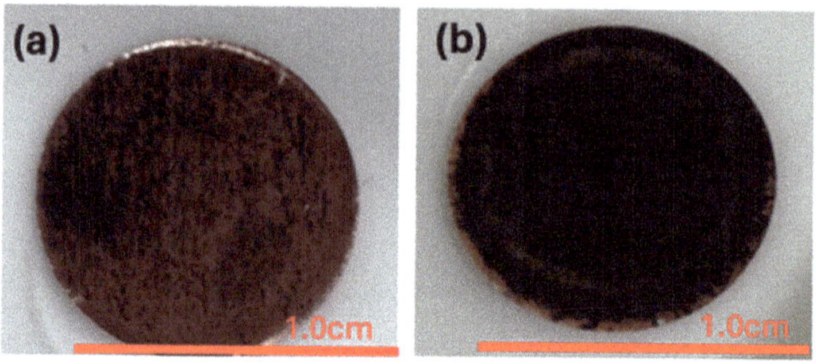

Figure 6. Optical microscope images of (**a**) an anode made of Si thin film and (**b**) an anode made of multi-layer Si/Ni/Si/Ni structure after 100 cycles of charge–discharge.

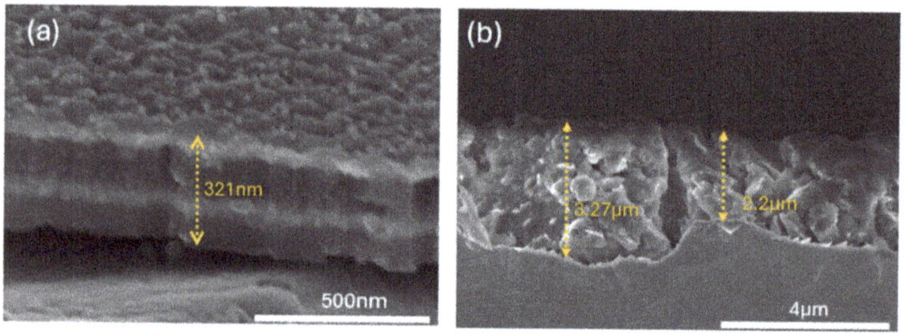

Figure 7. SEM cross-sectional images of Si/Ni/Si/Ni thin film anode (**a**) before and (**b**) after 100 cycles of discharge and charge operations under 1 mA/cm^2.

Figure 8a–c show that the surface of the Si thin-film anode developed significant cracks after 200 cycles. As depicted in the magnified view in Figure 8b,c, these cracks had widths of approximately 4–5 μm. Their formation is attributed to the substantial volume changes of silicon during alloying and de-alloying with lithium. The volume changes generate internal stress. These cracks increase the electrode's surface area, accelerating side reactions, causing the anode to lose electrical contact with the current collector and between neighboring silicon islands. This hinders electron transport as well as lithium-ion transport, leading to a declining capacity.

Figure 8d–f shows that the surface morphology of the Si/Ni/Si/Ni thin-film anode remained relatively smooth overall, with localized cracks approximately 1 μm wide. While the Ni layer effectively mitigates large-scale crack formation, nonuniform lithium-ion transport can still cause stress concentration in certain regions, leading to small cracks. These cracks were primarily confined to specific "hotspots", accounting for about 6% of the total surface area. Although limited in quantity, these cracks could accelerate electrolyte decomposition and gradually grow larger, with more cracks causing the long-term decline of the capacity.

X-ray photoelectron spectroscopy (XPS) analysis was used to examine the surface chemical composition. The protective role of the Ni layers and the composition of the solid electrolyte interphase (SEI) are emphasized in the analysis.

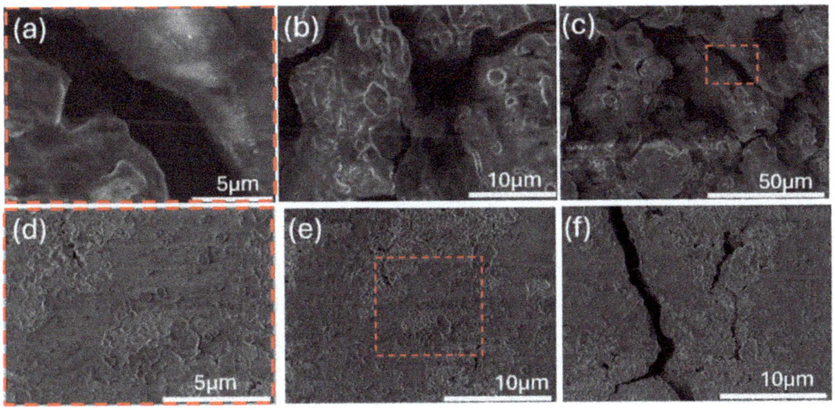

Figure 8. SEM images of a Si thin-film anode at (**a**–**c**) different magnifications after 200 cycles; SEM images of a Si/Ni/Si/Ni thin-film anode (**d**,**e**) displayed after different magnifications; (**f**) shows a different spot from that shown in (**e**) after 200 cycles.

Figure 9a,b show the C 1s XPS spectra, where the peak at 284.7 eV corresponds to C-C bonds, indicating conductive carbon, and the peak at 285.9 eV corresponds to CO_3, which is typically attributed to electrolyte decomposition. The Si/Ni/Si/Ni thin-film anode shows a notably lower peak intensity at 285.9 eV compared to the Si thin-film anode, highlighting the Ni layer's effectiveness in suppressing electrolyte decomposition and reinforcing its role as a protective barrier [30].

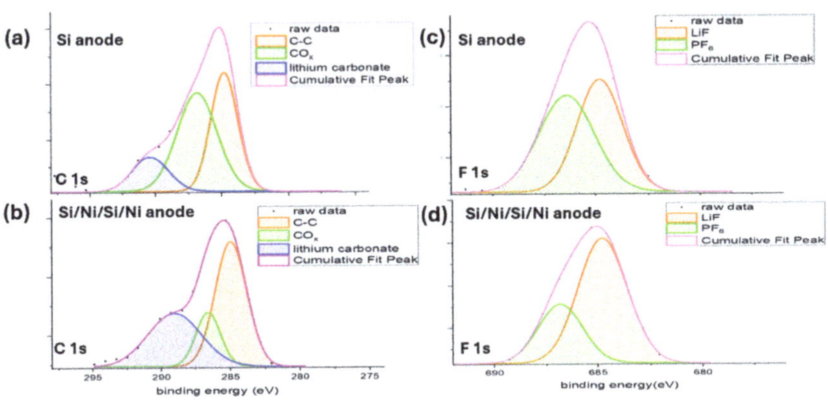

Figure 9. (**a**,**b**) C 1s and (**c**,**d**) F 1s X-ray photoelectron spectroscopy (XPS) spectra for two different electrodes after 200 cycles of charge and discharge.

Figure 9c,d show the F 1s XPS spectra, with a peak at 687 eV corresponding to PF_6^- and another at 685 eV corresponding to LiF [31,32]. The presence of LiF, a common inorganic SEI component, indicates SEI formation on the electrode surface. Known for its excellent mechanical stability, LiF suppresses electrolyte decomposition and enhances interfacial stability, thereby improving the cycling performance [33,34]. The higher LiF content observed in the Si/Ni/Si/Ni thin-film anode compared to the Si thin-film anode suggests that the Ni layer facilitates the formation of a more stable SEI, contributing to superior cycling stability.

Figure 8a illustrates significant cracks in the Si thin-film anode after cycling. The EIS analysis shown in Figure 10 indicates that these cracks contributed to an increase in both the internal resistance (R1) and charge transfer resistance (R2). As Table 1 shows, the Si thin-film anode had a larger R1 and R2. This is attributed to the enlarged electrode/electrolyte interfacial area caused by cracks, which accelerate the side reactions and additional growth of the SEI, and impede efficient Li-ion transport. Warburg Resistance (WR), which is the solid-state diffusion resistance for the silicon anode, was also larger than that of multi-layer Si/Ni/Si/Ni anode. This is attributed to the large surface area of the electrochemically inactive Ni layer, which results in a more stable SEI and enhances the electrode integrity.

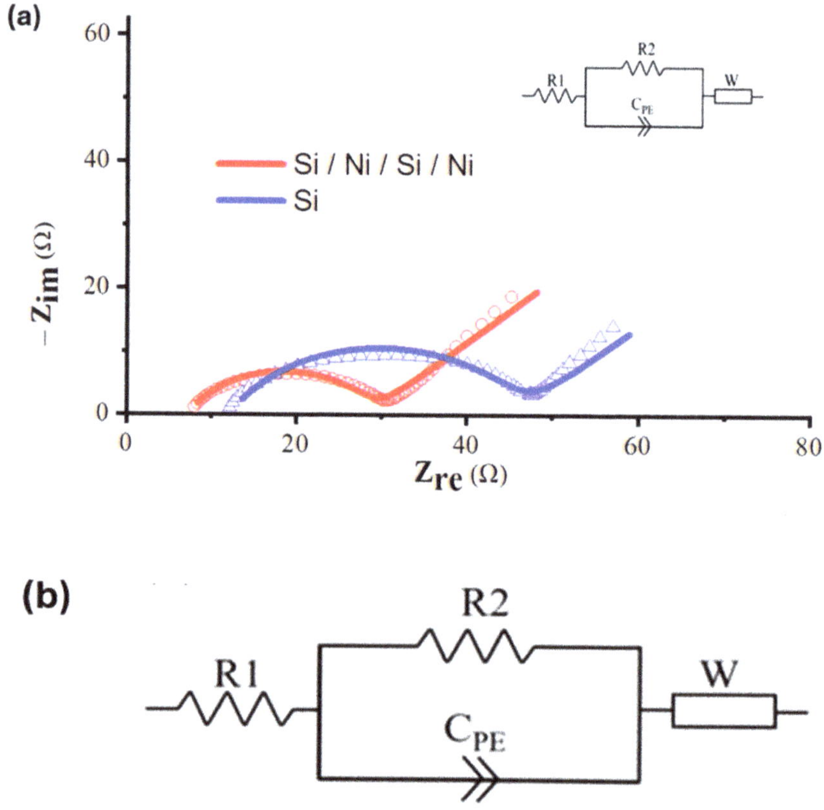

Figure 10. (**a**) The Nyquist plots comparing the impedance spectra of Si and Si/Ni/Si/Ni thin-film anode after 200 cycles; (**b**) the equivalent circuit model used to fit the impedance data.

Table 1. Comparative fitting results of EIS spectra for a Si thin-film anode and a Si/Ni/Si/Ni thin-film anode after 200 cycles of electrochemical cycling.

Sample/Resistance	R1(Ω)	R2(Ω)	WR(Ω)
Si/Ni/Si/Ni thin-film anode	7.5	21.3	32.2
Si thin-film anode	13	32.9	59.7

4. Conclusions

A Si/Ni/Si/Ni multi-layer anode was studied. The Ni top layer (Figure 3b) formed relatively uniform semispherical structures and numerous pores in between and, thus, facilitated more uniform lithium-ion transport and prevented a local high lithium-ion current density and internal stress due to the irregular volume changes of the silicon. Because nickel and its compound are more ductile than silicon, the nickel-based surface passivation layer may enhance the physical integrity of the anode. The electrochemically inert nickel layer and the compounds it forms during cycling act as a protective barrier, preventing direct contact between the electrolyte and the silicon while suppressing undesirable side reactions. Therefore, the multi-layer anode with an interlayer and a surface layer of nickel effectively accommodates silicon's substantial volume changes during charge and discharge cycling and enhances the battery's cycle life. The multi-layer anode demonstrated a significantly improved capacity retention of 82% after 200 cycles of charge–discharge. The Si/Ni/Si/Ni anode also demonstrated superior performance in C-rate tests compared to the Si anode under a high test current density.

Supplementary Materials: The following supporting information can be downloaded at: https://www.mdpi.com/article/10.3390/batteries11020046/s1, Table S1: Comparison of Si- and Sn-based material [35]. Table S2: Detailed peak values in CV curves; Figure S1: (a) SEM image and images of (b) Si EDS mapping and (c) Ni EDS mapping of a Si (100 nm)/Ni (20 nm)/Si (100 nm)/Ni (20 nm) anode; Figure S2: SEM cross-sectional image of Si (200 nm) anode before cycling; Figure S3: Cross-sectional EDS element content of Si (200 nm) anode after 100 cycles under 1 mA/cm^2. Figure S4: Cyclic performance of Si (200 nm), Si (200 nm)/Ni (20 nm), Si (100 nm)/Ni (20 nm)/Si (100 nm)/Ni (20 nm), Si (66 nm)/Ni (6.6 nm)/Si (66 nm)/Ni (6.6 nm)/Si (66 nm)/Ni (6.6 nm), and Si (100 nm)/Ni (10 nm)/Si (100 nm)/Ni (10 nm) under 0.05 mA/cm^2 test current. Figure S5: Cyclic performance of Si (100 nm)/Ni (10 nm)/Si (100 nm)/Ni (10 nm) with different Si thickness under 0.05 mA/cm^2.

Author Contributions: Y.T.: Conception, funding, data analysis, interpretation, final manuscript preparation; Y.-Y.C.: Material characterization, data analysis, device fabrication, data collection, initial data analysis, report preparation draft manuscript preparation; A.A.W.: Experimental assistance, data collection and verification, assistance to report preparation. All authors have read and agreed to the published version of the manuscript.

Funding: This research was funded by National Science and Technology Council, Taiwan under the grant number 113-2221-E-006-080-.

Data Availability Statement: The original contributions presented in this study are included in the article and Supplementary Materials. Further inquiries can be directed to the corresponding author.

Acknowledgments: The authors gratefully acknowledge the use of the scanning electron micro-scope (SEM, Hitachi-SU8000, Taipei, Taiwan) and X-ray photoelectron spectrometer (PHI Versa Probe 4, Taipei, Taiwan) belonging to the Core Facility Center of National Cheng Kung University.

Conflicts of Interest: The authors declare that they have no competing financial interests or personal relationships that appear to have influenced the work reported in this paper.

References

1. Miao, Y.; Hynan, P.; Von Jouanne, A.; Yokochi, A. Current Li-ion battery technologies in electric vehicles and opportunities for advancements. *Energies* **2019**, *12*, 1074. [CrossRef]
2. Gu, M.; He, Y.; Zheng, J.; Wang, C. Nanoscale silicon as anode for Li-ion batteries: The fundamentals, promises, and challenges. *Nano Energy* **2015**, *17*, 366–383. [CrossRef]
3. Ko, M.; Chae, S.; Cho, J. Challenges in accommodating volume change of Si anodes for Li-ion batteries. *ChemElectroChem* **2015**, *2*, 1645–1651. [CrossRef] [PubMed]
4. Hwang, T.H.; Lee, Y.M.; Kong, B.S.; Seo, J.S.; Choi, J.W. Electrospun core–shell fibers for robust silicon nanoparticle-based lithium ion battery anodes. *Nano Lett.* **2012**, *12*, 802–807. [CrossRef] [PubMed]
5. Ko, M.; Chae, S.; Jeong, S.; Oh, P.; Cho, J. Elastic a-silicon nanoparticle backboned graphene hybrid as a self-compacting anode for high-rate lithium ion batteries. *ACS Nano* **2014**, *8*, 8591–8599. [CrossRef]
6. Wen, Z.; Lu, G.; Mao, S.; Kim, H.; Cui, S.; Yu, K.; Huang, X.; Hurley, P.; Mao, O.; Chen, J. Silicon nanotube anode for lithium-ion batteries. *Electrochem. Commun.* **2013**, *29*, 67–70. [CrossRef]
7. Park, M.H.; Kim, M.G.; Joo, J.; Kim, K.; Kim, J.; Ahn, S.; Cui, Y.; Cho, J. Silicon nanotube battery anodes. *Nano Lett.* **2009**, *9*, 3844–3847. [CrossRef]

8. Ge, M.; Rong, J.; Fang, X.; Zhou, C. Porous doped silicon nanowires for lithium ion battery anode with long cycle life. *Nano Lett.* **2012**, *12*, 2318–2323. [CrossRef]
9. Chan, C.K.; Peng, H.; Liu, G.; McIlwrath, K.; Zhang, X.F.; Huggins, R.A.; Cui, Y. High-performance lithium battery anodes using silicon nanowires. *Nat. Nanotechnol.* **2008**, *3*, 31–35. [CrossRef]
10. Abel, P.R.; Lin, Y.M.; Celio, H.; Heller, A.; Mullins, C.B. Improving the stability of nanostructured silicon thin film lithium-ion battery anodes through their controlled oxidation. *ACS Nano* **2012**, *6*, 2506–2516. [CrossRef]
11. Bates, J.B.; Dudney, N.J.; Neudecker, B.; Ueda, A.; Evans, C.D. Thin-film lithium and lithium-ion batteries. *Solid State Ionics* **2000**, *135*, 33–45. [CrossRef]
12. Zhou, M.; Cai, T.; Pu, F.; Chen, H.; Wang, Z.; Zhang, H.; Guan, S. Graphene/carbon-coated Si nanoparticle hybrids as high-performance anode materials for Li-ion batteries. *ACS Appl. Mater. Interfaces* **2013**, *5*, 3449–3455. [CrossRef] [PubMed]
13. Luo, F.; Liu, B.; Zheng, J.; Chu, G.; Zhong, K.; Li, H.; Huang, X.; Chen, L. Nano-silicon/carbon composite anode materials towards practical application for next generation Li-ion batteries. *J. Electrochem. Soc.* **2015**, *162*, A2509. [CrossRef]
14. Tzeng, Y.; Jhan, C.Y.; Sung, S.H.; Chiou, Y.Y. Effects of Crystalline Diamond Nanoparticles on Silicon Thin Films as an Anode for a Lithium-Ion Battery. *Batteries* **2024**, *10*, 321. [CrossRef]
15. Sethuraman, V.A.; Kowolik, K.; Srinivasan, V. Increased cycling efficiency and rate capability of copper-coated silicon anodes in lithium-ion batteries. *J. Power Sources* **2011**, *196*, 393–398. [CrossRef]
16. Luo, W.; Chen, X.; Xia, Y.; Chen, M.; Wang, L.; Wang, Q.; Li, W.; Yang, J. Surface and interface engineering of silicon-based anode materials for lithium-ion batteries. *Adv. Energy Mater.* **2017**, *7*, 1701083. [CrossRef]
17. Chen, Z.; Soltani, A.; Chen, Y.; Zhang, Q.; Davoodi, A.; Hosseinpour, S.; Peukert, W.; Liu, W. Emerging organic surface chemistry for Si anodes in lithium-ion batteries: Advances, prospects, and beyond. *Adv. Energy Mater.* **2022**, *12*, 2200924. [CrossRef]
18. Li, H.; Bai, H.; Tao, Z.; Chen, J. Si–Y multi-layer thin films as anode materials of high-capacity lithium-ion batteries. *J. Power Sources* **2012**, *217*, 102–107. [CrossRef]
19. Kim, J.B.; Lee, H.Y.; Lee, K.S.; Lim, S.H.; Lee, S.M. Fe/Si multi-layer thin film anodes for lithium rechargeable thin film batteries. *Electrochem. Commun.* **2003**, *5*, 544–548. [CrossRef]
20. Tong, L.; Wang, P.; Chen, A.; Qiu, F.; Fang, W.; Yang, J.; Wang, C.; Yang, Y. Improved electrochemical performance of binder-free multi-layered silicon/carbon thin film electrode for lithium-ion batteries. *Carbon* **2019**, *153*, 592–601. [CrossRef]
21. Salah, M.; Hall, C.; Yap, P.L.; Fabretto, M. Silicon-tin thin-film anodes for low and high power-density lithium-ion batteries. *Thin Solid Films* **2024**, *796*, 140332. [CrossRef]
22. Fu, H.; Gu, F.; Niu, Y.; Liao, S.; Bu, Z.; Wang, H.; Yang, D.; Wang, X.; Li, Q. Spatially confined transition metals boost high initial coulombic efficiency in alloy anodes. *Chem. Sci.* **2024**, *16*, 418–424. [CrossRef] [PubMed]
23. Wang, K.-X.; Li, X.-H.; Chen, J.-S. Surface and interface engineering of electrode materials for lithium-ion batteries. *Adv. Mater.* **2015**, *27*, 527–545. [CrossRef] [PubMed]
24. Song, R.; Song, H.; Zhou, J.; Chen, X.; Wu, B.; Yang, H.Y. Hierarchical porous carbon nanosheets and their favorable high-rate performance in lithium-ion batteries. *J. Mater. Chem.* **2012**, *22*, 12369–12374. [CrossRef]
25. Deng, L.; Cui, Y.; Chen, J.; Wu, J.; Baker, A.P.; Li, Z.; Zhang, X. A core-shell Si@ NiSi2/Ni/C nanocomposite as an anode material for lithium-ion batteries. *Electrochim. Acta* **2016**, *192*, 303–309. [CrossRef]
26. Zhang, W.; Cai, T.H.; Sheldon, B.W. The impact of initial SEI formation conditions on strain-induced capacity losses in silicon electrodes. *Adv. Energy Mater.* **2019**, *9*, 1803066. [CrossRef]
27. Joshi, T.; Eom, K.; Yushin, G.; Fuller, T.F. Effects of dissolved transition metals on the electrochemical performance and SEI growth in lithium-ion batteries. *J. Electrochem. Soc.* **2014**, *161*, A1915. [CrossRef]
28. Tokranov, A.; Kumar, R.; Li, C.; Minne, S.; Xiao, X.; Sheldon, B.W. Control and optimization of the electrochemical and mechanical properties of the solid electrolyte interphase on silicon electrodes in lithium ion batteries. *Adv. Energy Mater.* **2016**, *6*, 1502302. [CrossRef]
29. Huang, X.; Pu, H.; Chang, J.; Cui, S.; Hallac, P.B.; Jiang, J.; Hurley, P.; Chen, J. Improved cyclic performance of Si anodes for lithium-ion batteries by forming intermetallic interphases between Si nanoparticles and metal microparticles. *ACS Appl. Mater. Interfaces* **2013**, *5*, 11965–11970. [CrossRef]
30. Jaumann, T.; Balach, J.; Klose, M.; Oswald, S.; Langklotz, U.; Michaelis, A.; Eckert, J.; Giebeler, L. SEI-component formation on sub 5 nm sized silicon nanoparticles in Li-ion batteries: The role of electrode preparation, FEC addition and binders. *Phys. Chem. Chem. Phys.* **2015**, *17*, 24956–24967. [CrossRef]
31. Etacheri, V.; Haik, O.; Goffer, Y.; Roberts, G.A.; Stefan, I.C.; Fasching, R.; Aurbach, D. Effect of fluoroethylene carbonate (FEC) on the performance and surface chemistry of Si-nanowire Li-ion battery anodes. *Langmuir* **2012**, *28*, 965–976. [CrossRef] [PubMed]
32. Philippe, B.; Dedryvère, R.; Allouche, J.; Lindgren, F.; Gorgoi, M.; Rensmo, H.; Gonbeau, D.; Edström, K. Nanosilicon electrodes for lithium-ion batteries: Interfacial mechanisms studied by hard and soft X-ray photoelectron spectroscopy. *Chem. Mater.* **2012**, *24*, 1107–1115. [CrossRef]

33. Tan, J.; Matz, J.; Dong, P.; Shen, J.; Ye, M. A growing appreciation for the role of LiF in the solid electrolyte interphase. *Adv. Energy Mater.* **2021**, *11*, 2100046. [CrossRef]
34. Kim, K.H.; Cho, J.H.; Hwang, J.U.; Im, J.S.; Lee, Y.S. A key strategy to form a LiF-based SEI layer for a lithium-ion battery anode with enhanced cycling stability by introducing a semi-ionic CF bond. *J. Ind. Eng. Chem.* **2021**, *99*, 48–54. [CrossRef]
35. Li, W.; Sun, X.; Yu, Y. Si-, Ge-, Sn-based anode materials for lithium-ion batteries: From structure design to electrochemical performance. *Small Methods.* **2017**, *1*, 1600037. [CrossRef]

batteries

MDPI

Article

Improved Self-Assembled Silicon-Based Graphite Composite Anodes for Commercially Viable High-Energy-Density Lithium-Ion Batteries

Ruye Cong [1], Da-Eun Jeong [2], Ye-Yeong Jung [2], Hyun-Ho Park [2], Jiyun Jeon [3], Hochun Lee [3] and Chang-Seop Lee [2,*]

[1] School of Chemical & Environmental Engineering, Liaoning University of Technology, Jinzhou 121001, China; hgcongruye@lnut.edu.cn

[2] Department of Chemistry, Keimyung University, Daegu 42601, Republic of Korea; de0602@ms.kmu.ac.kr (D.-E.J.); dpdud525@stu.kmu.ac.kr (Y.-Y.J.); rubchem@kmu.ac.kr (H.-H.P.)

[3] Department of Energy Science and Engineering, DGIST, Daegu 42988, Republic of Korea; jiyunn1234@dgist.ac.kr (J.J.); dukelee@dgist.ac.kr (H.L.)

* Correspondence: surfkm@kmu.ac.kr; Tel.: +82-53-580-5192; Fax: +82-53-580-5056

Abstract: Silicon-based anode materials are used to improve the performance of next-generation high-energy-density lithium-ion batteries (LIBs). However, the inherent limitations and cost of these materials are hindering their mass production. Commercial graphite can overcome the shortcomings of silicon-based materials and partially reduce their cost. In this study, a high-performance, low-cost, and environmentally friendly composite electrode material suitable for mass production was developed through optimizing the silicon content of commercial silicon–graphite composites and introducing a small amount of graphene and carbon nanofibers. This partially overcomes the inherent limitations of silicon, enhances the interface stability of silicon-based materials and the cycle stability of batteries, and reduces the irreversible capacity loss of the initial cycle. At a silicon content of 15 wt%, the initial Coulombic efficiency (ICE) of the battery was 65%. Reducing the silicon content in the composite electrode from 15% to 10% increased the ICE to 70% and improved the first lithiation and delithiation capacities. The battery exhibited excellent cycle stability at a current density of 0.1 A g^{-1}, retaining approximately 65% of its capacity after 100 cycles, good performance at various current densities (0.1–1 A g^{-1}), and an excellent reversible performance.

Keywords: lithium-ion batteries; silicon; graphite; anode materials; commercialization

1. Introduction

With the continuous advancement of science and technology and the increasing demand for clean energy, lithium-ion batteries (LIBs), which are efficient and environmentally friendly energy storage devices, have become an important part of the future energy field. LIB technology is continuously evolving to meet the demands of the growing electric vehicle and renewable energy markets [1,2]. However, LIBs that use traditional graphite anodes can no longer meet the increasing requirements of a high energy density, power density, and safety performance, owing to the low capacity of graphite (372 mAh g^{-1}). Therefore, it is important to develop high-capacity electrode materials. Silicon (Si) has attracted considerable attention as an electrode material owing to its high specific capacity (approximately 4200 mAh g^{-1}), low charge and discharge potential (<0.4 V vs. Li/Li^+), environmental friendliness, and abundant reserves. It is considered the most promising among the new generation of negative-electrode materials. Nevertheless, the alloying and dealloying reactions of Si materials during the charge–discharge process are accompanied by huge volume changes (>300%), causing damage to the electrode structure, the pulverization of Si particles, the repeated formation of the solid electrolyte interface (SEI), and the shedding of the active materials during the cycle, leading to a considerable decrease in the performance of the Si anode. In addition, the inherent low electrical conductivity of Si particles (10^{-5} S cm^{-1}) and slow lithium-ion (Li^+) diffusion kinetics (10–14 cm^{-2} S^{-1}) greatly hinder the widespread application of Si materials in high-capacity LIBs [3–5]. Therefore, reducing the internal stress caused by volume expansion and improving the ionic conductivity of the material to achieve long-term cycle stability present urgent research challenges.

To address the above challenges, new modified Si-based anodes were synthesized. These materials are constructed with (1) various nanostructures and voids in the Si main body to relieve internal stress and accommodate huge volume changes, and with (2) three-dimensional structures to form effective channels for Li^+ transmission, shortening the Li^+ diffusion length and improving the Li^+ diffusion kinetics [6,7]. Depositing Si layers on carbon surfaces to construct porous Si–C anodes with carbon nanotubes embedded into micron-size Si and Si-active materials incorporated into carbonaceous products has been extensively studied because carbon materials can minimize electrode deformation and considerably increase Si conductivity [8–10]. Researchers have extensively worked toward improving the performance of Si-based batteries. However, several challenges that hinder their use in practical applications arise when moving these batteries from the research phase to the production phase. For instance, high material costs and high technical difficulties have become major problems. The relatively high production cost needs to be effectively solved to meet the needs of large-scale commercial production, and the complexity of the preparation process requires us to continuously improve the structural and electrochemical stability of the material to overcome the technical difficulties. Previous studies have mainly focused on developing new materials and improving existing material structures. However, the battery cycle life, safety, and cost challenges still hinder the use of these materials in commercial applications. In the face of these challenges, future research could focus on reducing costs, optimizing processes, and improving stability, thereby promoting the industrialization of silicon–carbon negative-electrode materials. The development of a high-performance LIB-negative electrode that can be mass-produced remains a significant research challenge [11–13].

Herein, we consider the challenges hindering the use of Si-based batteries in practical applications, discussing, in particular, practical battery designs with improved electrode expansion. The development path of silicon-based negative-electrode materials mainly revolves around silicon–carbon and silicon–oxygen composite materials. The preparation of silicon–carbon negative-electrode materials is simple and can be achieved via a range of methods (chemical vapor deposition (CVD) method, mechanical ball-milling method, spray method, magnesium thermal reduction method, sol–gel method, and thermal decomposition method). This material has an excellent theoretical capacity, and could effectively improve the energy density of batteries, so it is widely used. Graphite is widely used in silicon–carbon negative-electrode materials. Graphite is a low-cost commercial carbon material with a high tap density, low surface area, and stable physical and chemical properties, and can considerably reduce the unnecessary decomposition of the electrolyte on the electrode surface while maintaining satisfactory anode integrity. M.G. et al. [14] investigated the relationship between the percentage of Si in a graphite anode and the realization of high-energy-density LIBs. They found that when the electrode composition was Si1$_5$Gr$_{75}$, the cell delivered a high ICE of approximately 82.9%, nearly equivalent to that achieved with a pure graphite anode. Furthermore, the $Si_{15}Gr_{75}$ Li cell exhibited excellent cyclic stability at a current rate of 0.5 C, retaining approximately 60% of its capacity after 215 cycles. This study paves the way for the development of high-energy-density LIBs by providing valuable insights into the optimization of Si–Gr composite anodes for commercial applications. Owing to the low expansion rate (≤20%) and small free space of graphite-based batteries, an acceptable energy density can be achieved despite their low theoretical specific capacity. However, owing to the larger electrode expansion of Si-based batteries, they require a larger free space than graphite batteries, which reduces the energy density and the advantage of the high specific capacity of Si-based anodes [15–17]. Therefore, designing a composite anode by introducing Si–carbon nanomaterials into a mature graphite backbone could be an effective method for mitigating the inherent shortcomings of Si-based and graphite-based anodes and obtaining a satisfactory specific capacity and cycle stability. Graphene is considered an efficient coating material for preparing LIBs owing to its unique structure, high electrical conductivity, high theoretical surface area, and other exceptional properties, providing it with significant potential for energy storage applications. Coating silicon with graphene can slow down the volume change of silicon and form a stable SEI film. However, the introduction of graphene presents another challenge. During the electrode preparation process, the graphene layers can be easily stacked, and it is difficult to ensure the uniform dispersion of the Si nanoparticles (SiNPs) on the graphene surface. The insufficient interpenetration of Li^+ between the graphene layers causes the diffusion distance of Li^+ through the graphene interlayer channel to increase with the superposition of the electrode size, reducing the Li^+ storage performance of the graphene electrode. We discovered that carbon nanofibers (CNFs) can be used to overcome this challenge. CNFs have sp^2 hybrid orbitals and a large specific surface area, and they exhibit excellent conductivity, chemical stability, heat resistance, and electrical conductivity, allowing their use in numerous fields [1,11,18–20]. Moreover, CNFs have excellent flexibility and structural stability, which can mitigate the Si–graphite negative-electrode problem. In contrast to previous studies, we are focusing on shifting the research on Si–graphite anodes toward practical applications to help achieve the widespread application of Si–graphite anodes in the next generation of mass-produced LIBs.

Herein, a low-cost, environmentally friendly, safe, stable, and high-performance Si–graphite composite electrode produced through a low-cost and simple synthesis process is proposed. In this study, SiNPs were evenly dispersed in an ethanol solution and combined with graphene oxide (GO) through simple physical processes and self-assembly. Next, a small amount of CNFs were dispersed in the mixture to form a multi-channel three-dimensional structure. Finally, the Si@carbon system was introduced into the graphite main chain to form a Si–graphite composite material. The final product was then obtained through multiple processes, such as drying and carbonization. This synthesis strategy does not involve an acid–base treatment, the use of various organic solvents, or a complex synthesis process. Thus, it reduces environmental pollution and experimental costs, avoids adverse effects on the human body, and improves the material performance.

2. Materials and Methods

2.1. Preparing Ni–Mo Binary Catalyst for the Synthesis of CNFs via Chemical Vapor Deposition

Ni–Mo (molar ratio = 6:4) bimetallic catalysts were prepared for the synthesis of CNFs using a coprecipitation method. Aqueous solutions A (nickel nitrate + aluminum nitrate) and B (ammonium molybdate) were thoroughly mixed. Next, the obtained mixture was combined with aqueous solution C (ammonium carbonate), added to distilled water in a dropwise manner, and stirred at room temperature and pH ≈ 9.0 until a precipitate was formed. The obtained solution containing the precipitate was vacuum-filtered and oven-dried at 100 °C for 24 h. The dried precipitate was then ground and collected as a powder to be used as a metal catalyst for the synthesis of CNFs. Details of the method used to prepare the catalyst are shown in Figure S1. Aluminum nitrate was used as a support for the transition metal catalyst, ammonium molybdate was used to inhibit the aggregation of the transition metal particles between particles during the reaction at high temperatures, and ammonium carbonate was used as a precipitant.

The CNF synthesis method is shown in Figure S1. Acetylene (C_2H_2), H_2, and Ar gases were used as the carbon source to synthesize the CNFs, promoting the production of gas and carrier gas, respectively. The prepared Ni–Mo catalyst powder was spread evenly on a quartz boat and placed in a tube furnace reactor. The temperature was then increased to 700 °C at a rate of 10 °C min^{-1} while maintaining the Ar gas flow. After reaching the target synthesis temperature, the Ar gas was replaced with H_2/Ar mixed gas (10%) to reduce the catalyst for 30 min. Next, C_2H_2 gas and the H_2/Ar mixed gas were passed to the reactor for 1 h. Finally, the reduced metal catalyst was slowly cooled to room temperature while eliminating the supply of other gases and maintaining the Ar gas flow.

2.2. Synthesis of the Si@G/CNF/Graphite Composites

The process used to synthesize the Si@G/CNF/graphite composites is shown in Figure 1. SiNPs (20 mg, APS ≤ 50 nm, 98%, Alfa Aesar, Inc., Ward Hill, MA, USA) were uniformly dispersed in ethanol (200 mL) by sonication for 2 h. Then, GO solution (40 mL, N002-PS, 0.5%, Angstron Materials, Dayton, OH, USA) was added to the above solution under vigorous magnetic stirring for 1 h. The mixed dispersion was then sonicated for 2 h to obtain the Si/GO mixture. Next, CNFs (0.2 g) were added to the mixture under stirring for 1 h and then sonicated for 2 h to obtain a stable Si/GO/CNFs dispersion. A certain amount of graphite (powder, <20 μm, synthetic, Sigma-Aldrich, St. Louis, MO, USA) was mixed into the dispersion under magnetic stirring, and the mixture was then sonicated for 2 h. The resulting composites were collected via centrifugation and dried at 80 °C for 24 h in a vacuum oven. The final product was transferred to a quartz tube furnace and heated to 700 °C at a rate of 10 °C min^{-1}, and the argon atmosphere flow was maintained for 5 h to obtain thermally reduced Si@G/CNF/graphite composites to be used as anode-active composite materials for LIBs (Figure 1).

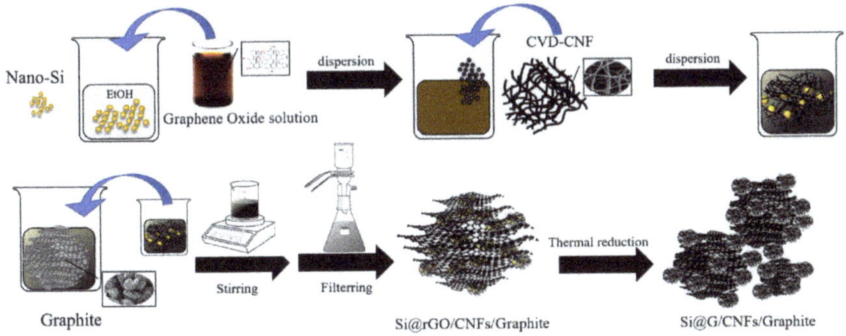

Figure 1. Schematic of the fabrication process of Si@G/CNF/graphite.

To optimize the electrochemical properties of the obtained materials, the mass ratio of graphite to the Si/G/CNF composite was changed, and the same method was used to prepare the Si@1-G/CNF/graphite (15:75, wt%) and Si@2-G/CNF/graphite (10:80, wt%) composite materials shown in Table 1.

Table 1. Description of the as-prepared Si@n-G/CNF/graphite samples.

No.	Sample Code Si@n-G/CNF/Graphite	Si (wt%)	GO (wt%)	CNF (wt%)	Graphite (wt%)	Conducting Carbon (wt%)	Binder (wt%)
1	Si@1-G/CNF/graphite	15	15	15	45	5	5
2	Si@2-G/CNF/graphite	10	10	10	60	5	5

2.3. Material Characterization

The surface morphologies and microstructures of the Si@G/CNF/graphite composites were characterized using field-emission scanning electron microscopy (FE-SEM, S-4800, Hitachi, Tokyo, Japan) and high-resolution transmission electron microscopy (HR-TEM, JEM-2100, JEOL, Tokyo, Japan). The qualitative and quantitative analyses of the elements in the prepared samples were performed using energy-dispersive X-ray spectroscopy (EDS) mapping (ARL-3460, Thermo Fisher Scientific, Waltham, MA, USA). The sample composition and crystal structure were characterized using powder X-ray diffraction (XRD, Ultima IV, Rigaku, Tokyo, Japan) with a 2 kW system and Cu-Kα radiation (K = 1.5418 Å) at a 2θ range of 2–90°. Raman spectroscopy was conducted on a LABRAM HR-800 (Horiba Jobin-Yvon, Paris, France) with a laser light (λ = 514 nm) in the wave number range of 100–3000 cm^{-1}. The Si, reduced GO (rGO), CNFs, and graphite contents in the composites were determined using thermogravimetric analysis (TGA, Diamond TG-DAT 8122 thermal analyzer system, Perkin Elmer, Waltham, MA, USA). In this test, the samples were heated from 25 °C to 800 °C at a rate of 10 °C min^{-1} under air atmosphere. Fourier-transform infrared spectroscopy (FTIR, Nicolet 6700 FTIR spectrophotometer, Nicolet, Madison, WI, USA) was used to analyze the changes in the surface functional groups of the samples, employing KBr pellets tested in the frequency range of 4000–500 cm^{-1}. The chemical bonding states were determined using X-ray photoelectron spectroscopy (XPS, Multilab-2000, Thermo Fisher Scientific, Waltham, MA, USA) with a twin anode and Al-Kα radiation as the X-ray source.

2.4. Cell Fabrication and Characterization

The anode slurry was prepared by dissolving the appropriate weight ratios of the active material (Si@G/CNF/graphite, 90 wt%), conductive carbon (Super P, 5 wt%), and the binder (polyvinylidene fluoride, 5 wt%) in N-methyl pyrrolidone solvent. To achieve a homogeneous mixture, the resulting viscous liquid was thoroughly stirred. Next, the slurry was cast onto a copper (Cu) current collector foil (thickness = 11 μm) and dried in a heating oven at 80–100 °C for 12 h. The dried electrode sheet was then punched into circular electrodes (diameter = 14 mm) for further processing. The cathode slurry was prepared in an N-methyl pyrrolidone solvent by incorporating the requisite quantities of the active material, conductive carbon (Super P), and the binder (PVDF). The slurry casting procedure and subsequent steps were the same as those used for the anode. Similarly, the dried cathode sheet was punched into electrodes (diameter = 14 mm) for cell assembly and electrochemical evaluation.

For the electrochemical characterization, coin-type (CR2032) button cells were assembled in an argon-filled glovebox with oxygen at a moisture level of less than 0.5 ppm. In the half-cells, Si@G/CNF/graphite acted as the working electrode, and lithium metal was used as both the counter and reference electrodes. A LiPF6 solution (1 M) dissolved in a mixture of ethylene carbonate and dimethyl carbonate was used as the electrolyte in all cell types. The electrodes were separated by a polypropylene membrane. Cyclic voltammetry (CV) and galvanostatic charge–discharge measurements were obtained at room temperature (25 °C) using an electrochemical workstation and a battery tester

(Neware Co., Ltd., Shenzhen, China) at a scan rate of 0.1 mV s^{-1} and a voltage range of 0.01–1.5 V (vs. Li/Li^+). The charge–discharge profiles of each composite electrode were measured for 100 cycles. Electrochemical impedance spectroscopy (EIS) measurements were conducted using a CHI 660D electrochemical analysis instrument (CH Instruments, Inc., Shanghai, China) at a frequency range of 100 kHz–0.01 Hz and an amplitude of 5 mV.

3. Results and Discussion

3.1. Structural and Morphological Characterizations of the Si@1-G/CNF/Graphite Composites

Figure 2 shows the SEM images of graphite, graphene, and the CNF/graphite Si@1-G/CNF/graphite composite. In addition, the atomic content ratio was determined based on the EDS spectrum of Si@1-G/CNF/graphite (Figure S2). Figure 2a shows the morphology of natural graphite. The image shows the clear flake-like layered structure characteristic of natural graphite, with numerous gaps between the flake structures. Figure 2b shows the morphological characteristics of the multilayer graphene, with a flaky structure and visible wrinkles on the surface. Figure 2c,d show the morphology of Si@1-G/CNF/graphite at different magnifications. Figure 2d shows smooth-surfaced slender fibers intricately entangled on the surfaces of graphite and graphene and tightly interspersed between the materials to form an intertwined three-dimensional network structure. These slender fibers are CNFs, and the round granular substances are SiNPs [21,22]. This structure can provide an effective buffer for the volume change in the electrode during the lithiation process, preventing electrode damage through volume expansion, which maintains the structural stability of the electrode material. Moreover, the three-dimensional network structure can reduce the transmission distance of the electrons and ions, effectively decreasing the transmission resistance during the cycle.

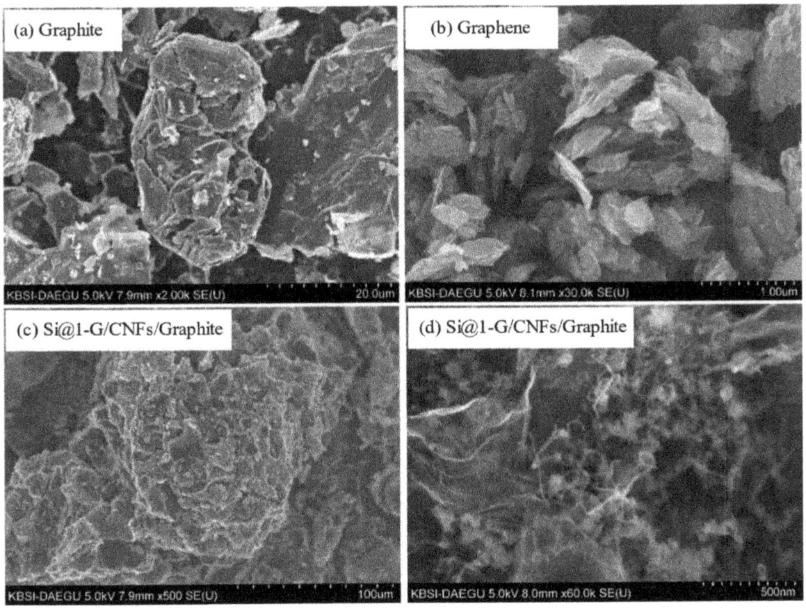

Figure 2. FE-SEM images of (**a**) graphite, (**b**) graphene, and (**c**,**d**) Si@1-G/CNF/graphite.

The HR-TEM images of the Si@1-G/CNF/graphite composites (Figure 3) were analyzed to further clarify their morphological and structural characteristics. Figure 3a,b show clear gaps between the layered structures. The lattice distances of graphite, which corresponded to the crystal plane spacing of (002) and (100) between two adjacent carbon atoms, were determined, based on Figure 3b, to be 0.351 and 0.2152 nm, respectively. The graphene image (Figure 3d) illustrates a similar lattice spacing between the carbon atoms. The morphological images of Si@1-G/CNF/graphite (Figure 3e,f) clearly show that the SiNPs were wrapped with wrinkled graphene and that the flaky graphite and CNFs were evenly distributed in the system. The encapsulation provided by the graphene layer can compensate for the volume change in the silicon particles and effectively prevent direct contact between silicon particles and the electrolyte solution, inhibiting excessive SEI film formation. The interwoven network structure formed among the CNFs, graphite, and graphene maintained the integrity of the electrode structure during the charge–discharge process, and the gaps formed by the network structure increased the specific surface area of the material, providing more effective channels for the transmission of ions, which improved the conductivity of the material. High conductivity, low resistivity, and good structural stability are key to maintaining the high specific capacity and excellent cycle performance and rate performance of anode materials [23]. In addition, Figure 3g,h show the typical lattice spacing between the carbon atoms as well as a lattice spacing of 0.313 nm, corresponding to the Si (111) of the SiNPs.

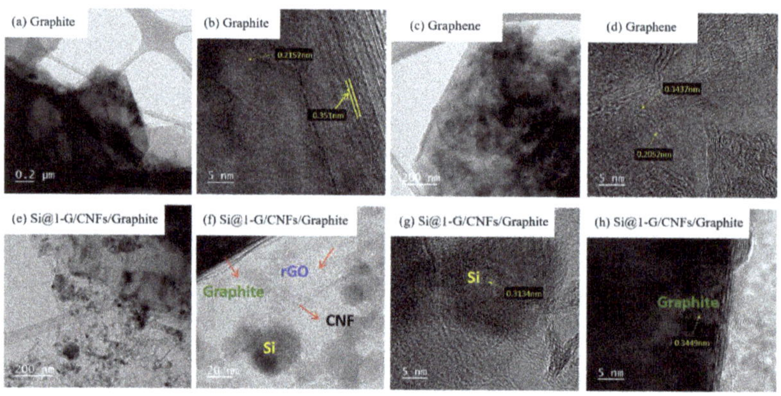

Figure 3. HR-TEM images of (**a**,**b**) graphite, (**c**,**d**) graphene, and (**e**–**h**) Si@1-G/CNF/graphite.

XRD was used to investigate the crystal structures of the individual components and the prepared composites (Figure 4a). In the diffraction curve of GO, a strong and broad representative diffraction peak was observed at 2θ = 12°. However, this representative peak disappeared in the diffraction pattern of the composite material, which can be attributed to the disappearance of the intrinsic –C–O–C–, C=O, –OH, and –COOH on the basal plane and the edges of GO owing to thermal reduction. Next, a strong peak was observed at 2θ = 26°, which is the characteristic peak of the (002) crystal plan of amorphous carbon and related to the layered structure of the crystal [24,25]. The removal of the oxygen-containing functional groups resulted in a decrease in the interlayer spacing between the stacked GO sheets, which changed the diffraction angle. This characteristic peak also appeared in the graphite and CNFs. In the diffraction results of the composite, the diffraction peak attributed to the (002) plan of graphite was observed at 2θ = 26° and the secondary peak of the (101) plan, which is related to the crystal orientation, was observed at 2θ = 44.6°, confirming the presence of carbon components in the composite. In addition, strong diffraction peaks were observed in the diffraction pattern of Si at 2θ = 28.4°, 47.3°, 56.1°, 69.1°, and 76.4°, corresponding to the typical (111), (220), (311), (400), and (331) planes of face-centered cubic Si crystals [23,26,27]. These representative Si diffraction peaks indicate that the synthesis process did not change the lattice structure of the composite components. Thus, these results confirm the presence of all components in the composite and that their structures were not changed by the synthesis process.

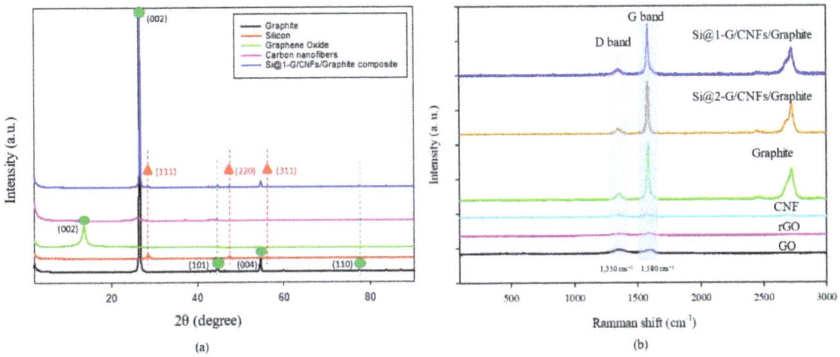

Figure 4. (**a**) XRD patterns of graphite, silicon, GO, CNFs, and Si@1-G/CNF/graphite; (**b**) Raman spectra of GO, rGO, CNFs, graphite, Si@1-G/CNF/graphite, and Si@2-G/CNF/graphite.

To further investigate the microstructure of the composites, the degree of graphitization was evaluated based on the material internal defects. In the Raman spectra (Figure 4b), all samples exhibited a characteristic sp3 hybrid disordered D band at approximately 1350 cm^{-1} and a G band associated with the sp2 hybrid ordered carbon structure at approximately 1580 cm^{-1} [28–30]. The relative intensity ratio of the D and G bands (ID/IG) reflects the degree of graphitization, the defect density, and the size of the graphitized area. The increase in the ID/IG value of GO from 0.84 to 0.85 after its thermal reduction to rGO is attributed to the increase in the number of O-terminated sp2 carbon edge atoms, resulting in a decrease in the in-plane sp2 graphitic domains [31]. The intensity of the G band in the composite was considerably higher than that of the D band. These results, combined with the curve characteristics of graphite, indicate the low defects and disorder of the composite material, suggesting that the material has all the graphite properties as well as good crystallinity.

The surface composition and chemical bonding state of each element in the Si@1-G/CNF/graphite composites were characterized using XPS (Figure 5). The figure shows the XPS high-resolution Si 2p, C 1s, and O 1s spectra of the composites. The XPS survey spectra of the composites (Figure 5a) showed strong O 1s and C 1s peaks, which were mainly due to the presence of highly graphitized lattices attributed to the graphite and graphene. The peak of Si 2p was relatively weak, indicating that the Si particles were encapsulated in the rGO sheets and that the surface was covered by the carbon layer. Figure 5b shows the high-resolution Si 2p spectra of the composites. The signal had two different peaks at 100.18 and 102.92 eV, which are attributed to bulk silicon (Si–Si) and oxygen-bonded silicon (Si–O), respectively [32,33], indicating that during the material preparation process, a small number of SiNPs exposed to the air underwent surface oxidation under certain conditions to produce a small amount of SiO_x. Because the surface was covered by SiO_x and carbon layers, the Si–Si peak of the Si particles was slightly lower than the Si–O peak. As shown in Figure 5c, the C 1s spectra of the composite showed two strong peaks at 284.2 and 285.73 eV, corresponding to the C–C/C=C in the aromatic ring and the C–O bonds in the carboxyl and epoxy groups, respectively [23,34]. In addition, the C 1s signals of the composites all had strong C–C/C=C peaks, whereas the peak intensity of the O-containing groups decreased considerably, which can be attributed to the removal of O atoms during the heat treatment (Figure S3). Two weak peaks at approximately 289.62 and 287.34 eV, which are attributed to the O–C=O bond in the –COOH groups in graphite and graphene, were also observed on the surface of the composite. Moreover, the high-resolution O 1s XPS spectra (Figure 5d) confirmed the presence of –COOH functional groups in the composite.

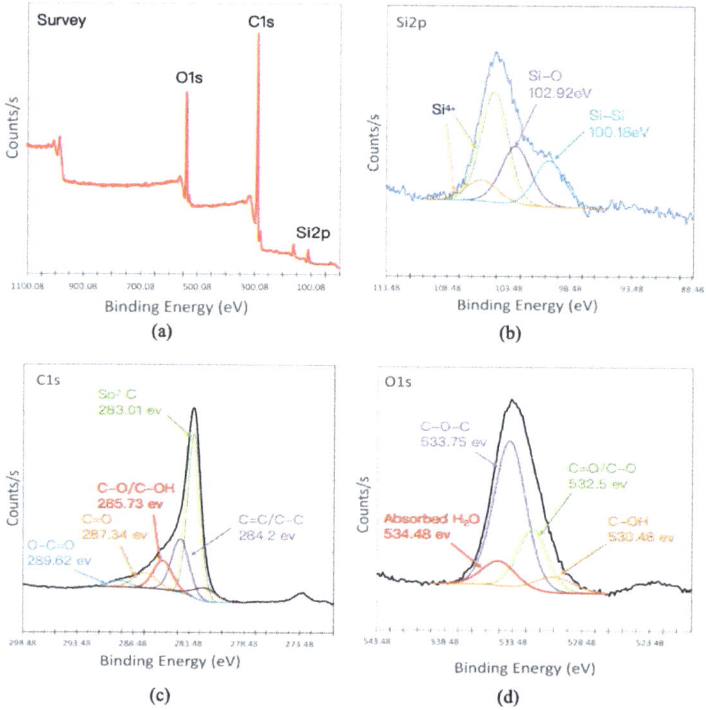

Figure 5. (**a**) XPS survey spectra of the Si@1-G/CNF/graphite composite; (**b**–**d**) high-resolution XPS of Si 2p, C 1s, and O 1s of the Si@1-G/CNF/graphite composite.

Figure 6a shows the thermogravimetry/differential thermal analysis (TG/DTA) of the two composite materials during continuous combustion under ambient conditions until the temperature cutoff point of 800 °C. The TG/DTA curves can be divided into four stages according to the different components of the composite materials, and the detailed analysis results are shown in Figure S4. The exothermic peaks of Si@1-G/CNF/graphite and Si@2-G/CNF/graphite started with the first weight loss at 400–450 °C, which corresponded to the decomposition of the graphene component in the composites [35]. In this stage, most O atoms in the composite material were eliminated from the GO layer. The second stage corresponded to the thermal decomposition of the CNF under an air atmosphere. The third stage was the rapid decomposition of the graphite component. Thus, the total C content in Si@1-G/CNF/graphite was approximately 79 wt%, whereas the total carbon content in Si@2-G/CNF/graphite was approximately 82 wt%, which corresponded to Si:C weight ratios of 15:75 and 1:80 during the composite manufacturing process, respectively. The slight difference in the data is attributed to a small amount of Si loss during the thermal reduction process and trace impurities on the surface of the composite. In addition, the weight of the fourth stage increased after the temperature reached 800 °C, which is attributed to the oxidation of the remaining Si component to SiO_x under the air atmosphere [23,36]. This was due to the direct contact between the carrier gas and the exposed Si after the decomposition of the surface carbon at high temperatures. This result is in good agreement with the XPS of Si 2p. Measuring the total C and Si contents accurately can be challenging because of the possible weight loss of the sample during annealing or thermal reduction.

Nevertheless, carbon materials, including graphene and CNFs, are typically combusted and become depleted at 400–800 °C, whereas Si remains stable. Therefore, the differences in the composite contents could be detected based on a rapid reduction in the weight of the sample.

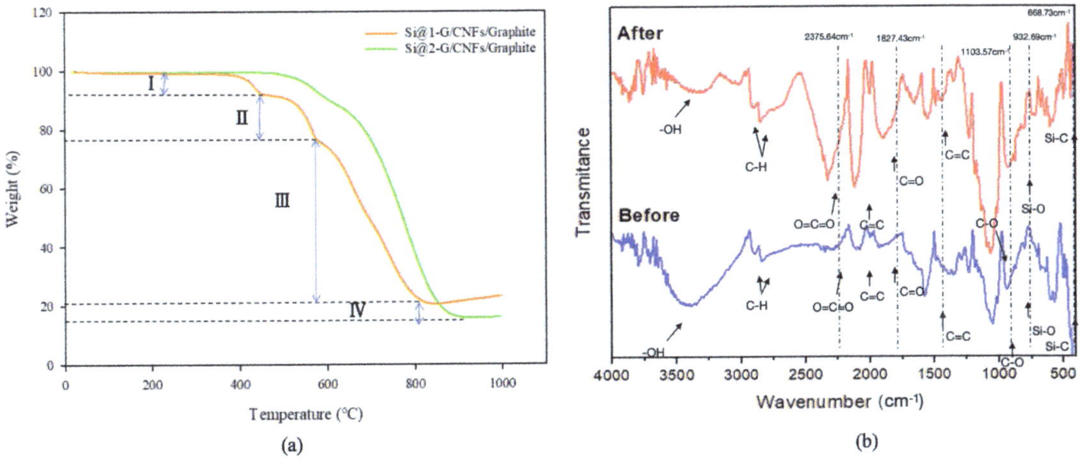

Figure 6. (**a**) TGA curves of Si@1-G/CNF/graphite and Si@2-G/CNF/graphite; (**b**) FTIR spectra of Si@1-G/CNF/graphite.

To further verify the surface chemical structure of the prepared samples, FTIR analysis was conducted, the results of which are presented in Figure 6b. The figure shows the FTIR spectra of the composite before and after the heat treatment and annealing. Through comparison, the peak intensities of the O=C=O and C=O oxygen-containing groups (2375.64 and 1870.0 cm^{-1}, respectively) were found to considerably decrease, owing to the reduction in some oxygen-containing groups after heat treatment [37,38]. The vibration peak at 932.69 cm^{-1} corresponded to the asymmetric stretching and bending of the siloxane group. The appearance of this peak was related to the oxidation of the surface of the SiNPs and the generation of a small amount of SiO_x, which is in good agreement with the previous XPS results. The multiple-vibration peak at approximately 800 cm^{-1} was the strongest characteristic peak of the Si–C bond, indicating an extremely tight bond between the silicon atoms and the carbon atoms, with high chemical and thermal stability. This suggests that the SiNP surface was successfully and tightly entangled with the carbon layer. Moreover, numerous peaks related to the O-containing functional groups were observed in the spectra. A broad peak was observed at approximately 3400 cm^{-1}, which corresponded to the stretching mode vibration of the O–H group superimposed on the O–H stretching of the carboxylic acid (R–COOH), owing to the presence of water molecules and alcohol groups. The peak at approximately 1827 cm^{-1} is attributed to the C=O stretching of the –COOH group, and the peak at 1103 cm^{-1} corresponds to the C–O stretching vibration of the C–O–C bond [39,40]. These results demonstrate the successful preparation of the Si@G/CNF/graphite composite.

3.2. Electrochemical Performance

Figure 7 demonstrates the CV curves of the fabricated coin cell using Si@1-G/CNF/graphite (Figure 7a) and Si@2-G/CNF/graphite (Figure 7b) during the initial five cycles at 0.01–1.5 V (vs. Li^+/Li) at a scan rate of 0.1 mV s^{-1}. During the first cathodic scan (lithiation process) of Si@1-G/CNF/graphite, two distinct peaks were observed at 0.32 and 0.70 V.

Similarly, two cathodic peaks were observed in the initial CV curves (at 0.32 and 0.75 V) of Si@2-G/CNF/graphite. The two reduction peaks in the cathode branch of the CV curve are attributed to the conversion of bulk Si into Li_xSi alloy and other lithiated precipitates (Li_xSiO_y, Li_2CO_3, and Li_2O) [23,41,42]. The sharp peak that appeared during the reduction process around 0.2V is a typical peak of graphite anode related to the embedding of lithium ions into the graphite interlayer. In addition, the broad peak related to SEI at around 0.8V overlapped with the broad peak caused by Si alloying. These materials were generated via a series of irreversible multi-step electrochemical reactions between Si and Li^+, which is in good agreement with previous electrochemical studies. In the first cathodic scan, a weak peak was observed at approximately 1.0 V for both electrodes, which can be attributed to the reaction between the electrode material and the electrolyte, resulting in the formation of an irreversible SEI layer on the electrode surface during the discharge process. However, this peak disappeared in the subsequent cycles, and the CV curve in the second cycle overlapped with the CV curve in the subsequent cycles, further confirming that the effective combination of SiNPs and carbon materials enhances the reversibility and cycle stability of electrode materials. In addition, during the anodic scan (delithiation process) in the first cycle, two broad oxidation peaks were observed at 0.52 and 0.33 V in Si@1-G/CNF/graphite and at 0.53 and 0.30 V in Si@2-G/CNF/graphite. These peaks corresponded to the partial decomposition of the highest lithiated phase and the complete delithiation of the Li_xSi alloy into amorphous Si, respectively. The strong reduction peak at 0.01–1.2 V in the cathodic scan corresponded to the amorphous Li_xSi alloy formed by amorphous SiNPs during the reversible Li^+ insertion–deinsertion process [23,43]. With the increase in the number of scans, a gradual and steady increase in the intensity of the oxidation peak was observed because the composite electrode material was gradually activated during the cycle, indicating the tendency of Li^+ to form alloys with SiNPs, which increases the ionic conductivity. During the oxidation process, the curve tended to be stable starting from the second cycle, and the splitting phenomenon of the oxidation peak weakened, indicating that the SEI film tended to be stable. The phase transformation results for Si and C during the first five cycles are summarized in Table S1.

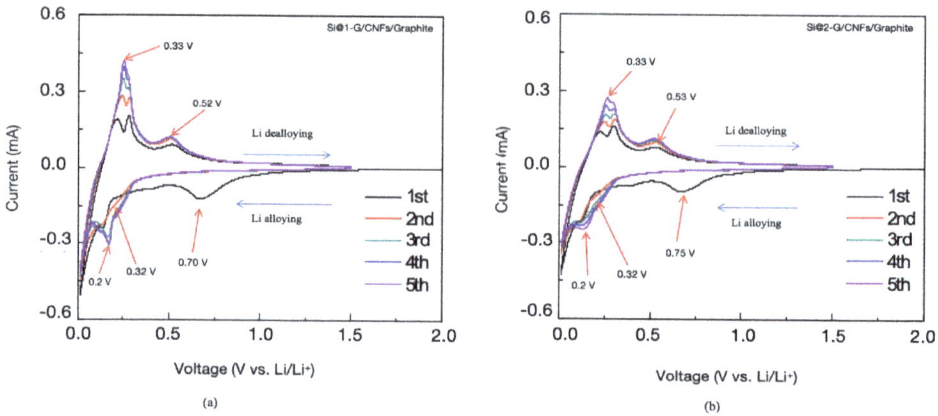

Figure 7. CV profiles of the (**a**) Si@1-G/CNF/graphite and (**b**) Si@2-G/CNF/graphite electrodes during the initial five cycles.

The electrochemical performance of Si@1-G/CNF/graphite and Si@2-G/CNF/graphite was evaluated for 100 cycles at a low current density (100 mA g^{-1}, Figure 8a), and the corresponding constant current curves are shown in Figure 8b. The discharge specific capacities of Si@1-G/CNF/graphite and Si@2-G/CNF/graphite were 521.1 and 558.7 mAh g^{-1}, respectively, in the initial cycle, after which they considerably decreased owing to the inability of the electrode to accommodate the volume change in the SiNPs during lithiation–delithiation and the formation of an SEI layer on the electrode surface when the SiNPs came into contact with the electrolyte. In addition to a low initial charge capacity, Si@1-G/CNF/graphite and Si@2-G/CNF/graphite exhibited low ICEs of 65.0% and 70.0%, respectively. This can be attributed to the irreversible SEI layer formed by the decomposition reaction at the electrode–electrolyte interface, during which Li^+ was consumed, leading to inevitable capacity loss. The Coulombic efficiency (CE) of Si@1-G/CNF/graphite and Si@2-G/CNF/graphite sharply increased to 92.1% and 93.3%, respectively, during the second cycle and stabilized in the subsequent cycles, reaching 99.0% after 100 cycles. After 100 cycles, the capacities of the Si@1-G/CNF/graphite and Si@2-G/CNF/graphite electrodes stabilized at 309.3 and 301.5 mAh g^{-1}, respectively, and their capacity retention rates reached 65% and 54%, respectively. To further verify the electrochemical performance of the composite materials, Figure 8c shows the rate performance of the Si@1-G/CNF/graphite and Si@2-G/CNF/graphite electrodes in the current density range of 0.1–1 A g^{-1}. The Si@1-G/CNF/graphite electrode exhibited a rate performance superior to that of the other electrodes at all current densities. The reversible capacities of the Si@1-G/CNF/graphite electrode were 568.5, 423.3, 373.3, 346.3, 336.2, and 332.3 mAh g^{-1} at current densities of 0.1, 0.2, 0.5, 1, 0.5, 0.2, and 0.1 A g^{-1}, respectively. When the current density was restored to 0.1 A g^{-1}, the reversible capacity shifted to 346.7 mAh g^{-1}, and the electrode exhibited a better cycle and rate performance overall, confirming the high cycle stability and reversibility of the Si@1-G/CNF/graphite electrode. With the increase in the silicon (wt%) content, the initial capacity loss increased, and the change in the lithiation and delithiation capacity considerably affected the ICE of the electrode, which is an important consideration for material commercialization. In previous studies, Si15 Gr75 batteries exhibited a capacity of approximately 70 mAh g^{-1} at a current density of 740 mA g^{-1}. However, our work demonstrated strong advantages in terms of the ICE and capacity retention after multiple cycles. These results indicate that graphite alone is not sufficient to compensate for the changes in silicon that occur during lithiation–delithiation processes. Moreover, doping with a small amount of graphene and CNFs effectively improved the conductivity of the electrode, resulting in the excellent cycle performance and rate performance of the electrode.

To understand the effect of the SEI and charge transfer resistance on the electrochemical performance of the materials, EIS spectra were acquired after 100 charge and discharge cycles (Figure 9). The inset in the Nyquist plot shows the equivalent circuit model of the assembled half-cell. The circuit model takes into account the resistance caused by the interaction between the electrolyte and the Si particles (R_S), the resistance caused by the migration of Li^+ between the electrodes through the SEI layer (R_{SEI}), the resistance during charge transfer (R_{CT}), and the Warburg impedance (W_Z) [23,44–46]. To observe the Nyquist plots of the two composites before cycling (Figure 9a) and after 100 cycles (Figure 9b), we obtained the Nyquist plots before cycling (Figure 9a) and after 100 cycles and evaluated the Li+ storage performance of the electrodes. The Nyquist plots of each composite material before cycling contain a semicircle in the mid-frequency region, which represents the charge transfer resistance (R_{CT}) between the electrode and the electrolyte, and an inclined line in the low-frequency region. The diameter of the semicircle was the sum of the resistance of Li^+ to passing through the insulating SEI layer on the surface of the active material (R_{SEI}) and R_{CT}, whereas the inclined line was directly related to the tortuosity of Li^+ diffusion. The Nyquist plots of the two composites before cycling (Figure 9a) show similar curve characteristics. Compared with the Si@2-G/CNF/graphite electrode (R_{CT} = 117.6 Ω), the Si@1-G/CNF/graphite electrode (R_{CT} = 78.5 Ω) exhibited a lower resistance. The resistance value of the composite electrode largely depends on the content of the SiNPs. To further evaluate the stability of the formed SEI film, the Nyquist plots were obtained after 100 cycles. The R_{SEI} values of Si@1-G/CNF/graphite and Si@2-G/CNF/graphite were 30.5 and 40.2 Ω, respectively. The lower resistance of Si@1-G/CNF/graphite is attributed to the formation of a stable SEI film on the surface of the SiNPs, which can prevent direct contact between the active materials and the electrolyte, thereby minimizing the decomposition of the electrolyte. After 100 cycles, the diameter of the Si@1-G/CNF/graphite semicircle was small, indicating that the transfer resistance of the Si@1-G/CNF/graphite electrode becomes smaller, so it can quickly transfer electrons and ions, effectively improving the ion–electron conductivity and reducing the charge transfer resistance. However, the electrode materials with more carbon content show a larger resistance, which is because the excessive carbon material considerably hinders the effective transmission of ions and electrons, resulting in higher resistance and a lower Li^+ diffusion rate. An appropriate silicon–carbon ratio induces the dispersion of carbon materials, minimizing the agglomeration of the SiNPs, providing an effective coating, and maintaining the structural integrity of the electrode during continuous cycling. The above results show that the Si@1-G/CNF/graphite electrode exhibited excellent performance, making it a promising anode material.

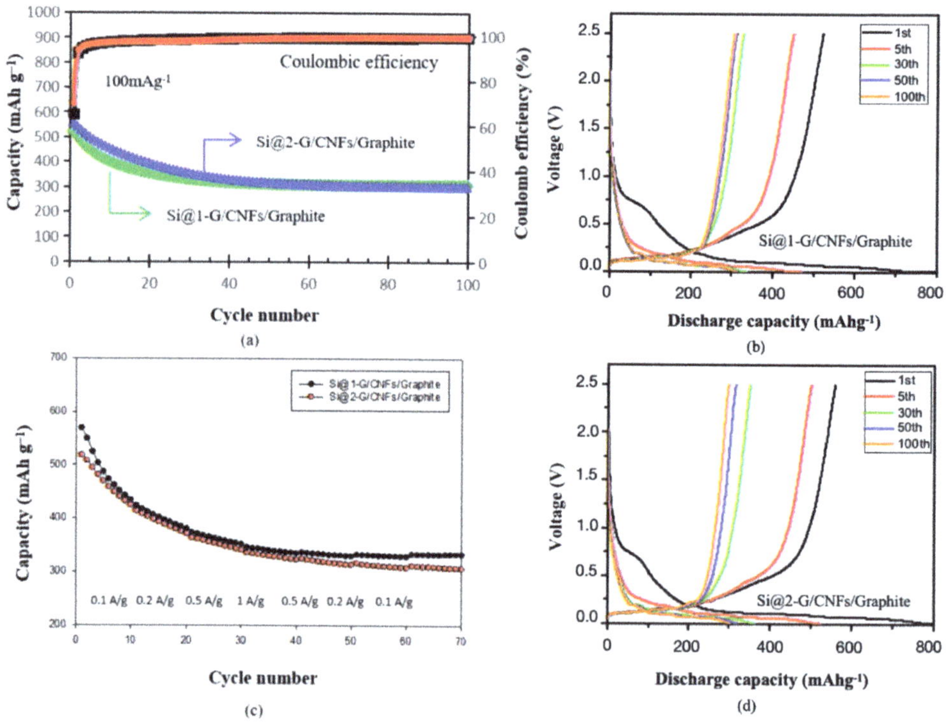

Figure 8. (**a**) Charge and discharge capacities of Si@1-G/CNF/graphite and Si@2-G/CNF/graphite with the corresponding CE at a current density of 100 mA g^{-1}; (**b**) cycling performance of Si@1-G/CNF/graphite and with the corresponding galvanostatic charge–discharge profiles at a current density of 100 mA g^{-1}; (**c**) rate performance of Si@1-G/CNF/graphite and Si@2-G/CNF/graphite at different current densities; (**d**) cycling performance of Si@2-G/CNF/graphite and with the corresponding galvanostatic charge–discharge profiles at a current density of 100 mA g^{-1}.

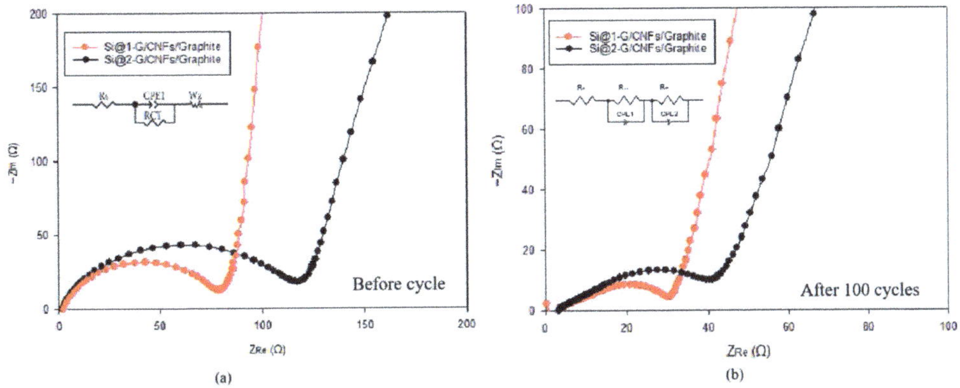

Figure 9. Nyquist plots and electrochemical impedance spectra of Si@1-G/CNF/graphite and Si@2-G/CNF/graphite (**a**) before cycling and (**b**) after the 100th cycle.

4. Conclusions

This study investigated the addition of appropriate amounts of silicon, graphene, and CNFs to existing commercial graphite anodes to improve their properties. The composite electrode was prepared using a simple physical mixing and low-temperature thermal reduction process. The reversible specific capacity was further improved without sacrificing the ICE and cycling stability. The developed multicomponent composite electrode exhibited superior performance at the half-cell level compared with pure graphite, pure silicon, and silicon/graphite binary composite electrodes. The results show that the composite electrode with a silicon content of 15 wt% exhibited a better capacity retention rate (60%) after 100 cycles. This can be attributed to the secondary carbon layer protection of the SINPs by graphene and CNFs, which mitigated the inherent volume change of the silicon, effectively reduced the contact between the silicon surface and the electrolyte, and stabilized the formation of the SEI layer. This study provides a low-cost and green electrode synthesis method that produces high-performance electrodes and has great potential in battery manufacturing. This synthesis method is promising for advancing effective theoretical support for the commercial use of silicon/carbon electrodes. Thus, this study paves the way for the development of high-energy-density LIBs and provides valuable technical support for optimizing silicon-based graphite composite anodes to achieve commercial feasibility. Future research work needs to focus on reducing costs, optimizing processes, and improving the production efficiency and structure and composition of silicon–carbon negative-electrode materials. The focus here should be on improving stability, strengthening the cooperation between industry, academia, and research teams, and jointly promoting the research and development of silicon–carbon anode materials, thereby accelerating its industrialization process from the laboratory to the market.

Supplementary Materials: The following supporting information can be downloaded at https://www.mdpi.com/article/10.3390/batteries11030115/s1. Figure S1: process for preparing the catalysts and the synthesis of carbon nanofibers; Figure S2: EDS spectrum of Si@1-G/CNF/graphite composite; Figure S3: high-resolution XPS of O 1s of the Si@1-G/CNF/graphite composite; Figure S4: TGA/DTA curves of Si@1-G/CNF/graphite and Si@2-G/CNF/graphite. Table S1: Summary of the phase changes of Si and C during the first five cycles; Table S2: Comparison of this work with other previous studies in terms of Discharge capacity, Coulombic efficiency, and Capacity retention.

Author Contributions: R.C. designed and performed the experiments, provided the concepts and methodology required for the experiments, organized the data, and wrote the main manuscript text. D.-E.J. and H.-H.P. prepared the samples, performed the experiments, and designed the figures using the relevant software. J.J., Y.-Y.J. and H.L. participated in the electrochemical measurements and validation of the relevant data. C.-S.L., as the corresponding author, was responsible for the review and submission of the manuscript. All authors reviewed and analyzed the manuscript. All authors have read and agreed to the published version of the manuscript.

Funding: This research was funded by a grant from the Korea Evaluation Institute of Industrial Technology (grant number: 20018434).

Data Availability Statement: The datasets generated during and/or analyzed during the current study are available from the corresponding author upon reasonable request.

Conflicts of Interest: The authors declare no conflicts of interest.

Abbreviations

The following abbreviations are used in this manuscript:

LIBs	Lithium-ion batteries
SEI	Solid electrolyte interface
CFTs	Carbon nanofibers
CVD	Chemical vapor deposition
GO	Graphene oxide
SEM	Scanning electron microscopy
TEM	Transmission electron microscopy
EDS	Energy-dispersive X-ray spectroscopy
XRD	X-ray diffraction
XPS	X-ray photoelectron spectroscopy
PVDF	Polyvinylidene fluoride
CV	Cyclic voltammetry
EIS	Electrochemical impedance spectroscopy
RCT	Charge transfer resistance
ICE	Initial Coulombic efficiency
CE	Coulombic efficiency

References

1. Mishra, G.K.; Gautam, M.; Bhawana, K.; Ghosh, J.; Mitra, S. High energy density lithium-ion pouch cell with modified high voltage lithium cobalt oxide cathode and graphite anode: Prototype stabilization, electrochemical and thermal study. *J. Power Sources* **2023**, *580*, 233395. [CrossRef]
2. Tubtimkuna, S.; Sawangphruk, M.; Duriyasart, F. Effect of electrolyte additives on cycling performance of 18650 graphite//nmc811 Li-ion batteries. *ECS Trans.* **2020**, *97*, 155–166. [CrossRef]
3. Duan, J.; Tang, X.; Dai, H.; Yang, Y.; Wu, W.; Wei, X.; Huang, Y. Building safe lithium-ion batteries for electric vehicles: A review. *Electrochem. Energ. Rev.* **2020**, *3*, 1–42. [CrossRef]
4. Asenbauer, J.; Eisenmann, T.; Kuenzel, M.; Kazzazi, A.; Chen, Z.; Bresser, D. The success story of graphite as a lithium-ion anode material—Fundamentals, remaining challenges, and recent developments including silicon (oxide) composites. *Sustain. Energy Fuels* **2020**, *4*, 5387–5416. [CrossRef]
5. Chae, S.; Ko, M.; Kim, K.; Ahn, K.; Cho, J. Confronting issues of the practical implementation of Si anode in high-energy lithium-ion batteries. *Joule* **2017**, *1*, 47–60. [CrossRef]
6. Chen, Y.; Kang, Y.; Zhao, Y.; Wang, L.; Liu, J.; Li, Y.; Liang, Z.; He, X.; Li, X.; Tavajohi, N. A review of lithium-ion battery safety concerns: The issues, strategies, and testing standards. *J. Energy Chem.* **2021**, *59*, 83–99. [CrossRef]
7. Lavigne Philippot, M.; Costa, D.; Cardellini, G.; De Sutter, L.; Smekens, J.; Van Mierlo, J.; Messagie, M. Life cycle assessment of a lithium-ion battery with a silicon anode for electric vehicles. *J. Energy Storage* **2023**, *60*, 106635. [CrossRef]
8. Yang, Z.; Wu, C.; Li, S.; Qiu, L.; Yang, Z.; Zhong, Y.; Zhong, B.; Song, Y.; Wang, G.; Liu, Y. A unique structure of highly stable interphase and self-consistent stress distribution radial-gradient porous for silicon anode. *Adv. Funct. Mater.* **2022**, *32*, 2107897. [CrossRef]

9. Wang, W.; Wang, Y.; Yuan, L.; You, C.; Wu, J.; Liu, L.; Ye, J.; Wu, Y.; Fu, L. Recent advances in modification strategies of silicon-based lithium-ion batteries. *Nano Res.* **2023**, *16*, 3781–3803. [CrossRef]
10. Nuhu, B.A.; Bamisile, O.; Adun, H.; Abu, U.O.; Cai, D. Effects of transition metals for silicon-based lithium-ion battery anodes: A comparative study in electrochemical applications. *J. Alloys* **2023**, *933*, 167737. [CrossRef]
11. Wang, J.; Huang, W.; Kim, Y.S.; Jeong, Y.K.; Kim, S.C.; Heo, J.; Lee, H.K.; Liu, B.; Nah, J.; Cui, Y. Scalable synthesis of nanoporous silicon microparticles for highly cyclable lithium-ion batteries. *Nano Res.* **2020**, *13*, 1558–1563. [CrossRef]
12. Zhao, Z.; Han, J.; Chen, F.; Xiao, J.; Zhao, Y.; Zhang, Y.; Kong, D.; Weng, Z.; Wu, S.; Yang, Q. Liquid metal remedies silicon microparticulates toward highly stable and superior volumetric lithium storage. *Adv. Energy Mater.* **2022**, *12*, 2103565. [CrossRef]
13. Sun, L.; Liu, Y.; Shao, R.; Wu, J.; Jiang, R.; Jin, Z. Recent progress and future perspective on practical silicon anode-based lithium ion batteries. *Energy Storage Mater.* **2022**, *46*, 482–502. [CrossRef]
14. Gautam, M.; Mishra, G.K.; Bhawana, K.; Kalwar, C.S.; Dwivedi, D.; Yadav, A.; Mitra, S. Relationship between silicon percentage in graphite anode to achieve high-energy-density lithium-ion batteries. *ACS Appl. Mater. Interfaces* **2024**, *16*, 45809–45820. [CrossRef]
15. Zhang, W.; Weng, Y.; Shen, W.; Lv, R.; Kang, F.; Huang, Z.-H. Scalable synthesis of lotus-seed-pod-like Si/siox@cnf: Applications in freestanding electrode and flexible full lithium-ion batteries. *Carbon* **2020**, *158*, 163–171. [CrossRef]
16. Yan, Z.; Yi, S.; Li, X.; Jiang, J.; Yang, D.; Du, N. A scalable silicon/graphite anode with high silicon content for high-energy lithium-ion batteries. *Mater. Today Energy* **2023**, *31*, 101225. [CrossRef]
17. Moyassari, E.; Roth, T.; Kücher, S.; Chang, C.-C.; Hou, S.-C.; Spingler, F.B.; Jossen, A. The role of silicon in silicon-graphite composite electrodes regarding specific capacity, cycle stability, and expansion. *J. Electrochem. Soc.* **2022**, *169*, 010504. [CrossRef]
18. Gautam, M.; Mishra, G.K.; Bhawana, K.; Kalwar, C.S.; Mitra, S. Enhancing electrochemical stability of silicon-carbon and nmc based lithium-ion batteries through the synergistic impact of charge balance and direct-contact pre-lithiation strategy. *J. Electrochem. Soc.* **2024**, *171*, 050526. [CrossRef]
19. Zhang, Y.; Wu, B.; Mu, G.; Ma, C.; Mu, D.; Wu, F. Recent progress and perspectives on silicon anode: Synthesis and prelithiation for libs energy storage. *J. Energy Chem.* **2022**, *64*, 615–650. [CrossRef]
20. Li, X.; Yan, P.; Xiao, X.; Woo, J.H.; Wang, C.; Liu, J.; Zhang, J.-G. Design of porous si/c–graphite electrodes with long cycle stability and controlled swelling. *Energy Environ. Sci.* **2017**, *10*, 1427–1434. [CrossRef]
21. Yu, J.; Zhang, C.; Wu, W.; Cai, Y.; Zhang, Y. Nodes-connected silicon-carbon nanofibrous hybrids anodes for lithium-ion batteries. *Appl. Surf. Sci.* **2021**, *548*, 148944. [CrossRef]
22. Wiggers, H.; Sehlleier, Y.H.; Kunze, F.; Xiao, L.; Schnurre, S.M.; Schulz, C. Self-assembled nano-silicon/graphite hybrid embedded in a conductive polyaniline matrix for the performance enhancement of industrial applicable lithium-ion battery anodes. *Solid State Ion.* **2020**, *344*, 115117. [CrossRef]
23. Cong, R.; Jo, M.; Martino, A.; Park, H.-H.; Lee, H.; Lee, C.-S. Three-dimensional network of nitrogen-doped carbon matrix-encapsulated si nanoparticles/carbon nanofibers hybrids for lithium-ion battery anodes with excellent capability. *Sci. Rep.* **2022**, *12*, 16002. [CrossRef]
24. Meng, W.-J.; Han, X.-Y.; Hou, Y.-L.; Xie, Y.; Zhang, J.; He, C.-J.; Zhao, D.-L. Defect-repaired reduced graphene oxide caging silicon nanoparticles for lithium-ion anodes with enhanced reversible capacity and cyclic performance. *Electrochim. Acta* **2021**, *382*, 138271. [CrossRef]
25. Zhang, Y.; Cheng, Y.; Song, J.; Zhang, Y.; Shi, Q.; Wang, J.; Tian, F.; Yuan, S.; Su, Z.; Zhou, C. Functionalization-assistant ball milling towards Si/graphene anodes in high performance Li-ion batteries. *Carbon* **2021**, *181*, 300–309. [CrossRef]
26. Liu, X.; Du, Y.; Hu, L.; Zhou, X.; Li, Y.; Dai, Z.; Bao, J. Understanding the effect of different polymeric surfactants on enhancing the silicon/reduced graphene oxide anode performance. *J. Phys. Chem. C* **2015**, *119*, 5848–5854. [CrossRef]
27. Zhu, J.; Ren, Y.; Yang, B.; Chen, W.; Ding, J. Embedded Si/graphene composite fabricated by magnesium-thermal reduction as anode material for lithium-ion batteries. *Nanoscale Res. Lett.* **2017**, *12*, 1186. [CrossRef]
28. Liu, W.; Xu, H.; Qin, H.; Lv, Y.; Zhu, G.; Lei, X.; Lin, F.; Zhang, Z.; Wang, L. Rapid coating of asphalt to prepare carbon-encapsulated composites of nano-silicon and graphite for lithium battery anodes. *J. Mater. Sci.* **2020**, *55*, 4382–4394. [CrossRef]
29. Kim, S.K.; Kim, C.; Chang, H.; Jang, H.D. Preparation of Silicon-carbon-graphene composites and their application to lithium ion secondary battery. *Aerosol Air Qual. Res.* **2022**, *22*, 220009. [CrossRef]
30. Duan, H.; Xu, H.; Wu, Q.; Zhu, L.; Zhang, Y.; Yin, B.; He, H. Silicon/graphite/amorphous carbon as anode materials for lithium secondary batteries. *Molecules* **2023**, *28*, 464. [CrossRef]
31. Lei, G.; Huaj, G.; Zhix, W. A facile synthesis of graphite/silicon/graphene spherical composite anode for lithium-ion batteries. *Electrochim. Acta* **2013**, *104*, 117–123.
32. Cabello, M.; Gucciardi, E.; Herrán, A.; Carriazo, D.; Villaverde, A.; Rojo, T. Towards a high-power Si@graphite anode for lithium ion batteries through a wet ball milling process. *Molecules* **2020**, *25*, 2494. [CrossRef] [PubMed]
33. Jo, A.H.; Kim, S.Y.; Kim, J.H.; Kim, Y.A.; Yang, C.-M. Robust core–shell carbon-coated silicon-based composite anode with electrically interconnected spherical framework for lithium-ion battery. *Int. J. Energy Res.* **2023**, *2023*, 6874429. [CrossRef]

34. Roland, A.; Fullenwarth, J.; Ledeuil, J.; Martinez, H.; Louvain, N.; Monconduit, L. How carbon coating or continuous carbon pitch matrix influence the silicon electrode/electrolyte interfaces and the performance in Li-ion batteries. *Battery Energy* **2022**, *1*, 1002. [CrossRef]
35. Su, M.; Wang, Z.; Guo, H.; Li, X.; Huang, S.; Xiao, W.; Gan, L. Enhancement of the cyclability of a Si/graphite@graphene composite as anode for lithium-ion batteries. *Electrochim. Acta* **2014**, *116*, 230–236. [CrossRef]
36. Nayak, S.K.; Mohanty, S.; Nayak, S.K. A new way synthesis of expanded graphite as a thermal filler to enhance the thermal conductivity of DGEBA resin as thermal interface material. *Adv. Polym. Sci.* **2020**, *32*, 506–523. [CrossRef]
37. Chu, Z.; Zhao, X.; Wang, Q.; Bao, T.; Li, H.; Cao, Y.; Zhang, B.; Cao, J.; Si, W. Preparation of a flexible reduced graphene oxide-si composite film and its application in high-performance lithium ion batteries. *Crystals* **2023**, *13*, 547. [CrossRef]
38. Xu, S.; Zhang, Z.; Liu, J.; Wang, Y.; Hu, J. Facile preparation of reduced graphene by optimizing oxidation condition and further reducing the exfoliated products. *J. Mater. Res.* **2017**, *32*, 383–391. [CrossRef]
39. Kareem, A.A. Enhanced thermal and electrical properties of epoxy/carbon fiber–silicon carbide composites. *Adv. Compos. Lett.* **2020**, *1*, 29. [CrossRef]
40. Sun, Z.G.; Wang, S.J.; Qiao, X.J.; Li, Y.; Zheng, W.H.; Bai, P.Y. Synthesis and microwave absorbing properties of SIC nanowires. *Appl. Phys. A* **2018**, *124*, 1007. [CrossRef]
41. Yuan, Y.; Li, H. Minimizing the volume expansion by a self-standing reduced graphene oxide/silicon nanoparticles/copper mesh hybrid electrodes for enhanced lithium-ion batteries. *J. Energy Storage* **2023**, *64*, 107202. [CrossRef]
42. Chae, C.; Choi, W.; Ji, S.; Lee, S.S.; Kim, J.-K.; Choi, S.; Kang, Y.; Choi, Y.; Kim, D.Y.; Jeong, S. Electrostatically assembled silicon–carbon composites employing amine-functionalized carbon intra-interconnections for lithium-ion battery anodes. *ACS Appl. Energy Mater.* **2019**, *2*, 1868–1875. [CrossRef]
43. Xin, L.; Kun, L.; Man, Y.; Jiapeng, Z.; Haiyan, L.; Ang, L.; Xiaohong, C.; Huaihe, S. Graphene-doped silicon-carbon materials with multi-interface structures for lithium-ion battery anodes. *J. Colloid Interface Sci.* **2024**, *667*, 470–477.
44. Woo Sung, C.; Heon-Cheol, S.; Ji Man, K.; Jae-Young, C.; Won-Sub, Y. Modeling and applications of electrochemical impedance spectroscopy (EIS) for lithium-ion batteries. *J. Electrochem. Sci. Technol.* **2020**, *11*, 1–13.
45. Haiping, S.; Xinrui, L.; Changwei, L.; Yazhuo, S.; Honglai, L. Scalable synthesis of micrometer-sized porous silicon/carbon composites for high-stability lithium-ion battery anodes. *Chem. Eng. J.* **2023**, *451*, 138394.
46. Angelica, M.; Ruye, C.; Minsang, J.; Hyun-Ho, P.; Hochun, L.; Chang-Seop, L. Characteristics and Electrochemical Performance of Hydroxyl-Functionalized Graphene Quantum Dot-Coated Si Nanoparticles/Reduced. *J. Nanomater.* **2023**, *6353894*, 23.

batteries

MDPI

Article

Do Silicon-Based Li-Ion Batteries Require a Time-Consuming Solid Electrolyte Interphase Formation Process?

Sheng S. Zhang

Battery Science Branch, Energy Sciences Division, DEVCOM Army Research Laboratory, Adelphi, MD 20783, USA; shengshui.zhang.civ@army.mil

Abstract: The solid electrolyte interphase (SEI) is a crucial component for ensuring the safe and long-term cycling of graphite-based Li-ion batteries. Traditionally, SEI formation requires a low current rate (0.05C–0.1C) and a moderate temperature (25–45 °C), and the same process has been widely applied in the manufacturing of silicon-based Li-ion batteries. However, silicon stores Li^+ ions through different mechanisms than graphite, raising the question of whether such a time-consuming SEI formation process is necessary. In this work, carbon-coated SiO_x is selected as a representative silicon material, and both Li/SiO_x half-cells and $SiO_x/LiNi_{0.8}Co_{0.1}Mn_{0.1}O_2$ (NCM811) full cells are assembled and cycled at varying current rates for the first 10 cycles, followed by identical cycling conditions for the subsequent cycles. The results show that the initial current rate has a minimal impact on the long-term cycling stability of both SiO_x-based Li metal half-cells and Li-ion cells. Notably, Li-ion cells formed at higher current rates exhibit lower overall impedance than those formed at lower current rates, consequently demonstrating better rate capability. These findings suggest that the time-consuming SEI formation process may not be necessary for the manufacturing of silicon-based Li-ion batteries, potentially simplifying production and reducing processing time.

Keywords: silicon anode; SEI formation; impedance; rate capability; Li-ion battery

1. Introduction

SEI formation is a critical process in the manufacture of graphite-based Li-ion batteries. During this process, a protective SEI layer forms on the surface of graphite particles, ensuring the safe and long-term cycling of the batteries. As suggested by its name, the SEI is an interphase that is highly conductive to Li^+ ions while being substantially electron-insulating. In batteries, the SEI functions to prevent solvated Li^+ ions from accessing the graphite surface, thereby avoiding solvent cointercalation into the graphite and the subsequent electrochemical reduction, which has been identified as the primary source for graphite structure exfoliation [1,2]. The SEI predominantly forms during the initial several cycles, especially the first charge, and it is primarily composed of the electrochemical reduction products of electrolyte components, mostly solvents. To form a high-quality SEI, it is essential that the reduction products of the electrolyte components are substantially insoluble and that the deposition kinetics are sufficiently slow to constitute a dense, stable, and highly ion-conductive structure. The latter is influenced largely by the current rate and temperature, with research showing that a slow current rate (0.05C–0.1C) [3] and a low-to-moderate temperature (25–45 °C) [4,5] favor the formation of a high-quality SEI. As such, SEI formation is a very time-consuming process, which significantly increases the manufacturing cost of Li-ion batteries. Additionally, the potential range was found to critically affect SEI properties. Generally, an SEI formed at higher potentials (>0.5 V vs. Li/Li^+) exhibits high ionic conductivity but inferior structural stability, whereas an SEI formed at lower potentials (<0.5 V vs. Li/Li^+) offers better structural stability with slightly lower ionic conductivity [6]. An SEI formed during the intercalation of Li^+ ions into graphite, namely at low potentials, has a particularly stable structure [7], indicating that the reduction kinetics of the solvents along with the dissociation of the solvated Li^+ ions is most favorable for forming a stable SEI. Based on these findings, various charging protocols have been explored to improve SEI stability and reduce formation time, such as multiple-step current [6,8] or voltage [9,10] charge formation, raising formation temperature [11–13], narrowing formation potential range [14,15], and pulse formation [16,17].

To date, the SEI formation process for graphite-based Li-ion batteries has matured significantly, and the same process has been widely applied to the manufacture of silicon-based Li-ion batteries [18–22]. However, silicon stores Li^+ ions through a mechanism entirely different from graphite. In graphite, the primary function of the SEI is to prevent solvent cointercalation into graphite together with Li^+ ions and the resulting reduction between the graphene planes of the graphite structure. This solvent cointercalation does not occur with silicon anodes, raising the question of whether the time-consuming SEI formation process is necessary for the manufacture of silicon-based Li-ion batteries. To address this question, in this work, we study the effect of the formation current rate on the cycling performance, particularly the cycling stability, of silicon by using SiO_x as the anode material and NCM811 as the cathode material. Surprisingly, we observed from both the Li/SiO_x half-cells and SiO_x/NCM811 full cells that the formation current rate has a minimal effect on the cycling performance of Si anode materials. Moreover, it was found that the cells formed at higher current rates exhibit lower overall impedance compared to those formed at lower current rates. These findings suggest that the time-consuming SEI formation process may not be necessary for Si-based Li-ion batteries, potentially reducing the manufacturing cost of these batteries significantly.

2. Materials and Methods

Carbon-coated SiO_x powder, with a D50 of 5.0 ± 1.0 μm and a D90 of 10.0 ± 1.0 μm, was purchased from MSE Supplies LLC (Tucson, AZ, USA) and coated onto a copper foil in a composition of 60% SiO_x, 20% Super-P carbon, and 20% poly(acrylic acid) binder by weight. The resulting SiO_x electrode had SiO_x loading of 1.51 ± 0.1 mg cm^{-2} and was punched into 1.27 cm^2 discs ($^1/_2$ inch diameter). A single-side-coated $LiNi_{0.8}Co_{0.1}Mn_{0.1}O_2$ (NCM811) cathode, with an NCM811 loading of 10.8 mg cm^{-2}, was also purchased from MSE Supplies LLC (Tucson, AZ, USA) and punched into 0.97 cm^2 discs ($^7/_{16}$-inch diameter). Both the SiO_x and NCM811 electrode discs were dried at 110 °C under vacuum overnight and then transferred to an argon-filled glove box. In the glove box, an electrolyte consisting of 1.0 m (molality) $LiPF_6$ dissolved in a 1:2:7 (wt.) mixture of fluorinated ethylene carbonate, ethylene carbonate, and ethylmethyl carbonate was prepared. Using the electrodes described above and a piece of Celgard 2350 membrane as the separator, CR2032 Li/SiO_x and SiO_x/NCM811 coin cells were assembled and filled with a fixed amount (40 μL) of electrolyte. The negative-to-positive (N/P) capacity ratio in SiO_x/NCM811 cells was controlled at 1.05 ± 0.01, based on specific capacities of 1500 mAh g^{-1} for SiO_x and 200 mAh g^{-1} for NCM811. The cells were tested on a Maccor Series 4000 cycler (Maccor, Tulsa, OK, USA) using two expressions for C rates: * C and * C/** h. The * C represents constant-current charging or discharging with a voltage cutoff, while * C/** h indicates constant-current charging or discharging followed by a voltage hold until the total charge/discharge time reaches ** h. The detailed testing conditions are provided in the discussion or figure captions. The AC impedance of the SiO_x/NCM811 coin cells before and after SEI formation was measured at 20 °C with a 10 mV perturbation in a frequency range from 100,000 Hz to 0.01 Hz, using SI 1260 Impedance/Gain-Phase Analyzer in combination with a Solartron SI 1287 Electrochemical Interface (AMETEK, Inc., Berwyn, PA, USA).

3. Results and Discussion

3.1. Li/SiO_x Half-Cell

The effect of the formation current rate on cycling performance was first evaluated in Li/SiO_x cells. The voltage profiles and differential capacity versus voltage plots for various current rates during the first cycle are shown in Figure 1a and Figure 1b, respectively. At a slow current rate (0.1C), the Li/SiO_x cell exhibits a specific capacity of 1452 mAh g^{-1} with a coulombic efficiency (CE) of 70.4%. The low CE can be attributed to two factors: (1) the irreversible reduction of SiO_x to Si accompanied by the formation of Li_2SiO_3 and the similar compounds, as representatively described by Equation (1), and (2) the irreversible reduction of electrolyte solvents, both contributing to the formation of the SEI on the surface of SiO_x particles.

$$3SiO_x + 2xLi^+ + 2xe^- \rightarrow (3-x)Si + xLi_2SiO_3 \quad (1)$$

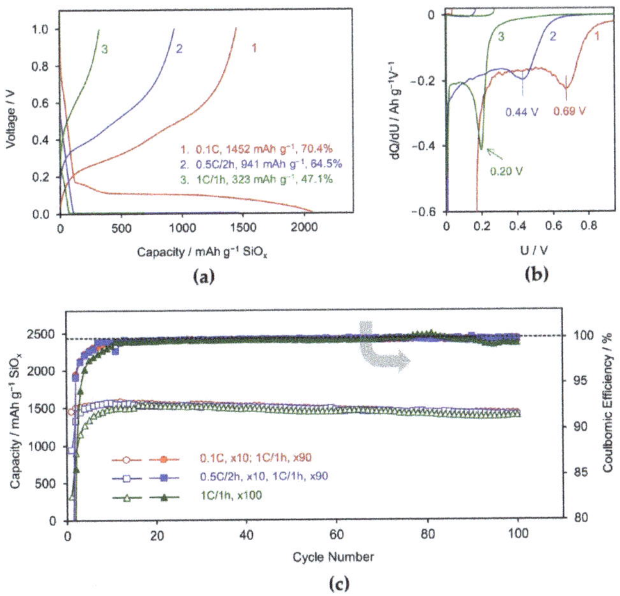

Figure 1. Performance comparison of Li/SiO_x half-cells formed at varying formation currents. (**a**) Voltage profile of the first cycle, (**b**) differential capacity vs. voltage plot for the first cycle, and (**c**) cycling performance, where from the 11th cycle, all cells were cycled under identical conditions with lithiation at 1C for 1 h and delithiation at 1C, cycling between 0.005 V and 1.0 V. The grey arrow indicates that these lines correspond to the right vertical axis.

In Equation (1), the resulting Si instantly forms Li–Si alloy through the electrochemical reduction of Li^+ ions. As the current rate increases, both the specific capacity and CE decrease. For instance, when the lithiation process is conducted at 1C for 1 h (1C/1 h), the specific capacity drops to 323 mAh g^{-1} and the CE decreases to 47.1%, as noted in the inset of Figure 1a. This decline is attributed to electric polarization resulting from the increased current rate. As observed in Figure 1a, the cell's voltage rapidly drops to the lithiation cutoff voltage (0.005 V), with most capacities achieved by the constant-voltage (CV) discharge step at 0.005 V. This effect is more clearly seen in the differential capacity plot (Figure 1b), where the differential capacity peak of the first reduction shifts from 0.69 V at 0.1C to 0.20 V at 1C due to significant polarization, leading to a substantial capacity reduction. It is worth noting that the large polarization at high current rates arises not only from the SiO_x electrode but also from the Li counter electrode [23]. The observed decrease in CE with increasing lithiation current rate can be attributed to both the reductions of SiO_x and the electrolyte solvents occurring at higher potentials than Li–Si alloying. Their contribution to the low CE becomes more pronounced at higher currents.

After 10 formation cycles at varying current rates, the cells were cycled using an identical current rate: constant current–constant voltage (CC–CV) lithiation at 1C for 1 h, followed by constant current (CC) delithiation at 1C. The cycling performances of the cells are compared in Figure 1c. The results indicate that the current rate during formation cycles has a minor effect on the capacity retention of Li/SiO_x cells, although it significantly affects the specific capacity and CE during the initial several formation cycles. These findings reveal that the current rate in formation cycles has minimal impact on the cycling stability of Li/SiO_x cells, differing significantly from observations in Li/graphite cells. This difference can be attributed to the distinct Li storage mechanisms between graphite and silicon. In graphite, Li^+ ions intercalate between the graphene layers, during which graphite is reduced, becoming negatively charged to accommodate the positively charged Li^+ ions. In contrast, in silicon, Li^+ ions are first reduced to Li atoms on the surface of the silicon particles; the resulting Li atoms then react with silicon to form a Li–Si alloy. As the process continues, the Li–Si alloy layer gradually moves inward toward the core of the silicon particles. Therefore, the solvent cointercalation related reduction does not occur with the Si electrodes. In addition to the observations mentioned above, Figure 1c reveals that during the initial cycles, both the cells' specific capacity and CE gradually increase until they reach a plateau. This behavior can be attributed to the slow conversion kinetics of SiO_x to Si, preventing the full capacity of SiO_x from being accessed within a single cycle.

3.2. *SiO_x/NCM811 Full Cells*

3.2.1. Formation Cycle

The effect of the formation current rate on cycling performance was further investigated in SiO_x/NCM811 cells. Figure 2 compares the voltage profiles and differential capacity versus voltage plots of the first two cycles for SiO_x/NCM811 cells formed by charging and discharging at 0.1C and charging at 1C for 1 h while discharging at 1C, respectively. Unlike the results observed from Li/SiO_x half-cells, both cells exhibit nearly identical specific capacities and CE. The specific capacity slightly decreases from the first to the second cycle, while the CE increases significantly, as noted in the insets of Figure 2a,c. The primary difference between the two formation protocols is that the cell charged at 1C/1 h experiences higher polarization due to the high current applied. During discharge, however, this increased polarization is only evident in the high-voltage region, as shown in Figure 2a,c. Since the irreversible conversion of SiO_x to Si in the first cycle consumes a large number of Li^+ ions from the NCM811 cathode, both the specific capacity and CE of NCM811 of the SiO_x/NCM811 cells in the first cycle are significantly lower than the normal values generally obtained in Li/NCM811 cells. As shown in the inset of Figure 2a, the SiO_x/NCM811 cells exhibit a first-cycle specific capacity of only 141.5–145.3 mAh g^{-1} and a CE of 63.4–63.9%. Although the CE increases to 94.9% in the second cycle, its value is far below the threshold required for high-performance Li-ion batteries. The low initial CE and its gradual increase over subsequent cycles can be attributed to the slow and incomplete conversion of SiO_x to Si during the early cycling stages. This slow conversion along with the initial low CE presents a significant challenge for the practical application of SiO_x as an anode material in Li-ion batteries. Additionally, it should be noted that the 1C/1 h formation used in this work is only an example, demonstrating that a slow SEI formation process is not necessarily required. The practical formation rate may vary depending on cell design, for instance, an energy-focused vs. a power-focused design. As a rule of thumb, the formation rate can be the same as those used in normal cycling.

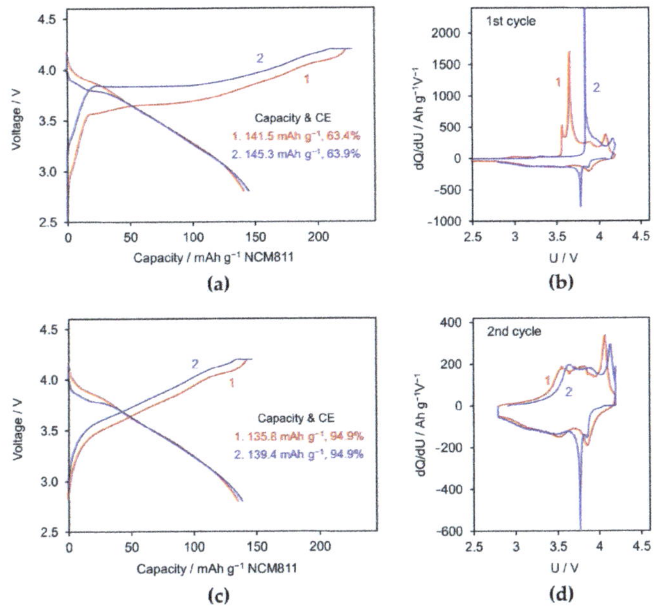

Figure 2. Voltage profile and differential capacity vs. voltage plot of SiO_x/NCM811 cells formed at varying currents, where Cell 1 was formed at 0.1C and Cell 2 at 1C for 1 h. (**a**,**b**) First cycle and (**c**,**d**) second cycle.

The impedances of two SiO_x/NCM811 cells subjected to different formation protocols were measured before and after the formation process, and the results are compared in Figure 3. Before the formation process (i.e., the freshly assembled cells), the two cells had similar overall impedance. After two formation cycles, the cells ended at the discharged state, i.e., at 0% state of charge (SOC). At the discharged state, the cell formed using the 1C/1 h charge and 1C discharge protocol displayed significantly lower overall impedance than the cell formed using the 0.1C cycling protocol. Subsequently, the cells were charged to 100% SOC by charging at 0.5C to 4.2 V, followed by a voltage hold at 4.2 V for a total charge time of 2 h. At 100% SOC, the similar impedance difference between the two cells persisted. This discrepancy can be attributed to the fact that the cell formed under the 1C/1 h charge protocol had much lower SEI resistance (R_{SEI}) than the cell formed under the 0.1C protocol, as indicated by the smaller semicircle in the high-frequency region of the Nyquist plot in Figure 3. Furthermore, a similar trend extended the charge-transfer resistance (R_{ct}), represented by the second semicircle in the low-frequency region. The cell formed under the 1C/1 h protocol demonstrated substantially lower R_{ct}, contributing to its overall lower impedance compared to the 0.1C formed cell. The above results suggest that the SEI formed under high current rates is more porous, facilitating greater electrolyte penetration and enhancing ionic conductivity.

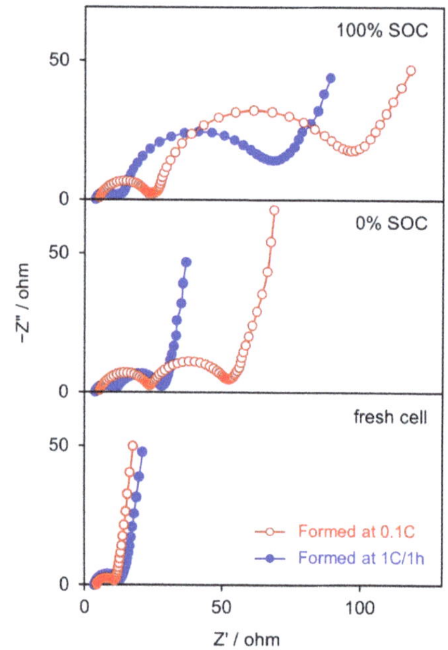

Figure 3. Nyquist plot of SiO_x/NCM811 cells at various SOCs after two formation cycles at different formation conditions.

3.2.2. Rate Capability

After completing the impedance analysis, the cells underwent rate capability testing. Figure 4 compares the discharge rate capability of two cells. Both cells were charged at 0.5C to 4.2 V, followed by a voltage hold at 4.2 V for a total charge time of 2 h, and they were subsequently discharged at various current rates. As shown in Figure 4a, at low discharge rates (<3C), both cells formed using 0.1C and 1C/1 h formation protocols, respectively, exhibited similar specific capacities. However, at high discharge rates (5C and 6C), the cell formed using the 1C/1 h protocol displayed superior capacity. This trend aligns with the impedance results and is attributed to the lower overall impedance of the cell formed at the higher current rate. Figure 4b presents the voltage profiles of the two cells at representative discharge rates, illustrating that differences in both specific capacity and polarization become more pronounced as the discharge current increases.

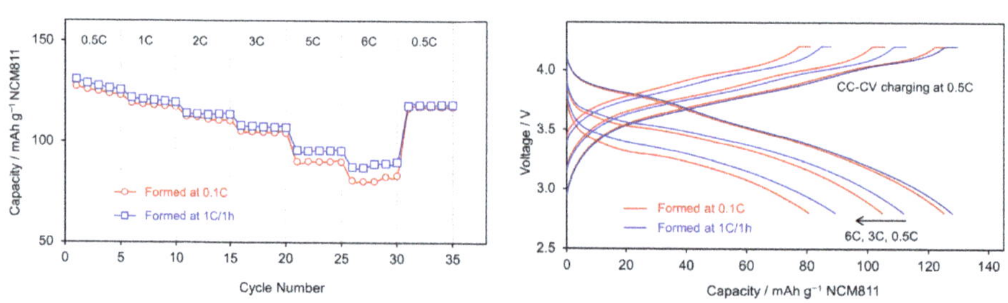

(a) (b)

Figure 4. Discharge capability of SiO_x/NCM811 cells formed at different current rates. The cells were charged using a CC–CV protocol at 0.5C to 4.2 V, followed by a voltage hold at 4.2 V for a total charge time of 2 h, and then discharged at various rates. (**a**) Specific capacity and (**b**) voltage profile.

Next, the charge rate capability of the cells was evaluated using a CC–CV charging protocol. The cells were charged at a specific rate to 4.2 V and then held at 4.2 V for a total time corresponding to the specific rate, followed by discharge at 0.5C to 2.8 V. The specific capacities of the cells are compared in Figure 5a, while their voltage profiles at representative charging current rates are shown in Figure 5b. Similar to the discharge rate capability results, the difference in specific capacity between the two cells becomes more pronounced as the charge current rate increases. The cell formed using the higher current rate (1C/1 h protocol) consistently outperforms the one formed at the lower current rate (0.1C). As illustrated in Figure 5b, the differences in charge capacity between the two cells at 3C and 6C primarily exist during the CC charging stage. Specifically, the cell formed using the 1C/1 h protocol exhibits a larger CC charge capacity compared to the cell formed at 0.1C, which correlates with its lower overall impedance.

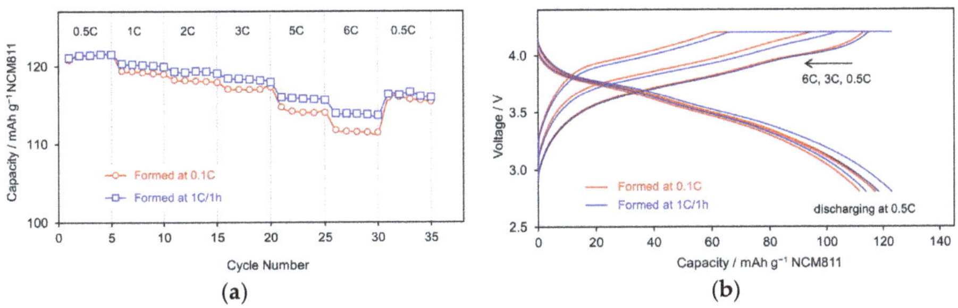

Figure 5. Charge capability of SiO_x/NCM811 cells formed at different current rates. The cells were CC-charged at a specific current rate to 4.2 V, followed by a voltage hold at 4.2 V until the designated time was reached, and then galvanostatically discharged at 0.5C to 2.8 V. (**a**) Specific capacity and (**b**) voltage profile.

Comparing Figures 4a and 5a shows that, with an increase in the current rate, the charge rate capability significantly outperforms the discharge rate capability, with the former exhibiting much higher capacities than the latter. This discrepancy arises from differences in the testing protocols. For the charge rate capability, a CC–CV charging step was applied, whereas the discharge rate capability testing utilized only a CC discharging step. The results are expected to align if a CV discharging step, i.e., a voltage hold at 2.8 V for a certain duration, is incorporated into the latter.

3.2.3. Cycling Stability

Another set of cells was tested for long-term cycling stability, as shown in Figure 6. In this experiment, Cell 1 underwent formation by cycling at 0.1C for 10 cycles, followed by a 1C/1 h charge and a 1C discharge. Cell 2 was formed at 1C for 10 cycles without a voltage hold at 4.2 V, followed by the same 1C/1 h charge and 1C discharge. As observed in Figure 6, during the initial 10 formation cycles, Cell 2 exhibited lower capacities than Cell 1 due to the absence of the CV charging step at 4.2 V, which limited full-capacity access. However, in the subsequent cycles, both cells demonstrated nearly identical specific capacities and capacity retention. This observation indicates that the formation current rate has minimal impact on the cycling performance, in terms of specific capacity and capacity retention, of the SiO_x-based Li-ion cells. Therefore, the traditionally time-consuming formation process may not be necessary for manufacturing Si-based Li-ion batteries. Furthermore, the Li-ion cells formed at higher current rates exhibit lower overall impedance compared to those formed using the more time-intensive, low-current rate process.

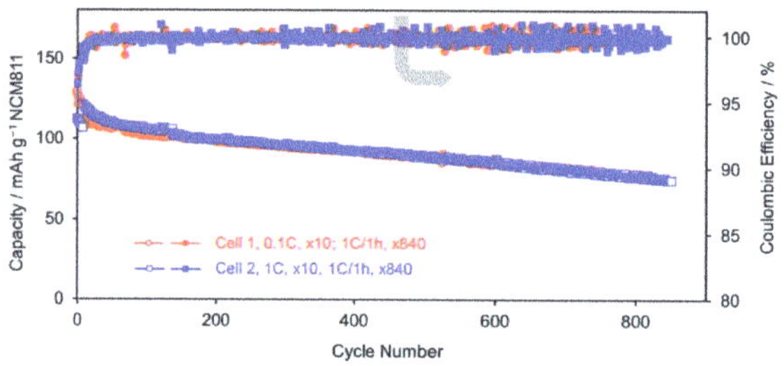

Figure 6. Cycling stability of SiO_x/NCM811 cells cycled at 1C between 2.8 V and 4.2 V, of which Cell 1 was formed at 0.1C for 10 cycles and Cell 2 was formed at 1C without a CV charging step for 10 cycles. The grey arrow indicates that these lines correspond to the right vertical axis.

4. Conclusions

In summary, we studied the effect of formation current rates on the cycling performance of Si-based Li-ion batteries, using carbon-coated SiO_x as a representative silicon material. It is uncovered that, in both Li/SiO_x and SiO_x/NCM811 cells, the formation current rate has minimal impact on cycling performance, specifically regarding both the specific capacity and capacity retention of the Si electrodes. This behavior contrasts with that observed in traditional graphite-based Li-ion batteries and can be attributed to the different Li^+ storage mechanisms between silicon and graphite. Silicon stores Li^+ ions through Li–Si alloying, eliminating issues such as solvent cointercalation and associated reductions, which occurs with graphite. Moreover, SiO_x/NCM811 cells formed at higher current rates exhibit lower overall impedance compared to those formed at lower current rates, enhancing their rate capabilities. The results of this work suggest that the traditionally time-consuming SEI formation process may not be necessary for the production of Si-based Li-ion batteries. This advantage, rooted in the fundamental differences between the Li^+ storage mechanisms of silicon and graphite, could significantly reduce the manufacturing costs of these batteries.

Funding: DEVCOM Army Research Laboratory.

Data Availability Statement: The data presented in this study are available on request from the corresponding author.

Acknowledgments: The author acknowledges the support of the DEVCOM Army Research Laboratory.

Conflicts of Interest: The author declares no conflicts of interest.

References

1. Wood, D.L.; Li, J.; An, S.J. Formation Challenges of Lithium-Ion Battery Manufacturing. *Joule* **2019**, *3*, 2884–2888. [CrossRef]
2. Schomburg, F.; Heidrich, B.; Wennemar, S.; Drees, R.; Roth, T.; Kurrat, M.; Heimes, H.; Jossen, A.; Winter, M.; Cheong, J.Y.; et al. Lithium-ion battery cell formation: Status and future directions towards a knowledge-based process design. *Energy Environ. Sci.* **2024**, *17*, 2686–2733. [CrossRef]
3. Oh, S.M.; Song, J.; Lee, S.; Jang, I.-C. Effect of current rate on the formation of the solid electrolyte interphase layer at the graphite anode in lithium-ion batteries. *Electrochim. Acta* **2021**, *397*, 139269. [CrossRef]
4. Zhang, S.S.; Xu, K.; Jow, T.R. Optimization of the forming conditions of the solid-state interface in the Li-ion batteries. *J. Power Sources* **2004**, *130*, 281–285. [CrossRef]
5. He, Y.; Wang, J.; Wang, L.; He, X. Thermodynamic Understanding of Formation and Evolution of Solid Electrolyte Interface in Li-Ion Batteries. *Batter. Supercaps* **2024**, *7*, e202400059. [CrossRef]
6. Antonopoulos, B.K.; Stock, C.; Maglia, F.; Hoster, H.E. Solid electrolyte interphase: Can faster formation at lower potentials yield better performance? *Electrochim. Acta* **2018**, *269*, 331–339. [CrossRef]
7. Zhang, S.S.; Xu, K.; Jow, T.R. EIS study on the formation of solid electrolyte interface in Li-ion battery. *Electrochim. Acta* **2006**, *51*, 1636–1640. [CrossRef]

8. Chiang, P.-C.J.; Wu, M.-S.; Lin, J.-C. A Novel Dual-Current Formation Process for Advanced Lithium-Ion Batteries. *Electrochem. Solid-State Lett.* **2005**, *8*, A423. [CrossRef]
9. Pathan, T.S.; Rashid, M.; Walker, M.; Widanage, W.D.; Kendrick, E. Active formation of Li-ion batteries and its effect on cycle life. *J. Phys. Energy* **2019**, *1*, 044003. [CrossRef]
10. Li, Y.; Guo, J.; Pedersen, K.; Gurevich, L.; Stroe, D.-I. Investigation of multi-step fast charging protocol and aging mechanism for commercial NMC/graphite lithium-ion batteries. *J. Energy Chem.* **2023**, *80*, 237–246. [CrossRef]
11. Ellis, L.D.; Allen, J.P.; Hill, I.G.; Dahn, J.R. High-Precision Coulometry Studies of the Impact of Temperature and Time on SEI Formation in Li-Ion Cells. *J. Electrochem. Soc.* **2018**, *165*, A1529. [CrossRef]
12. Moretti, A.; Sharova, V.; Carvalho, D.V.; Boulineau, A.; Porcher, W.; de Meatza, I.; Passerini, S. A Comparison of Formation Methods for Graphite//$LiFePO_4$ Cells. *Batter. Supercaps* **2019**, *2*, 240–247. [CrossRef]
13. Yan, C.; Yao, Y.-X.; Cai, W.-L.; Xu, L.; Kaskel, S.; Park, H.S.; Huang, J.-Q. The influence of formation temperature on the solid electrolyte interphase of graphite in lithium ion batteries. *J. Energy Chem.* **2020**, *49*, 335–338. [CrossRef]
14. Lee, H.-H.; Wang, Y.-Y.; Wan, C.-C.; Yang, M.-H.; Wu, H.-C.; Shieh, D.-T. A fast formation process for lithium batteries. *J. Power Sources* **2004**, *134*, 118–123. [CrossRef]
15. An, S.J.; Li, J.; Du, Z.; Daniel, C.; Wood, D.L. Fast formation cycling for lithium ion batteries. *J. Power Sources* **2017**, *342*, 846–852. [CrossRef]
16. Wang, F.-M.; Wang, H.-Y.; Yu, M.-H.; Hsiao, Y.-J.; Tsai, Y. Differential pulse effects of solid electrolyte interface formation for improving performance on high-power lithium ion battery. *J. Power Sources* **2011**, *196*, 10395–10400. [CrossRef]
17. Wang, F.-M.; Wang, J.-C.; Rick, J. Forward and reverse differential-pulse effects applied in the formation of a solid electrolyte interface to enhance the performance of lithium batteries. *Electrochim. Acta* **2014**, *147*, 582–588. [CrossRef]
18. Arreaga-Salas, D.E.; Sra, A.K.; Roodenko, K.; Chabal, Y.J.; Hinkle, C.L. Progression of Solid Electrolyte Interphase Formation on Hydrogenated Amorphous Silicon Anodes for Lithium-Ion Batteries. *J. Phys. Chem. C* **2012**, *116*, 9072–9077. [CrossRef]
19. Horowitz, Y.; Ben-Barak, I.; Schneier, D.; Goor-Dar, M.; Kasnatscheew, J.; Meister, P.; Grünebaum, M.; Wiemhöfer, H.-D.; Winter, M.; Golodnitsky, D.; et al. Study of the Formation of a Solid Electrolyte Interphase (SEI) on a Silicon Nanowire Anode in Liquid Disiloxane Electrolyte with Nitrile End Groups for Lithium-Ion Batteries. *Batter. Supercaps* **2019**, *2*, 213–222. [CrossRef]
20. Kim, J.; Chae, O.B.; Lucht, B.L. Perspective—Structure and Stability of the Solid Electrolyte Interphase on Silicon Anodes of Lithium-ion Batteries. *J. Electrochem. Soc.* **2021**, *168*, 030521. [CrossRef]
21. Yang, G.; Frisco, S.; Tao, R.; Philip, N.; Bennett, T.H.; Stetson, C.; Zhang, J.-G.; Han, S.-D.; Teeter, G.; Harvey, S.P.; et al. Robust Solid/Electrolyte Interphase (SEI) Formation on Si Anodes Using Glyme-Based Electrolytes. *ACS Energy Lett.* **2021**, *6*, 1684–1693. [CrossRef]
22. Chen, B.; Xu, D.; Chai, S.; Chang, Z.; Pan, A. Enhanced Silicon Anodes with Robust SEI Formation Enabled by Functional Conductive Binder. *Adv. Funct. Mater.* **2024**, *34*, 2401794. [CrossRef]
23. Zhang, S.S. Is Li/Graphite Half-Cell Suitable for Evaluating Lithiation Rate Capability of Graphite Electrode? *J. Electrochem. Soc.* **2020**, *167*, 100510. [CrossRef]

batteries

MDPI

Article

Structural Design of Dry-Processed Lithium-Rich Mn-Based Materials with High Loading for Enhanced Energy Density

Yujie Ma, Haojin Guo, Tai Yang * and Zhifeng Wang *

School of Materials Science and Engineering, Hebei University of Technology, Tianjin 300401, China; 202231802015@stu.hebut.edu.cn (Y.M.); 202221801135@stu.hebut.edu.cn (H.G.)
* Correspondence: yangtai@hebut.edu.cn (T.Y.); wangzf@hebut.edu.cn (Z.W.); Tel.: +86-22-60204129 (T.Y.); +86-22-60202006 (Z.W.)

Abstract: With the growing demand for electric vehicles and consumer electronics, lithium-ion batteries with a high energy density are urgently needed. Lithium-rich manganese-based materials (LRMs) are known for their high theoretical specific capacity, rapid electron/ion transfer, and high output voltage. Constructing electrodes with a substantial amount of active materials is a viable method for enhancing the energy density of batteries. In this study, we prepare thick LRM electrodes through a dry process method of binder fibrillation. A point-to-line-to-surface three-dimensional conductive network is designed by carbon agents with various morphologies. This structural design improves conductivity and facilitates efficient ion and electron transport due to close particle contact and tight packing. A high-loading cathode (35 mg cm^{-2}) is fabricated, achieving an impressive areal capacity of up to 7.9 mAh cm^{-2}. Moreover, the pouch cell paired with a lithium metal anode exhibits a remarkable energy density of 949 Wh kg^{-1}. Compared with the cathodes prepared by the wet process, the dry process optimizes the pathways for e^-/Li^+ transport, leading to reduced resistance, superior coulombic efficiency, retention over cycling, and minimized side reaction. Therefore, the novel structural adoption of the dry process represents a promising avenue for driving innovation and pushing the boundaries for enhanced energy density for batteries.

Keywords: Li-ion battery; Li-rich Mn-based materials; cathode; energy density

1. Introduction

Lithium-ion batteries (LIBs) have become the leading energy storage technology due to their outstanding properties, such as a high energy density, superior power output, extended cycle life, and exceptional reliability [1–3]. As global demand continues to surge, increasing energy density has become a paramount challenge for the battery industry [4]. However, the energy density of commercial cathode materials has largely plateaued, highlighting the urgent need for innovations in manufacturing processes. By advancing process engineering, it is possible to achieve breakthroughs in efficiency, scalability, and performance, paving the way for developing batteries with even greater capacity and power [5]. One promising approach to enhancing energy density is the development of electrodes with a higher proportion of active materials. This strategy reduces the reliance on non-active components, such as separators and conductive additives, thereby maximizing the energy storage potential of LIBs [6,7].

The conventional wet processing technique frequently faces challenges in handling thick electrodes, especially during the solvent evaporation stage. This process may lead to several issues, like the non-uniform distribution of binders and conductive carbon materials [8]. N-methyl-2-pyrrolidone (NMP), widely employed as a solvent in the wet processing of mixing cathode composites, presents environmental and economic challenges due to its evaporation and recycling [9,10]. Dry electrode technology signifies a transformative advancement in energy storage manufacturing, delivering substantial advantages that extend far beyond process simplification and environmental footprint reduction. By obviating the requirement for solvents, dry electrode fabrication substantially mitigates operational hazards and ecological risks. This solvent-free approach contributes to safer and more sustainable battery production. Additionally, the inherent scalability and cost-effectiveness of dry processing techniques make them well suited for large-scale production, facilitating the widespread adoption of energy storage solutions across various industries [11,12].

Dry processing techniques are particularly well suited for constructing electrodes with a high loading of active materials, which has been widely studied by researchers [13]. While Tesla has successfully implemented dry processing for anode fabrication, technical hurdles persist in cathode adaptation due to PTFE fibrillation challenges. During the process of binder fibrillation, the active materials, conductive carbon, and binder polytetrafluoroethylene (PTFE) are mixed to achieve uniform dispersion. During shear processing, PTFE undergoes microstructural transformation from spherical particles to interconnected fibrils that form percolating networks. These nanofibrils mechanically interlock active material particles with conductive carbon additives, achieving greater improvements [14–16].

Layered lithium-rich manganese-based materials (LRMs) have emerged as promising next-generation cathode candidates for LIBs due to their high specific capacity (>250 mAh g^{-1}) and low cost [17]. This study employs LRM as a cathode active material, synergistically integrating three conductive additives: Super C65 carbon black, vapor-grown carbon fibers (VGCFs), and graphene nanosheets. By constructing a point-to-line-to-surface three-dimensional conductive network, the intimate contact between the active material particles and the conductive carbon particles is significantly enhanced. This intricate network architecture facilitates efficient charge transfer pathways, minimizes electron transport resistance, and maximizes the utilization of active materials. The LRM cathodes prepared by the binder fibrillation process were adjusted to various thickness and mass loadings ranging from 15 mg cm^{-2} to 35 mg cm^{-2}. These cathodes exhibited a high areal density of 7.9 mAh cm^{-2} and a remarkable energy density of 949 Wh kg^{-1}. Compared with the wet process, the dry-processed cathodes optimize the pathways for e^-/Li^+ transport. This optimization leads to reduced resistance, superior Coulombic efficiency, enhanced retention over cycling, and minimized side reactions.

2. Materials and Methods

2.1. Preparation of Electrode

The synthetic route of the LRM can be found in Supporting S1 [18]. The cathode composite was formulated by integrating lithium-rich manganese-based oxide with a ternary conductive system (C65:graphene:VGCF = 1:1:4 *w*/*w*) and polytetrafluoroethylene (PTFE) binder at precisely controlled mass ratios of 90:6:4, and then 200 °C drying for 1 h. The mixed material was rolled into film by a roller several times until it became a dense self-supporting film. The areal loadings of the self-supporting films with a thickness of 60 μm, 100 μm and 130 μm are 15 mg cm^{-2}, 25 mg cm^{-2} and 35 mg cm^{-2}, respectively. For comparison, the electrode prepared via the wet process was composed of LRM, a composite conductive agent (C65:graphene:VGCF = 1:1:4), and polyvinylidene fluoride (PVDF) according to the mass ratio of 90:6:4.

2.2. Material Characterization

The electrode morphology was characterized using a scanning electron microscope (SEM, Zeiss Merlin, Bochum, Germany) at an acceleration voltage of 5 kV. Surface chemistry analysis was performed by X-ray photoelectron spectra (XPS, ESCALAB 250Xi, Waltham, MA, USA) featuring a monochromatic Al K_α (1486.6 eV) X-ray source, focused on a 400 μm spot under a vacuum pressure lower than 1×10^{-7} mbar. Contact angle measurements were performed with a KRUSS DSA100 (Hamburg, Germany). The structural changes of the cathode, both before and after cycling, were analyzed by X-ray diffraction (XRD, Rigaku D8 Discover, Karlsruhe, Germany) using a PANalytical Empyrean Series 2 equipped with a Cu K_α radiation source (45 kV, 40 mA). The electrical resistivity of the electrodes was measured using an IEST BER2500 (Xiamen, China). For large-area cross-sectional milling and subsequent SEM imaging, a Thermo Scientific Helios 5 Hydra Plasma Focused Ion Beam (PFIB) DualBeam system (Göteborg, Sweden) was utilized. Cross-sections of various electrode samples were milled and cleaned using a Xe^+ ion source. High-resolution transmission electron microscopy (HRTEM) images of both pristine and cycled samples were acquired using a ThermoFisher Talos F200X (Waltham, MA, USA), equipped with a Ceta camera and operated at 200 kV. BET analysis of various particles was conducted using a gas adsorption/surface area analyzer (Micromeritic ASAP 2020, Atlanta, GA, USA). The TOF-SIMS (time-of-flight secondary ion mass spectrometry) measurements were performed using TOF-SIMS 5-100 (IONTOF, Ingolstadt, Germany) with a primary ion beam of Bi^{3+} and sputtering ion beam of Cs^+.

2.3. Electrochemical Measurements

Electrochemical performance was evaluated in an argon-filled glove box using 2032-type coin cells with a Li metal anode, polyethylene (PE) separator, and Shanshan 003 liquid electrolyte. Cell performance was tested within the 2.0–4.8 V vs. Li/Li^+ range. The cycling test was carried out under various charging/discharging conditions. EIS and CV were performed using an electrochemical workstation (Interface 1010E, Gamry Instruments, Philadelphia, PA, USA). The lithium metal pouch cell was assembled with a 100 μm lithium metal anode in a dry room at a −50 °C dew point.

3. Results and Discussion

Figure 1 illustrates a schematic diagram of the binder fibrillation process. During the process, LRM, conductive carbon, and PTFE are mixed, followed by the induction of fibrillation of PTFE under shear force. The resulting fibers, with diameters of only a few nanometers and lengths of tens of micrometers, form a three-dimensional network structure that aggregates the active material and carbon black (Supplementary Materials, Figure S1). Subsequently, this mixture is compressed between two rollers to form films of varying thickness. The conductive agent plays a critical role in electrode performance. Its type and content can influence the flexibility and mechanical strength of the electrode, thereby affecting the battery's flexibility and durability. The conductive agent also possesses excellent electrical conductivity, facilitating electron transport within the electrode, which reduces internal resistance and improves overall conductivity. Additionally, the conductive agent helps to form a conductive network, establishing effective electron transfer pathways between the active material and the electrode substrate, thereby enhancing the electrode's structure, stability, and cycling life. Therefore, when designing electrodes, we selected different carbon and carbon composites to optimize performance. As shown in Table S1, we performed a series of control tests with different carbon ratios, and the four ratios were determined based on the integrity and flexibility of film during the preparation process. As shown in Figure S2, the carbon composites of VGCF, C65, and graphene with the mass ratio of 4:1:1 demonstrated higher conductivity than single VGCF, VGCF and C65 (mass ratio of 5:1), and VGCF and graphene (mass ratio of 5:1). Therefore, the point-to-line-to-surface three-dimensional conductive network facilitated efficient electron transport throughout the cathodes and enhanced overall conductivity.

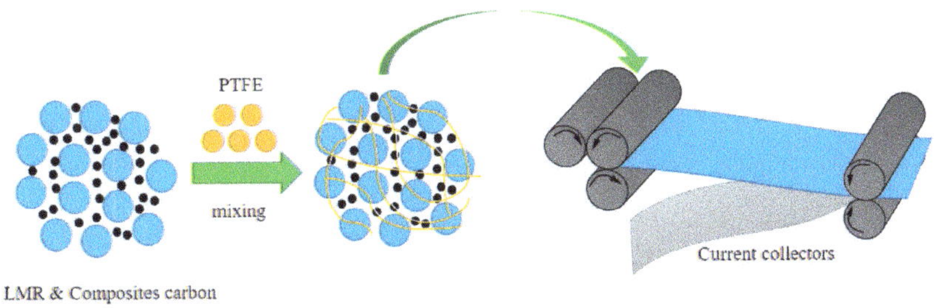

Figure 1. The schematic diagram of the binder fibrillation process.

Figure 2 illustrates the cross-section topography of LRM electrodes with varying thicknesses (60, 100, 130 μm). It is evident that LRM particles, carbon, and PTFE are closely intertwined, forming a cohesive and conductive framework. As shown in Figure S3, the XRD patterns of the LRM raw materials and dry-processed electrodes are consistent, indicating that the macroscopic crystal structure is maintained. This suggests that the dry-processing method has a minimal impact on the material's structure and may offer advantages in terms of structural stability. This fibrous structure provides excellent mechanical support and conductive pathways for the electrode, enhancing its stability and electrochemical performance. As the electrode thickness decreases, a finer and more interconnected network of fibers is observed throughout the electrode. This tighter integration improves electrical conductivity and facilitates efficient charge transfer. Additionally, the reduced thickness brings the electrode components closer together, maximizing their interaction and synergistic effects. The rolling process is effective in controlling the porosity and compaction density of the cathodes. As shown in Figure S4, increasing thickness results in lower compaction density and higher porosity. Contact angle results (Figure 2c,f,i) indicate that, as the thickness decreases, the contact angle increases, reducing wettability. To optimize wettability, we proactively increased the volume of electrolytes during the battery assembly process. Firstly, we completely flooded the cathodes with electrolytes for 2h before assembling the cell, and then we added 150 μl of electrolytes to assemble the cell to ensure a sufficient amount of electrolytes during the measurement. By doing so, we aimed to improve the penetration and coverage of the electrode surface, enhancing the electrolyte–electrode interaction. This adjustment not only promotes efficient ion transport within the electrode but also improves electrochemical performance by minimizing electrolyte depletion and ensuring adequate contact between the electrolyte and active materials. Additionally, optimizing electrolyte volume helps to address reduced wettability due to higher electrode packing densities, ensuring consistent and reliable battery performance under various operating conditions.

The rate and cycle performances of the Li-rich manganese material were tested, with the results shown in Figure 3 (1 C = 200 mAh g^{-1}). At a current rate of 0.1 C, the cathode with a mass loading of 15 mg cm^{-2} achieved an initial discharge capacity of 270 mAh g^{-1}. Increasing the mass loading to 25 and 35 mg cm^{-2} resulted in initial discharge capacities of 250 mAh g^{-1} and 226 mAh g^{-1}, respectively (Figure 3a). Consequently, a high areal capacity of 7.9 mAh cm^{-2} was achieved with a mass loading of 35 mg cm^{-2}. During rate testing from 0.1 C to 2 C, the electrode with a mass loading of 15 mg cm^{-2} demonstrated significantly superior rate capability compared to the others (Figure 3b). As shown in Figure 3c, after 20 cycles, the electrode with a 15 mg cm^{-2} loading exhibited a capacity retention rate of 87.8%, outperforming the electrodes with 25 mg cm^{-2} (81.1%) and 35 mg cm^{-2} (63.3%) loadings. The cycling stability exhibited an inverse correlation with electrode mass loading, as evidenced by a more pronounced capacity degradation in higher-loading samples. Notably, the cathode fabricated with a mass loading of 15 mg cm^{-2} demonstrated 65% capacity retention after 100 charge/discharge cycles (Figure S5). It was found that, under low-rate discharge conditions, electrode thickness has a minimal effect on battery efficiency. However, as the electrode thickness increases and discharge rates rise, battery efficiency declines. The rate capability of the battery decreases with both increasing electrode thickness and higher discharge rates. A Li||LRM pouch cell was assembled using Li metal (100 μm) and LRM (25 mg cm^{-2}) (Figure S6).

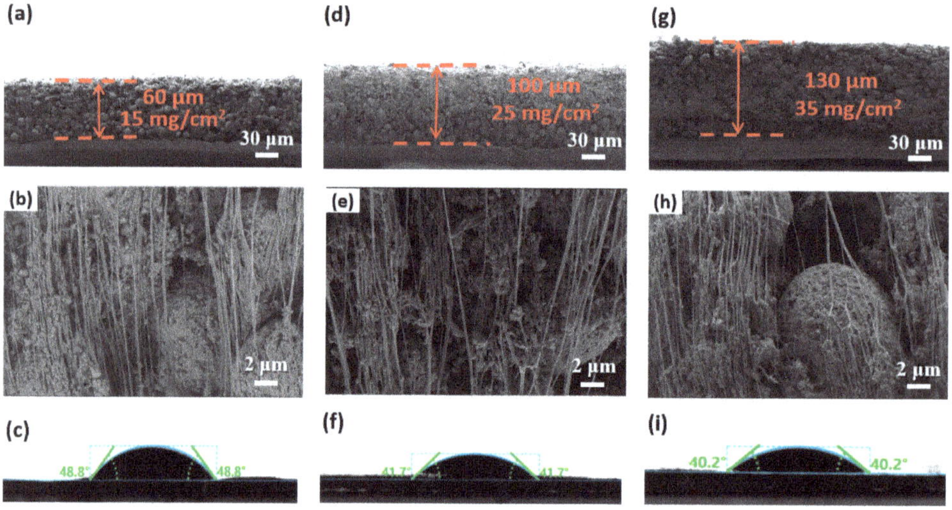

Figure 2. Cross-section SEM images (**a**,**d**,**g**), SEM images (**b**,**e**,**h**) and Contact angles (**c**,**f**,**i**) of LRM with various mass loading.

The pouch cell exhibited an initial discharge capacity of 300 mAh g^{-1} at 0.1 C, corresponding to an energy density of 949 Wh kg^{-1}. We investigated the impact of thickness on efficiency and rate capability using time-of-flight secondary ion mass spectrometry (TOF-SIMS) analysis of cycled cathodes (Figure 3d). Interactions with organic electrolytes lead to surface structural degradation inside the cathodes [19,20]. The quantities of F^-, PO^{2-}, and C_2HO^- species identified in the cathodes with 15 mg cm^{-2} were significantly lower compared to those of cathodes with 25 and 35 mg cm^{-2} [21].

SEM images of the electrodes after cycling reveal a clear trend: as the thickness of the electrodes increases, the presence of surface cracks becomes more pronounced (Figure 4). This provides additional evidence of the heightened occurrence of side reactions in thicker electrodes. Cracks on the electrode surface provide penetration paths for the electrolyte, allowing the electrolyte to more readily access the interior of the electrode material, thus exacerbating side reactions at the electrode/electrolyte interface [22–24]. The cracks can compromise the structural integrity of the electrode, contributing to higher impedance and reduced cycling stability, as confirmed by the tests.

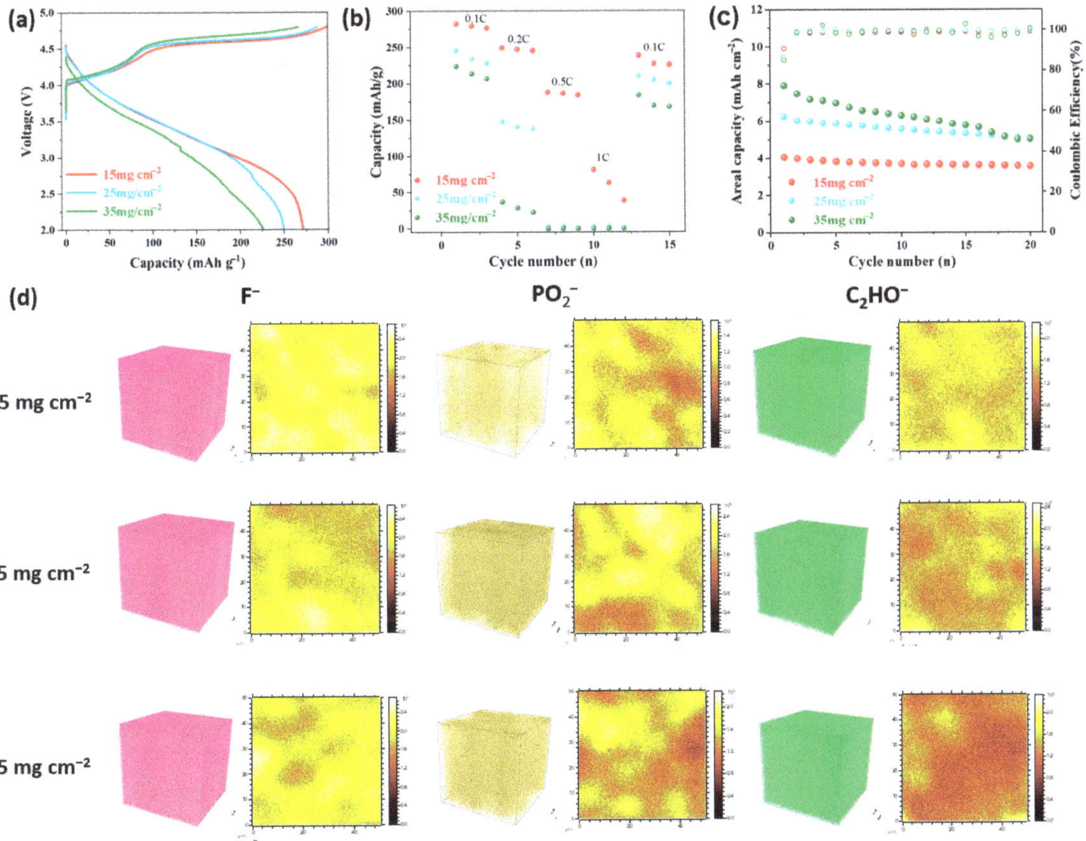

Figure 3. (**a**) Charge/discharge voltage profiles, (**b**) rate capability performance, (**c**) cycling performance of cathodes with various mass loadings at 0.1C and a voltage range of 2.0–4.8 V. (**d**) The 3D views and chemical imaging for F^-, PO^{2-}, and CH2O$^-$ of cycled cathodes.

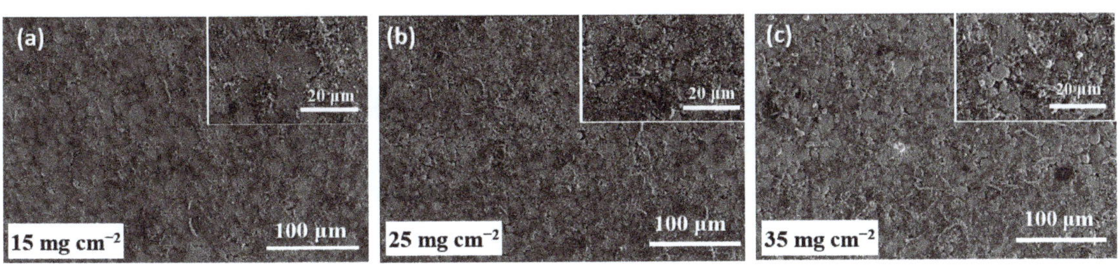

Figure 4. SEM of cycled cathodes with various mass loadings: (**a**) 15 mg cm^{-2}, (**b**) 25 mg cm^{-2}, (**c**) 35 mg cm^{-2}.

As electrode thickness increases, there is enhanced infiltration of the electrolyte into the electrode structure; this suggests that the infiltration may lead to an increase in secondary reactions within the electrode, which could negatively impact battery performance and cycling stability. For thick electrode batteries, transport limitations of the electrolyte and ohmic polarization are the primary factors limiting efficiency at high discharge rates [25]. An et al. demonstrated the spatial distribution of Li^+ at the end of discharge in cathodes with various thicknesses using a granular stacking model [26]. The results showed that the gradient of Li^+ concentration becomes more pronounced as areal density increases. In thick cathodes with high areal density, the active material particles exhibit uneven sizes. Li^+ tends to diffuse more rapidly within smaller particles, leading to significant irregularities in the distribution of Li^+ concentration. The uneven distribution of Li^+ results in inconsistent discharge depths among active particles, further affecting the performance and efficiency of the battery.

We compared the performance of cathodes prepared by the dry process and wet process, as shown in Figure 5. The electrochemical impedance spectroscopy (EIS) results (Figure 5a) further show that the dry-process cathode exhibits lower impedance compared to the wet-process cathode, resulting in tighter interfacial contact between particles. This forms more efficient ion transport pathways, thereby reducing interfacial contact resistance and minimizing energy loss during ion transport. In contrast, in wet-processed electrodes, the presence of solvent may lead to uneven distribution between the active material and conductive agent. After solvent evaporation, interfacial defects are formed, leaving discontinuous ion transport pathways, which ultimately reduce ion transport efficiency [27]. The cycling performance (Figure 5b) shows that the discharge specific capacity of dry film is slightly lower than that of the wet film. However, the dry film demonstrates a higher initial Coulombic efficiency, indicating more efficient utilization of the active material. Additionally, the dry film shows a capacity retention rate of 84.5%, which is higher than the wet film (83.6%). The comparison of dQ/dV curves between the dry film (Figure 5c) and wet film (Figure 5d) reveals that the dry film exhibits a more stable performance [28,29], consistent with the cycling test results.

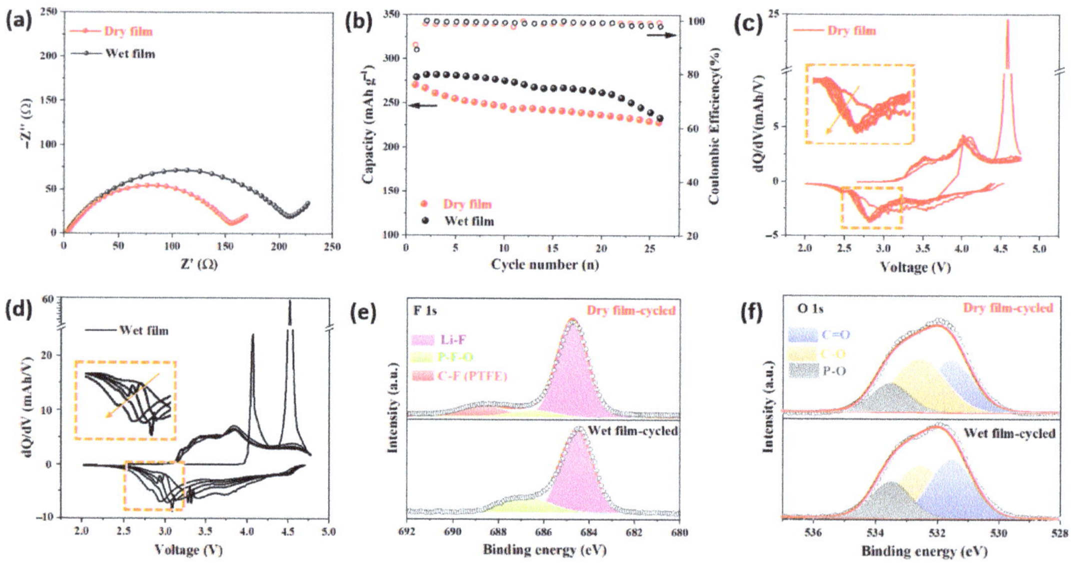

Figure 5. (**a**) EIS results, (**b**) cycling performance of dry film and wet film. dQ/dV curves of (**c**) dry film and (**d**) wet film. (**e**) F 1s, and (**f**) O 1s core spectra of cycled dry film and wet film.

To gain insight into the structural improvements that contribute to the dry film's enhanced battery performance, we analyzed the composition of the cathode electrolyte interphase (CEI). We retrieved the cathodes of both dry-film and wet-film cells after formation for XPS characterization. The F 1s spectrum (Figure 5e) reveals a higher presence of LiF (79.7% vs. 77.4%) on the surface of the dry film, indicating the formation of a robust LiF-rich CEI as a protective layer that efficiently inhibits unwanted side reactions between the cathode and the electrolyte [30]. Additionally, the O 1s spectrum (Figure 5f) shows a higher proportion of P-O (22.08% vs. 17.10%) on the surface of the wet film, suggesting more side reactions with organic electrolytes, which leads to surface structural degradation. These results indicate that the network structure of the dry film optimizes ion transport and electron conduction pathways, resulting in reduced resistance and superior Columbic efficiency and retention over cycling. This highlights the superior utilization of active materials in the dry film, potentially due to its more effective structural integrity and minimized side reactions during operation [31,32]. In contrast, the wet film exhibited higher impedance due to the formation of clusters or less uniform distribution of active materials and binders, which hinders efficient e^-/Li^+ transfer and contributes to increased side reactions with electrolytes.

The SEM image (Figure S7) of the wet film reveals that the active material, carbon materials, and PVDF binder aggregate into clusters. In contrast, the dry film forms a three-dimensional conductive network that reduces polarization effects and enhances charge transfer efficiency (Figure S1). To ensure the accuracy of the comparison results, we selected dry and wet cathodes with similar mass loadings (Figure 6). According to the porosity formula [33], dry films have lower porosity than wet films (30% vs. 43%). It is noteworthy that PTFE is typically used in dry-processed electrodes, whereas PVDF is commonly employed in wet-processed electrodes, as this pairing is dictated by their respective processing requirements. The choice of binder significantly influences the SEI composition, the electrode structure, and, consequently, the electrochemical cycling performance. This inherent constraint reveals the intricate relationship between binder chemistry and electrode fabrication methods in determining overall battery performance.

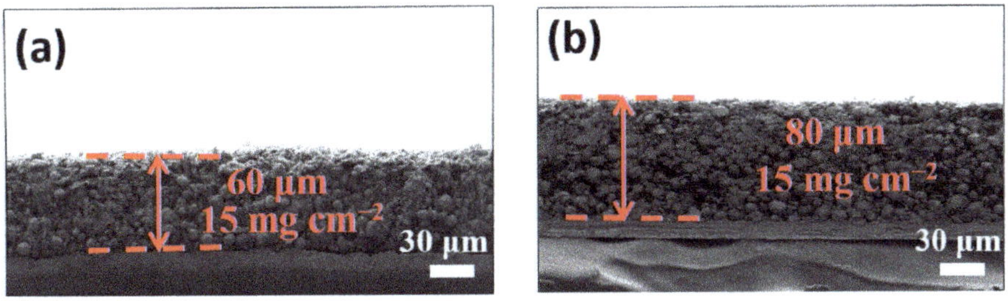

Figure 6. Cross-section SEM images of LRM with (**a**) dry film and (**b**) wet film.

SEM images of the cycled wet film section (Figure 7) showed obvious large voids, whereas SEM images of the dry film section did not have excessive pores, confirming the above illustration. Therefore, the choice of manufacturing process not only affects the structural integrity of the electrode but also plays a crucial role in determining the electrochemical performance and efficiency of the battery.

The comparative performance analysis between dry-processed and wet-processed electrodes conclusively demonstrates that LRM is highly prone to surface degradation (e.g., $Li_2CO_3/LiOH$ formation) and irreversible oxygen release when exposed to moisture or polar solvents during traditional slurry processing [34]. Therefore, dry processing, which eliminates solvent use, is suitable for the preparation of LRM electrodes. Furthermore, compared with commercially dominant cathode materials, such as lithium iron phosphate (LFP) and nickel–cobalt–manganese (NCM), as shown in Table S2 [27,35], LRMs with a higher capacity (>250 mAh/g) demonstrate the potential of dry processing in enabling next-generation high-energy-density cathodes.

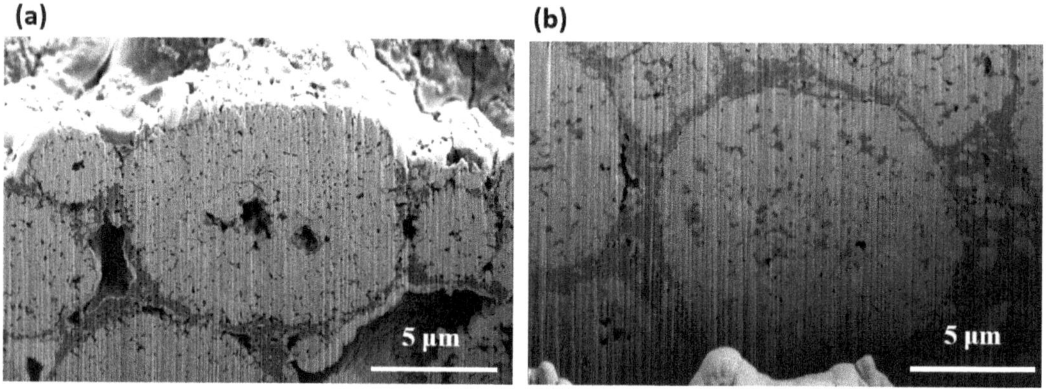

Figure 7. SEM image of cycled (**a**) wet film and (**b**) dry film.

4. Conclusions

This study successfully demonstrates the fabrication of lithium-rich manganese oxide (LRM) cathodes through an innovative dry binder fibrillation process. Through various morphologies of conductive carbon combinations, we established an optimized three-dimensional structure connected by points, lines, and surfaces for the cathodes. The structural design with close contact and tight packing improves the conductivity and facilitates more efficient e^-/Li^+ transport, ultimately resulting in enhanced battery performance. Comparing the electrochemical performance of dry electrodes with various thickness, a high areal capacity of 7.9 mAh cm^{-2} and a pouch cell with an energy density of 949 Wh kg^{-1} are achieved. The choice of manufacturing process also plays a crucial role in determining the performance. The dry film with a network structure optimizes the pathways for ion transport and electron conduction, leading to reduced resistance and superior coulombic efficiency and retention over cycling, potentially due to the more effective structural integrity and minimized side reactions during operation. In contrast, the wet film exhibits higher impedance due to clusters or a less uniform distribution of active materials and binders, which can hinder efficient e^-/Li^+ transfer and lead to increased side reactions with electrolytes. Therefore, this study proposed a practical strategy for designing lithium-rich manganese cathodes with high mass loading, aiming to address issues related to manufacturing processes. The design concept of building continuous interfaces and optimizing electrode structures can be applied to various cathodes to prepare high-mass-loading, long-life, and high-energy-density lithium-ion batteries, and can be further extended to sodium-ion batteries and beyond.

Supplementary Materials: The following supporting information can be downloaded at https://www.mdpi.com/article/10.3390/batteries11040146/s1, Supporting S1: Material preparation route; Figure S1: SEM image of internal structure of dry film; Figure S2: Electrical conductivity with different ratios of carbon; Figure S3: Comparison of XRD patterns between LRM raw materials and dry-processed electrodes; Figure S4: Compaction density and porosity of cathodes with different thicknesses; Figure S5: Cycling performance of cathodes with various mass loading at 0.1 C; Figure S6: Charge/discharge voltage profile of the Li metal pouch cell at 0.1 C; Figure S7: SEM image of wet film; Table S1: Film-forming properties and flexibility of electrodes with different material ratios. Table S2: Comparison of properties of different materials.

Author Contributions: Conceptualization, T.Y. and Z.W.; methodology, T.Y. and Z.W.; investigation, Y.M., H.G., T.Y. and Z.W.; data curation, Y.M., H.G., T.Y. and Z.W.; writing—original draft preparation, Y.M.; writing—review and editing, Y.M., T.Y. and Z.W.; visualization, Y.M., T.Y. and Z.W.; supervision,

Y.M., H.G., T.Y. and Z.W.; project administration, T.Y. and Z.W.; funding acquisition, Z.W. All authors have read and agreed to the published version of the manuscript.

Funding: This study was financially supported by the Science and Technology Program of Tianjin, China (24YDTPJC00140) and the Natural Science Foundation of Hebei Province, China (E2023202253).

Data Availability Statement: The data presented in this study are available upon request from the corresponding author.

Conflicts of Interest: The authors declare no conflicts of interest.

References

1. Sun, X.; Qin, C.L.; Zhao, B.Y.; Jia, S.F.; Wang, Z.F.; Yang, T.Z.; Liu, X.C.; Pan, L.N.; Zheng, L.L.; Luo, D.; et al. A cation and anion dual-doping strategy in novel Li-rich Mn-based cathode materials for high-performance Li metal batteries. *Energy Storage Mater.* **2024**, *70*, 103559. [CrossRef]
2. Xu, J.J.; Cai, X.Y.; Cai, S.M.; Shao, Y.X.; Hu, C.; Lu, S.R.; Ding, S.J. High-energy lithium-ion batteries: Recent progress and a promising future in applications. *Energy Environ. Mater.* **2023**, *6*, e12450. [CrossRef]
3. Wang, Z.F.; Yan, Y.J.; Zhang, Y.G.; Chen, Y.X.; Peng, X.Y.; Wang, X.; Zhao, W.M.; Qin, C.L.; Liu, Q.; Liu, X.J.; et al. Single-atomic Co-B_2N_2 sites anchored on carbon nanotube arrays promote lithium polysulfide conversion in lithium-sulfur batteries. *Carbon Energy* **2023**, *5*, e306. [CrossRef]
4. Yonaga, A.; Kawauchi, S.; Mori, Y.; Xuanchen, L.; Ishikawa, S.; Nunoshita, K.; Inoue, G.; Matsunaga, T. Effects of dry powder mixing on electrochemical performance of lithium-ion battery electrode using solvent-free dry forming process. *J. Power Sources* **2023**, *581*, 233466. [CrossRef]
5. Wang, Z.F.; Wang, H.Y.; Liu, X.L.; Chen, Y.X.; Zhao, Y.; Zhang, Y.G.; Han, Q.Q.; Qin, C.L.; Bakenov, Z.; Wang, Y.C.; et al. Single Zn atoms anchored on hollow carbon nanofiber network for dendrite-free lithium metal anode of flexible Li-S full cell. *Rare Met.* **2023**, *42*, 3705–3717. [CrossRef]
6. Arnot, D.J.; Mayilvahanan, K.S.; Hui, Z.; Takeuchi, K.J.; Marschilok, A.C.; Bock, D.C.; Wang, L.; West, A.C.; Takeuchi, E.S. Thick electrode design for facile electron and ion transport: Architectures, advanced characterization, and modeling. *Acc. Mater. Res.* **2022**, *3*, 472–483. [CrossRef]
7. Kuang, Y.; Chen, C.; Kirsch, D.; Hu, L. Thick electrode batteries: Principles, opportunities, and challenges. *Adv. Energy Mater.* **2019**, *9*, 1901457. [CrossRef]
8. Yao, W.; Chouchane, M.; Li, W.; Bai, S.; Liu, Z.; Li, L.; Chen, A.X.; Sayahpour, B.; Shimizu, R.; Raghavendran, G. A 5 V-class cobalt-free battery cathode with high loading enabled by dry coating. *Energy Environ. Sci.* **2023**, *16*, 1620–1630. [CrossRef]
9. Ryu, M.; Hong, Y.K.; Lee, S.Y.; Park, J.H. Ultrahigh loading dry-process for solvent-free lithium-ion battery electrode fabrication. *Nat. Commun.* **2023**, *14*, 1316. [CrossRef]
10. Kim, H.; Lim, J.H.; Lee, T.; An, J.; Kim, H.; Song, H.; Lee, H.; Choi, J.W.; Kang, J.H. Ozone-treated carbon nanotube as a conductive agent for dry-processed lithium-ion battery cathode. *ACS Energy Lett.* **2023**, *8*, 3460–3466. [CrossRef]
11. Tao, R.M.; Steinhoff, B.; Sun, X.G.; Sardo, K.; Skelly, B.; Meyer, H.M., III; Sawicki, C.; Polizos, G.; Lyu, X.; Du, Z.J.; et al. High-throughput and high-performance lithium-ion batteries via dry processing. *Chem. Eng. J.* **2023**, *471*, 144300. [CrossRef]
12. Sul, H.; Lee, D.; Manthiram, A. High-loading lithium-sulfur batteries with solvent-free dry-electrode processing. *Small* **2024**, *20*, 2400728. [CrossRef] [PubMed]
13. Gao, Y.N.; Yang, Y.; Yang, T.Z.; Zhang, Z.; Tang, L.; Mao, Z.Y.; Zhang, Y.G.; Luo, D.; Chen, Z.W. Design lithium exchanged zeolite based multifunctional electrode additive for ultra-high loading electrode toward high energy density lithium metal battery. *Adv. Energy Mater.* **2024**, *15*, 2403063. [CrossRef]
14. Lv, Z.W.; Liu, J.; Li, C.; Peng, J.X.; Zheng, C.X.; Zheng, X.F.; Wu, Y.Q.; Xia, M.; Zhong, H.Y.; Gong, Z.L.; et al. High-areal-capacity all-solid-state Li-S battery enabled by dry process technology. *Etransportation* **2024**, *19*, 100298. [CrossRef]
15. Fu, J.Z.; Gong, X.T.; Jin, W.T.; Podder, C.; Liu, Y.T.; Yang, Z.Z.; Sultanov, M.; Pan, H.; Wang, Y. Enable superior performance of ultra-high loading electrodes through the cost-efficient solvent-free electrode manufacturing technology. *Energy Storage Mater.* **2024**, *69*, 103423. [CrossRef]
16. Lu, Y.; Zhao, C.Z.; Yuan, H.; Hu, J.K.; Huang, J.Q.; Zhang, Q. Dry electrode technology, the rising star in solid-state battery industrialization. *Matter* **2022**, *5*, 876–898. [CrossRef]
17. Tang, W.H.; Zhu, J.P.; Chen, C.; Ye, Q.; Zeng, F.H.; Ma, Z.P. Modification strategies and challenges of high-performance lithium-rich manganese-based cathode materials. *Energy Technol.* **2024**, *12*, 2301254. [CrossRef]
18. Li, Q.; Yang, Y.; Yu, X.Q.; L, H. A 700 $W{\cdot}h{\cdot}kg^{-1}$ rechargeable pouch type lithium battery. *Chin. Phys. Lett.* **2023**, *40*, 048201. [CrossRef]

19. Yuan, X.D.; Dong, T.; Liu, J.X.; Cui, Y.Y.; Dong, H.T.; Yuan, D.; Zhang, H.T. Biaffinity electrolyte optimizing high-voltage lithium-rich manganese oxide battery via interface modulation strategy. *Angew. Chem. Int. Ed.* **2023**, *62*, e202304121. [CrossRef]
20. Zhang, B.D.; Wu, X.H.; Luo, H.Y.; Yan, H.; Chen, Y.L.; Zhou, S.Y.; Yin, J.H.; Zhang, K.; Liao, H.G.; Wang, Q.S.; et al. Gradient interphase engineering enabled by anionic redox for high-voltage and long-life Li-ion batteries. *J. Am. Chem. Soc.* **2024**, *146*, 4557–4569. [CrossRef]
21. Ge, B.C.; Deng, J.J.; Wang, Z.J.; Liang, Q.H.; Hu, L.; Ren, X.Y.; Li, R.M.; Lin, Y.X.; Li, Y.S.; Wang, Q.R.; et al. Aggregate-dominated dilute electrolytes with low-temperature-resistant ion-conducting channels for highly reversible Na plating/stripping. *Adv. Mater.* **2024**, *36*, 2409161. [CrossRef] [PubMed]
22. Steven, L.; Su, L.S.; Alex, M.; Cui, Z.H.; Arumugam, M. Cracking vs. surface reactivity in high-nickel cathodes for lithium-ion batteries. *Joule* **2023**, *7*, 2430–2444.
23. Wade, A.; Llewellyn, A.V.; Heenan, T.M.M.; Tan, C.; Brett, D.J.L.; Jervis, R.; Shearing, P.R. First cycle cracking behaviour within Ni-rich cathodes during high-voltage charging. *J. Electrochem. Soc.* **2023**, *170*, 070513. [CrossRef]
24. Gao, T.H.; Andrew, K.; Lu, W. Modeling electrode-level crack and quantifying its effect on battery performance and impedance. *Electrochim. Acta* **2020**, *363*, 137197. [CrossRef]
25. Xu, M.; Reichman, B.; Wang, X. Modeling the effect of electrode thickness on the performance of lithium-ion batteries with experimental validation. *Energy* **2019**, *186*, 115864. [CrossRef]
26. An, F.Q.; Zhou, W.N.; Li, P. A comparison of model prediction from P2D and particle packing with experiment. *Electrochim. Acta* **2021**, *370*, 137775.
27. Kwon, K.; Kim, K.; Han, S.; Lee, J.; Lee, H.; Kwon, J.; Lee, J.; Seo, J.; Kim, P.J.; Song, T.; et al. Low-resistance $LiFePO_4$ thick film electrode processed with dry electrode technology for high-energy-density lithium-ion batteries. *Small Sci.* **2024**, *4*, 2300302. [CrossRef]
28. Liang, J.W.; Zhu, Y.M.; Li, X.N.; Luo, J.; Deng, S.X.; Zhao, Y.; Sun, Y.P.; Wu, D.J.; Hu, Y.F.; Li, W.H.; et al. A gradient oxy-thiophosphate-coated Ni-rich layered oxide cathode for stable all-solid-state Li-ion batteries. *Nat. Commun.* **2023**, *14*, 146. [CrossRef]
29. Chen, D.C.; Kan, W.H.; Chen, G.Y. Understanding performance degradation in cation-disordered rock-salt oxide cathodes. *Adv. Energy Mater.* **2019**, *9*, 1901255.
30. Bai, P.X.; Ji, X.; Zhang, J.X.; Zhang, W.R.; Hou, S.; Su, H.; Li, M.J.; Deng, T.; Cao, L.S.; Liu, S.F.; et al. Formation of LiF-rich cathode-electrolyte interphase by electrolyte reduction. *Angew. Chem. Int. Ed.* **2022**, *61*, e202202731. [CrossRef]
31. Feng, Z.R.; Guo, L.Y.; Liu, X.F.; Li, W.W.; Zhang, R.P.; Wang, D.; Zhang, W.; Zheng, W.T. Stabilizing the lithium-rich manganese-based oxide cathode via regulating a CEI film. *ACS Appl. Energy Mater.* **2024**, *7*, 2791–2799. [CrossRef]
32. Li, A.L.; Li, G.H.; Lu, S.G.; Ren, Z.M.; Wang, J.T.; Zhuo, H.X.; Quan, W.; Zhang, G.N.; Han, F.J.; Xia, Y.M.; et al. Interface stabilization of 1,1,2,2-tetrafluoroethyl-2,2,3,3-tetrafluoropropyl ether to high-voltage Li-rich Mn-based layered cathode materials. *Rare Met.* **2022**, *41*, 822–829. [CrossRef]
33. Beuse, T.; Fingerle, M.; Wagner, C.; Winter, M.; Boerner, M. Comprehensive Insights into the Porosity of Lithium-Ion Battery Electrodes: A Comparative Study on Positive Electrodes Based on $LiNi_{0.6}Mn_{0.2}Co_{0.2}O_2$ (NMC622). *Batteries* **2021**, *7*, 70. [CrossRef]
34. Li, Q.H.; Wang, Y.; Wang, X.L.; Sun, X.R.; Zhang, J.N.; Yu, X.Q.; Li, H. Investigations on the fundamental process of cathode electrolyte interphase formation and evolution of high-voltage cathode. *ACS Appl. Mater. Interfaces* **2019**, *12*, 2319–2326. [CrossRef]
35. Wu, F.; Shi, Q.; Chen, L.; Dong, J.; Zhao, J.; Wang, H.; Gao, F.; Liu, J.; Zhang, H.; Li, N.; et al. New insights into dry-coating-processed surface engineering enabling structurally and thermally stable high-performance Ni-rich cathode materials for lithium ion batteries. *Chem. Eng. J.* **2023**, *470*, 144045. [CrossRef]

MDPI AG
Grosspeteranlage 5
4052 Basel
Switzerland
Tel.: +41 61 683 77 34

Batteries Editorial Office
E-mail: batteries@mdpi.com
www.mdpi.com/journal/batteries

www.ingramcontent.com/pod-product-compliance
Lightning Source LLC
Chambersburg PA
CBHW061604210926
51850CB00009B/2212
* 9 7 8 3 7 2 5 8 6 1 1 9 4 *